AF837285

Suburban Books

40 East Scenic Ave, Richmond, CA 94801

Molecular Evolution and Protobiology [Hardcover] [1984]
Matsuno, Koichiro

.

Order ID: 112-7787570-7338626 (000WU646)
Purchased: 2026-09-22T08:14:21-07:00

Shipping Address:
AMZ ORDERS APTP
800 AVONDALE AVE
COLUMBUS, OH 43212-3473

Shipping Method: Standard

Thank you,

Tyson

P.S. If you have any questions or comments, you can contact me at: tyson@bichonbooks.com

MOLECULAR EVOLUTION AND PROTOBIOLOGY

MOLECULAR EVOLUTION AND PROTOBIOLOGY

Edited by
Koichiro Matsuno
Technological University of Nagaoka
Nagaoka, Japan
Klaus Dose
J. Gutenberg University
Mainz, Federal Republic of Germany
Kaoru Harada
University of Tsukuba
Ibaraki, Japan
and
Duane L. Rohlfing
University of South Carolina
Columbia, South Carolina

PLENUM PRESS • NEW YORK AND LONDON

Library of Congress Cataloging in Publication Data

Main entry under title:

Molecular evolution and protobiology.

Includes bibliographical references and index.
1. Chemical evolution. 2. Life—Origin. I. Matsuno, Koichiro.
QH325.M64 1984 577 83-24465
ISBN 0-306-41509-7

Cover: Photo of proteinoid microspheres, courtesy of Mr. Steven Brooke

A Division of Plenum Publishing Corporation
233 Spring Street, New York, N.Y. 10013

Printed in the United States of America

PREFACE

In recent years, an ever-increasing amount of research has been conducted on the physico-chemical basis of the origin and evolution of life, or protobiology. Many questions are raised in this endeavor: What research methodology should be employed? What sort of dependable facts are available as a firm frame of reference upon which the physico-chemical origin of life or protolife could be examined? Is the origin due exclusively to chance events? If not, what is then responsible for the origin? What physical reality underlies the evolutionarily selective process leading to the origin? What role does variation assume and how is it generated in the course of evolution?

Many research workers have pursued various avenues toward answering the stated questions. Among them, we believe Sidney W. Fox has been playing a very unique and pivotal role over the past quarter of a century, presiding over 240 man-years or more of laboratory work. His laboratory syntheses of thermal proteins called proteinoids and proteinoid microspheres have emphasized the principle of the self-sequencing of amino acids as a key concept of protobiological synthesis. The significance of his contribution is seen in presenting the experimental evidence that the origin of life is largely due to nonrandom events. This discovery marks a new epoch in the conceptual development of studying the origin of life by focusing on the molecular processes that underlied the emergence and evolution of protobiological information.

Laboratory experiments on simulated protobiogenesis seek the origin of biological information either in the initial condition, in the external boundary condition, or within the matter that itself formed the first biological systems. What makes thermal proteins or proteinoids significant especially from an evolutionary perspective is their experimental indication of the protobiological information originating within matter. The capacity of matter for self-assembly plays a most fundamental role.

Questions become sharper as the research proceeds. We thus thought that the time is ripe for inviting scientists in the field to express their own ideas and to see how their problems evolve. In particular, the year 1983 appears to be a good opportunity because we also observe the 25th anniversary of the first report on the thermal synthesis of heteropolyamino acids (proteinoid). The research programs initiated by the conceptualization of protoproteins, among whose laboratory models are thermal proteins or proteinoids, have continued to gain momentum both experimentally and theoretically.

The central problems cover the assembly of prebiotic molecules and protocells, as well as their structural and functional analysis. For instance, we imagine that a deeper study of precursors of protein and protocells could awaken us to much sharper and more penetrating questions. Protobiology is providing a bridge over the chasm between physics and chemistry on the one hand and biology on the other. It is our hope to see where we are now and to where we should go in the scientific investigation of the origin or, rather, emergence of life, even though we may not command a precise definition or understanding of life itself.

This volume is comprised of five parts and 32 chapters. In the introductory chapter the status of protobiology is briefly reviewed. The historical and personal reviews in Part I focus on the concept of self-assembly, with emphasis on the proteinoid theory. In Part II are discussed several models of molecular self-assembly, such as carbonaceous small molecules, pyrophosphate and thermal polyamino acids. Part III deals with the spontaneous emergence of protocells and related systems. Part IV refers to the origin of the modern genetic code and protein synthesis. Part V comprises theoretical concepts on the emergence of protolife and life. In particular, the last chapter attempts to integrate the major ideas presented in this volume. The views expressed in these chapters are not necessarily consistent with each other. Some of the conflicts, we believe, serve as a source of the new problems to be tackled in the future.

We offer this volume as a tribute commemorating the 25th anniversary of the pioneering thermal heteropolycondensation of amino acids and as a dedication to Sidney W. Fox on the occasion of his seventieth birthday.

We would like to acknowledge Michael Conrad, James C. Lacey, Jr., Paul Melius, and Robert M. Syren for their generous time offered to help us.

Koichiro Matsuno
Klaus Dose
Kaoru Harada
Duane L. Rohlfing

CONTENTS

PROTOCELLS

THE ORIGIN OF THE GENETIC CODE AND PROTEIN SYNTHESIS

THEORETICAL SYNTHESIS

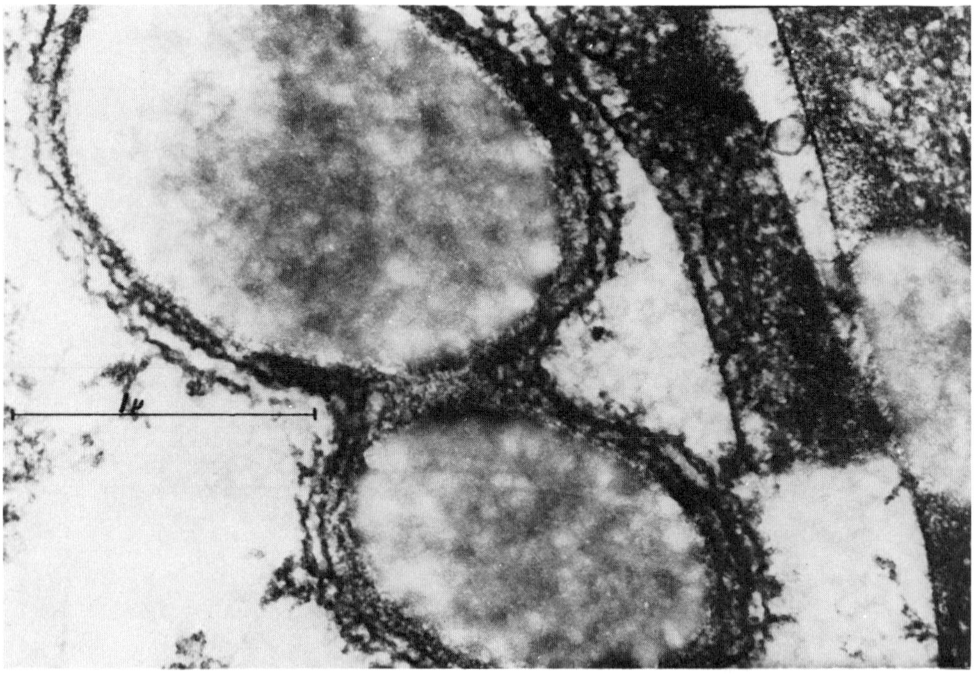

Proteinoid Microsphere (Courtesy of Dr. Walther Stoeckenius)

MOLECULAR EVOLUTION AND PROTOBIOLOGY: AN OVERVIEW

Klaus Dose

Institute for Biochemistry
J. Gutenberg University
Mainz, Federal Republic of Germany

INTRODUCTION

During the winter semester 1865/66 the German biologist Ernst Haeckel (1834-1919) proposed in his lectures on Darwinism and evolutionary sequence in order to explain the origin of the first cells by self-assembly and selection. These lectures were published in 1868 (Haeckel, 1868). Haeckel's pioneering ideas constituted a departure from the then prevailing view that the origin of life could not be a subject of scientific research.

This view is best expressed by statements of C. Darwin made at about the same time. In 1859 (C. Darwin, 1859) C. Darwin suggested that all forms of life that ever existed on Earth evolved from a common ancestor that was created by a divine act. Again in 1863 Darwin rejected the idea that the question of the origin of life could be rationally analyzed. In a letter to Hooker (edited by F. Darwin, 1896) he wrote, "It is mere rubbish thinking at the presence of the origin of life--one might as well think of the origin of matter." But later, in 1871, perhaps impressed by Haeckel's thesis, Darwin no longer rejected thinking of an evolutionary origin of life (Darwin, 1959, posthumous).

Haeckel's concept on a stepwise self-organization of matter into a living system was certainly a controversial issue in his time because it was not backed by appropriately designed experimentation. Our modern era of the field was initiated when Oparin in 1924 (Oparin, 1924) for the first time published his concept on the origin of life. But Oparin actually revived Haeckel's nineteenth-century concept. Oparin could certainly be more specific and more fact-based than Haeckel whose detailed views today sound quite naive. The gap between the idealized concept and the known facts, however, was also tremendous in 1924. It took more than 30 years until some of Oparin's (and Haeckel's) ideas on the spontaneous formation of biologically significant molecules from "inorganic" precursors could be verified in the laboratory. The history of this era of experimentation has been detailed elsewhere (see e.g., Kenyon and Steinman, 1969; Miller and Orgel, 1974; Fox and Dose, 1977). Numerous experiments in the field have since 1955 dealt with the abiotic or prebiotic origin of small molecules on the primitive Earth. This area of research is often referred to as "chemical evolution."

The number of experiments on the origin of the first biologically relevant macromolecules and their self-organization into prebiotic systems including models for protocells has also been impressive. This field is often referred to as "molecular evolution", a term generally accepted in analogy to "molecular biology", in order to express the additional involvement of molecular (and supramolecular) interactions beyond the level of typical (bio)chemical reactions. In spite of the many intriguing experiments on the origin of the first cells, pioneered in particular by S. W. Fox and his associates during the past 30 years, we cannot as yet unambiguously conclude that a protocell has been constructed in the laboratory. Experiments with prebiotic molecules, however, convincingly demonstrate the central role of self-organization also for the prebiotic realm. The various cell-like or cell organelle-like structures that have been produced in simulation experiments can be regarded as precells or preorganelles. It has to be established, however, whether some of these precells, for instance, are capable of evolving to primitive cells (or primordial cells), that is, to the evolutionary precursors of contempory cells. Only these precellular structures are protocells. The form of "life" they represent is thus protolife. The science concerned is protobiology.

Dealing with protobiology or molecular evolution requires a mode of reasoning usually different from that in biology: Since nobody was there when life arose the mode of reasoning is largely constructionistic in protobiology, it is largely deductionistic, however, in biology.

The various evolutionary sequences from the beginning of the Universe to the appearance of contemporary cells have often been summarized in a single scheme as shown in Fig. 1.

THE ABUNDANCE OF BIOELEMENTS IN PRESOLAR CLOUDS

When we now view the abundances of elements in interstellar clouds, that is, in clouds that contain the starting materials for the formation of new stars, including those of the kind of our Sun

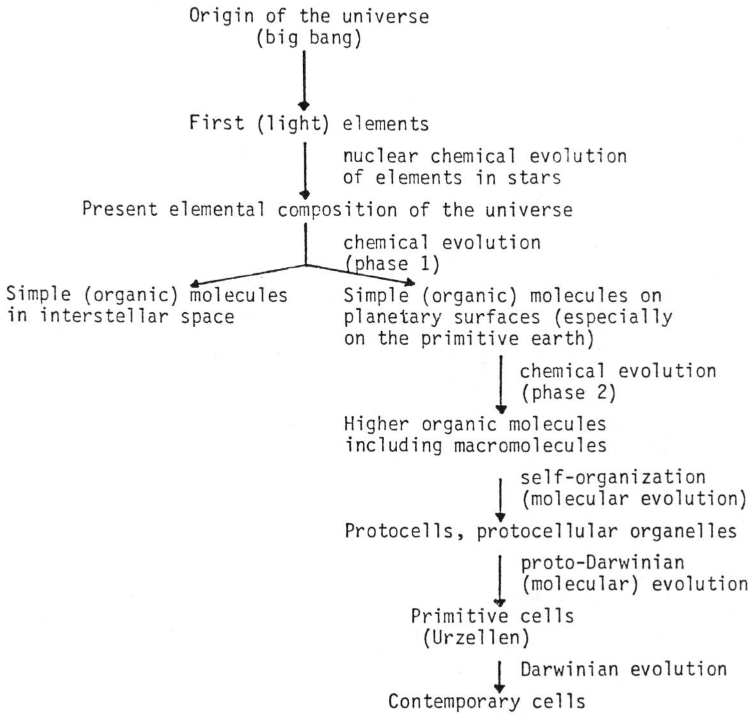

Fig. 1. Sequences of nuclear chemical and molecular evolution. The scheme has been compiled on the basis of experimentation, observation, and theoretical reasoning.

with its planets, we realize that the four most important bioelements hydrogen, carbon, oxygen, and nitrogen are also the most abundant elements besides the noble helium and neon (Fig. 2). The production of these elements by nuclear chemical processes that occur in the interior of stars is directly controlled by the physical properties of neutrons, protons, and other particles of which the nuclei of these elements are made. The laws of physics and the properties of matter have predetermined our Universe to produce predominantly the four bioelements besides helium and neon.

All chemical elements (with the general exception of the noble gases) readily interact chemically whenever the conditions are favorable. Spontaneous formation and interconversion of molecules occur in a variety of cosmological settings, even at the extremely low temperature and pressure of interstellar clouds. Most of these chemical compounds, so far observed and identified, are carbon compounds (Table 1). The most abundant molecular species, however, is water. The available data on the chemistry of the Universe

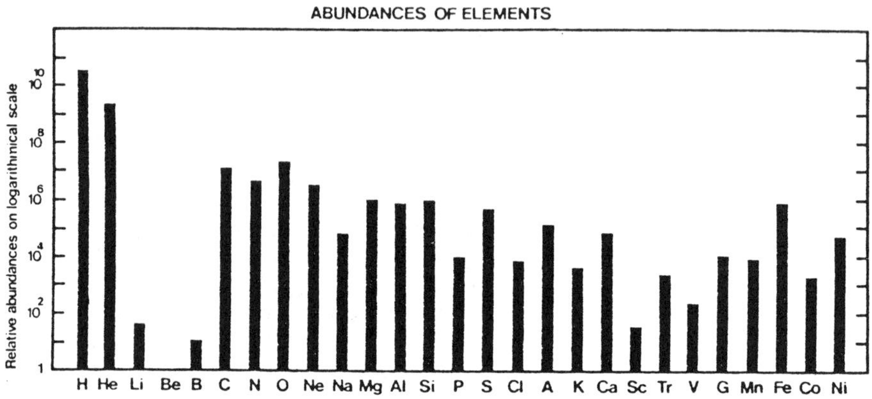

Fig. 2. Interstellar abundances of elements. Source: Cameron (1970).

suggest that any thinkable form of life is most likely based on the chemistry of carbon compounds with water as the solvent.

PREBIOTIC MOLECULES: PREBIOTIC AND BIOTIC FITNESS OF AMINO ACIDS AND THEIR POLYMERS

Simulation experiments on the prebiotic origin of biomolecules from sources containing the elements C, H, O, and N generally show that amino acids (or precursors that are readily hydrolyzed to amino acids) are formed with significant yields (see e.g, review by Fox and Dose, 1977). Mostly these yields are substantially higher than those of other classes of organic compounds. Besides glycine, alanine, glutamic acid, and aspartic acid other α-amino acids appear not to be most abundant prebiotic amino acids, and certainly amino acids are formed in racemic mixtures. As yet we cannot satisfactorily explain why amino acids are the predominant reaction products and why they are so fit to survive the stresses of various harsh geochemical environments. We know, of course, that their geochemical fitness is closely related to their zwitter ion character, their high melting (decomposition) points, their low vapor pressure, their generally good solubility and stability in water and their selective binding to minerals with cation and anion exchange capacity.

Another remarkable property of amino acids is their ability to undergo intermolecular condensation reactions to yield relatively stable polymers such as polypeptides, proteins, and other polyamino acids. Due to this process and the multitude of biochemical pathways involving amino acids these classes of compounds also exhibit a high degree of biological fitness.

Table 1. Observed Interstellar Molecules in dense molecular clouds

Inorganic		Organic
H_2 hydrogen	DIATOMIC	CH methylidyne
OH hydroxyl		CH^+ methylidyne ion
SiO silicon monoxide		CN cyanogen
SiS silicon sulfide		CO carbon monoxide
NS nitrogen sulfide		CS carbon monosulfide
SO sulfur monoxide		
H_2O water	TRIATOMIC	CCH ethynal
H_2H^+		HCN hydrogen cyanide
H_2S hydrogen sulfide		HCN hydrogen isocyanide
SO_2 sulfur dioxide		HCO^+ formyl ion
		HCO formyl
		OCS carbonyl sulfide
NH_3 ammonia	4-ATOMIC	H_2CO formaldehyde
		HNCO isocyanic acid
		H_2CS thioformaldehyde
	5-ATOMIC	H_2CHN methanimine
		H_2NCN cyanamide
		HCOOH formic acid
		HC_3N cyanoacetylene
	6-ATOMIC	CH_3OH methanol
		CH_3CN cyanomethane
		$HCONH_2$ formamide
	7-ATOMIC	CH_3NH_2 methylamine
		CH_3C_2H methylacetylene
		$HCOCH_3$ acetaldehyde
		H_2CCHCN vinyl cyanide
		HC_5N cyanodiacetylene
	8-ATOMIC	$HCOOCH_3$ methyl formate
	9-ATOMIC	$(CH_3)_2O$ dimethyl ether
		C_2H_5OH ethanol
		HC_7N cyanotriacetylene

Laboratory experiments, in particular those by S. W. Fox and his associates (Fox et al., 1959a,b; see also the review by Fox and Dose, 1977), have shown that mixtures of amino acids can be polymerized under simulated geological conditions to a variety of individual, nonrandom polymers. These polymers have in common with biologically produced proteins many significant properties (Fox and Dose, 1977). They are therefore often referred to as proteinoids. Their significance as prebiotic precursors of proteins, that is, as protoproteins, has been discussed (Fox, 1981). Because of their geochemical origins, proteinoids (protoproteins) also are to be included in the group of geologically fit molecules.

THE POWER OF SELF-ASSEMBLY: THE ORIGINS OF PROTOCELLS AND OTHER PROTOBIOLOGICAL STRUCTURES

For the first time Fox et al. (1959a) have described the formation of "spherules" during cooling of saturated proteinoid (aqueous) solutions. Later, the spherical, cell-like structures have been called (proteinoid) microspheres. Proteinoids rich in acidic amino acids (aspartic acid, glutamic acid) are generally good starting material for microsphere production. The process is simple, quick, and geologically relevant. To some degree proteinoid microspheres exhibit structures similar to the "organized elements" in carbonaceous chondrites or in ancient sediments. However, the structures in carbonaceous chondrites have very likely a non-aqueous origin. No other model for the origin of the first protocells is as much based on the experimental results of prebiotic evolution as that involving microsphere self-assembly from proteinoids alone or from proteinoids, lipids and related compounds found also in the membranes of contemporary cells.

The properties of proteinoid microspheres and related structures as well as their significance in protobiological evolution have been detailed elsewhere (Fox and Dose, 1977) and they will be discussed and referred to in several articles of this volume.

OPEN QUESTIONS: THE ORIGIN OF THE GENETIC CODE, PROTEIN-BIOSYNTHESIS AND OPTICAL ACTIVITY

The experimentation with proteinoid has lead to a number of principal questions. Three of these are:

1. Do the proteinoids and models for protocells made from proteinoids in the laboratory contain within their nonrandom sequences of amino acids the appropriate information required for evolution towards first living (primordial) cells?

2. Is this information a protogenetic information (an information from which the contemporary genetic information in nucleic acid evolved) and how could information stored in proteinoids (protoproteins) become translated into the sequence of nucleotides in nucleic acids, especially DNA?

3. How did the first nucleic acids evolve (including their stereospecific structures required for their replication)?

Some molecular biologists and theoreticians have postulated that the genetic information of living systems has ab initio evolved in polynucleotides. The problem is that polynucleotides are not self-replicating. They need enzymes for "self-replication". The chances for a practically simultaneous formation of stereochemically appropriate polynucleotides and specific polynucleotide replicases are supposedly extremely small, they are definitely zero according to the experimental data as yet available.

The following two obvious reasons have therefore been put forward in devaluation of the "gene-first" view:

1. As already stated, there is no self-replication of nucleic acids in contemporary cells or in cell-free systems. Catalysts are required. In contemporary cells the catalysts are highly specific nucleic acid replicases. Like the synthesis of all other enzymes the synthesis of these enzymes is coded by DNA and RNA.

2. Even if there exists an (as yet only suggested) non-enzymic, but sufficiently specific catalysis of polynucleotide replication, where did the first replicable polynucleotide come from? To be replicable the polynucleotide must have a highly specific structure as exhibited by contemporary DNA (D-2-deoxyribose in 3'- and 5'-position esterified with phosphoric acid and in 1'-position β-glycosidially linked to a specific N-atom of a purine or pyrimidine base).

Only if these two objections are neglected does the "genes-first" view (see also Fig. 3, right) become a realistic theory. Even then, the prebiotic selection and conservation of optically pure D-ribose and the spontaneous formation of an appropriately structured oligonucleotide, perhaps a decanucleotide, appear to be extremely unlikely on the basis of experimental data available.

In summary, prebiotic simulation experiments back the protoproteins-first view (Fig. 3, left). The step from informative amino acid sequences in polypeptides to nucleotide sequences coding for amino acid sequences does not require the actual reversal of what is called translation in molecular biology. The notorious "central dogma of molecular biology" only forbids direct reversal of translation in contemporary cells. But this dogma does not generally rule out that proteins can catalyze and direct the synthesis of

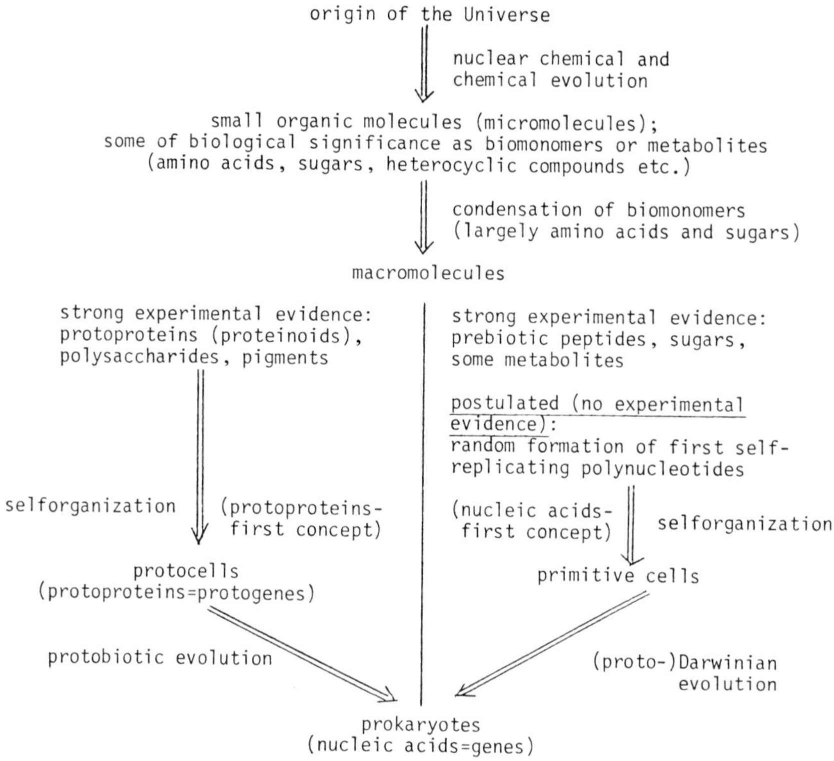

Fig. 3. Scheme of evolution of prokaryotes: According to the "proteins-first concept" (left) proteinoids (protoproteins-protogenes) were the first informative biopolymers. During further evolution their information became translated into specific nucleotide sequences of first (informative) nucleic acids. According to the "nucleic-acids-first concept" (right) biological information has evolved ab initio from spontaneously formed polynucleotides (protogenes). The latter concept is not supported by simulation experiments on prebiotic evolution.

specific nucleotide sequences in the absence of a nucleic acid acting as a primer. Such a flow of information, e.g., has been verified in the case of Qβ-replicase (Sumper and Luce, 1975). We therefore emphasize that biological information can flow from protein to nucleic acid. Mechanistically there exist various ways to achieve such a flow of information by direct interaction of nucleotides with specific active sites of a nucleotide polymerase. It will be shown by a number of articles in this volume that there is no need to violate the "central dogma" in order to demonstrate or postulate that the first nucleic acid-based genes were produced by protoproteins (proteinoids).

REFERENCES

Cameron, A. G. W., 1970, Space Science Reviews, 15:121.
Darwin, C., 1859, "On the Origin of Species by Means of Natural Selection," Murray, London.
Darwin, C., 1959 (posthumous), Some Unpublished Letters (1871) Sir Gavin de Beer, ed., Notes Rec. Roy. Soc. London, 14:1.
Darwin, F., 1896, "The Life and Letters of Charles Darwin," Vol. 2, p. 202, Appleton and Co., New York.
Fox, S. W., 1981, Amer. Biol. Teach., 43:127.
Fox, S. W., and Dose, K., 1977, "Molecular Evolution and the Origin of Life," 2nd edn., Marcel Dekker, New York.
Fox, S. W., Harada, K., and Kendrick, J., 1959a, Science, 129:1221.
Fox, S. W., Harada, K., and Vegotsky, A., 1959b, Experientia, 15:81.
Haeckel, E., 1968, "Natürliche Schöpfungsgeschichte," G. Riemer, Berlin.
Kenyon, D. H., and Steinman, G., 1969, "Biochemical Predestination," McGraw-Hill, New York.
Miller, S. L., and Orgel, L. E., 1974, "The Origins of Life on the Earth," Prentice Hall, Englewood Cliffs, New Jersey.
Oparin, A. I., 1924, "The Origin of Life," 1st edn., (Russ., Proiskhozdenic Zhizny), Moskovski Rabochii, Moscow.
Sumper, M., and Luce, R., 1975, Proc. Nat. Acad. Sci. U.S., 72:162.

CONCEPTUAL ORIGIN OF THERMAL HETEROPOLYCONDENSATION OF AMINO ACIDS

Koichiro Matsuno

Technological University of Nagaoka

Nagaoka 949-54, Japan

Any experimental breakthrough presupposes fledgling new ideas as Popper (1959) rightly pointed out. Thermal heteropolycondensation of amino acids is no exception to this rule.

Professor Harada's review of history of the development of the proteinoid model in the following chapter properly covers, as he points out, the period 1957-1960. That is the period in which he worked most directly in the program. It has become apparent to me that thermal heteropolycondensation reported in 1958 (Fox and Harada, 1958) had prior conceptual incubation of its own. Significant seminal events occurred before 1957. Since I believe that the research program both culminating in and originating in thermal copolymerization of amino acids will benefit from an exposition of the emergence of concepts *ab initio*, I have consulted with Professor Fox and other sources in tracing essential concepts from their beginning.

Laboratory synthesis of the proteinoid model for organic evolution began with several hints from prior studies. One was the inferred primordial role of aspartic acid and glutamic acid; this inference arose from studies of comparative composition of protein molecules (Fox, 1956; Vegotsky and Fox, 1962). Another was an indication of self-ordering of amino acids into peptide derivatives formed by the action of proteases. This suggested that amino acids combining without aid of enzymes might order themselves during the process (Fox, 1956). Yet another essential hint was found in a review by Katchalski (1951), which stated that the early chemists, Schiff and Schaal, made polyaspartic acid by heating aspartic acid but, as Katchalski stated, the product was of indefinite constitution.

The general prospect for producing useful polymers of amino acid was for such reasons a clouded one. Indeed, Carothers of Du Pont chose to polymerize ε-amino acids rather than α-amino acids, and thereby initiated the nylons. Most α-amino acids decompose on heating (Katchalski, 1951). The workers at Iowa State learned that they would not do so if they are copolymerized with aspartic acid (Fox and Middlebrook, 1954; Maier, 1956; Fox, 1956).

At about this time Kovacs et al. (1953) showed that thermal polyaspartic is a polyimide that easily hydrolyzes to peptide-type polyaspartic acid. This observation was extended by Vegotsky et al. (1958). At Florida State, the proteinoid investigators learned also that glutamic acid, which does not homopolymerize like aspartic acid (Harada and Fox, 1958), does form copolymers much as does aspartic acid. _The key to production of protein-like polymers by heating is especially in the copolymerization._ This principle applied, as shown at about the same time, to all of the α-amino acids simultaneously (Fox and Harada, 1958).

The synthesis of polymers of twenty amino acids simultaneously thus did not require organisms. Nor did this result signify that the first proteins contained 20 types of α-amino acid. Rather the copolymerization reaction with (a) aspartic acid, (b) glutamic acid, and then (c) all the amino acids documented the wide scope of simple thermal copolymerization of α-amino acids. The research with aspartic acid, including the clarification of information on polyaspartic acid, was part of a crucial conceptual bridge to thermal copolymerization of α-amino acids, including proteinoid as one of the many products.

REFERENCES

Fox, S. W., 1956, _Amer. Sci._, 44:347.
Fox, S. W., and Harada, K., 1958, _Science_, 128:1214.
Fox, S. W., and Middlebrook, M., 1954, _Federation Proc._, 13:211.
Harada, K., and Fox, S. W., 1958, _J. Amer. Chem. Soc._, 80:2694.
Katchalski, E., 1951, _Advances Protein Chem._, 6:123.
Kovacs, J., Konyves, I., and Pusztai, A., 1953, _Experientia_, 9:459.
Maier, G. D., 1956, The pyropolymerization of aspartic acid and leucine. M.S. thesis, Iowa State College.
Popper, K., 1959, "The Logic of Scientific Discoveries," Hutchinson, London.
Vegotsky, A., 1957, Thermal synthesis and characterization of amino acid polymers, M.S. thesis, Florida State University, p. 72.
Vegotsky, A., 1961, Thermal copolymers of amino acids. Ph.D. dissertation, Florida State University, p. 108.
Vegotsky, A., Harada, K., and Fox, S. W., 1958, _J. Amer. Chem. Soc._, 80:3361.

Vegotsky, A., and Fox, S. W., 1962, in: "Comparative Biochemistry," vol. IV, M. Florkin, and H. S. Mason, eds., pp. 185-244, Academic Press, New York.

SOME EARLY HISTORICAL ASPECTS OF THE THERMAL POLYCONDENSATION OF AMINO ACIDS

Kaoru Harada

Institute of Chemistry
University of Tsukuba
Sakura-mura, Ibaraki 305, Japan

The main objective of this review is to track a part of the very early history of Professor Fox's proteinoid theory, which was carried out in the Florida State University in Tallahassee during 1957-1960.

INTRODUCTION

There are two kinds of carbon chemistry in natural science. One is organic chemistry and the other is biochemistry, and these two chemistries are closely related with each other. Contemporary organic chemistry is largely a carbon chemistry in nonaqueous solutions, and the reactions are usually controlled by the use of various solvents, catalysts, temperatures and pressures. Therefore, organic chemistry comprises a systematized knowledge of carbon chemistry under man-made reaction conditions. On the other hand, biochemistry is a carbon chemistry in aqueous solutions, and biochemical reactions are controlled mainly by enzymes.

Prebiological chemistry is the chemistry of the formation and the organization of bio-organic compounds under assumed primitive geological and astronomical conditions. In the study of chemical evolution of bioorganic compounds, we have to choose the conditions which may have occurred on the primitive Earth. The reaction conditions are inevitably related to water molecules, which are among the most abundant molecules in the universe. Therefore, prebiological chemistry is in part also a water-oriented chemistry as are biochemistry and organic geochemistry.

We now know several organic chemical methods of synthesizing peptides or polypeptides from amino acids. These are:

a) step-by-step synthesis by organic chemical techniques,
b) polycondensation of activated amino acid derivatives (e.g., Leuchs anhydride),
c) aminolysis of amino acid esters or peptide esters.

Method (a) was originated by E. Fischer, T. Curtius and M. Bergmann et al. and the recent extension of this method has resulted in the total syntheses of peptide hormones and enzymes such as ribonuclease. In method (b), especially by using the Leuchs anhydride, many investigations have been conducted to synthesize high molecular weight homopolymers of amino acids as chemical and physical models of proteins. Method (c) could also be utilized to prepare homopolymers of several amino acids.

However, these synthetic methods (a), (b) and (c) are based on organic chemical techniques, and the starting compounds, reagents, solvents and the conditions used in these syntheses are usually man-made and have almost never occurred in nature. Therefore, in order to investigate the prebiological origin of protein or protein-like compounds, we have to find some other chemical pathway.

HISTORICAL BACKGROUND

One way to the prebiotic formation of polyamino acid was seen in the polycondensation of free amino acids or dipeptides by heating.

Several attempts to synthesize polyamino acids by heating were reported at the end of the last century and also in the beginning of this century. However, the results obtained by these reactions were not satisfactory because of the lack of knowledge and techniques about high polymers. Amino acids were usually decomposed by heating or converted into diketopiperazines. However, according to the early literature, glycine and aspartic acids were converted into horny or colloidal products.

Polyglycine

Balbiano and Trasciatti (1900) heated glycine in glycerol and obtained diketopiperazine and a small amount of horny material. The material did not dissolve in water and glycine was obtained upon hydrolysis. The heating of alanine, however, did not give any horny material. Maillard (1911, 1914) also heated glycine in glycerol and confirmed the earlier results. Abderhalden and Komm (1924) heated glycine in a sealed tube at 160°C and obtained diketopiperazine as the major product together with a small amount of polyglycine. Shibata (1925) heated diketopiperazine in glycerol in a sealed tube at 170°C and obtained a colloidal product which did not pass through a dialysis membrane.

Later in 1949, Polyakova and Vereschagin (1949) heated diketopiperazine in water at 170°C under high pressure (200-4,000 atm) and reported the formation of diketopiperazine as the major product together with water insoluble polyglycine (25%). Meggy (1953) studied the polycondensation of glycine, glycylglycine and diketopiperazine and found that the yield of polyglycine was dependent on the heating time and the amount of water in the reaction system. In the case of the polymerization of diketopiperazine, the yield of polyglycine was the highest when the ratio of diketopiperazine to water was 2:1, prolonged heating decreased the polymer yield. Meggy also studied the effects of mineral acid, alkali and inorganic salts on polyglycine formation. Watanabe and Kosai (1963) studied the effects of water, acids, organic bases and salts on the polycondensation of glycine. Okamoto and Nakahara (1964) also studied the thermal polycondensation of glycine and alanine.

Oligo- and Polyaspartic Acid

Schaal (1871) heated aspartic acid or aspargine under a carbon dioxide atmosphere. Grimaux (1882) carried out a similar study and obtained certain reaction products. Schiff (1897, 1899) described the isolation of tetra- and octa-aspartic anhydride from heated aspartic acid. These compounds were converted to the corresponding oligoaspartic acids by treatment with boiling water. The resulting acids formed salts with potassium hydroxide. Free octa-aspartic acid was formed to be a clear amorphous glass and was reconverted to the anhydride by heating at 190-200°C. The copper and silver complexes of octa-aspartic acid were prepared and the reactions with ammonia, aniline and phenylhydrazine were studied.

More recently, the thermal polycondensation of aspartic acid was reinvestigated by Kovács et al. (Kovács et al., 1953; Kovács and Könyves, 1954; Kovács et al., 1954; Kovács et al., 1961). They prepared polyaspartic acid anhydride by heating DL-aspartic acid at about 180-200°C. The polymer was easily hydrolyzed to polyaspartic acid. The chain length of the polyaspartic acid was estimated to be 71 by Van Shlyke amino nitrogen analysis. The polyaspartic acid was reconverted to anhydropolyaspartic acid by heating at 200°C or by heating in tetralin. Kovács et al. postulated the structure of anhydropolyaspartic acid as shown below (structure A).

Polyaspartic acid obtained by alkaline treatment is readily soluble in water, but its solubility is different from that of polyaspartic acid synthesized from the Leuch's anhydride of aspartic acid. Therefore, the structure of the hydrolyzed thermal polymer was thought to be a β-linkaged polyaspartic acid (B). Both a and b bonds in the anhydropolyaspartic acid could be hydrolyzed to form α- and β-aspartyl residues, respectively. Therefore, the solubility of the hydrolyzed polyaspartic acid yields no solid evidence for the presence of β-linkages. Later Kovács et al. determined the ratio of α- and β-aspartyl residues of the polyaspartic acid by chemical methods using the Hofmann reaction. The ratio of α- and β-aspartyl residues in the polymer was 1:1.3, and the structure of polyaspartic acid was determined as structure (C).

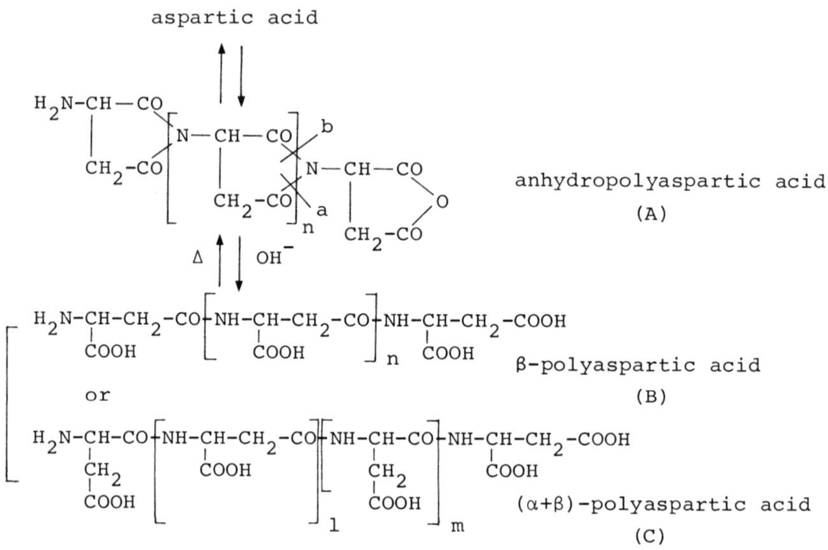

EARLY STUDY OF THE THERMAL SYNTHESES OF POLYAMINO ACIDS

In the middle of the 19th century, Dessaignes (1850a, 1850b) synthesized a compound that was later identified as aspartic acid by heating of ammonium salts of malic acid, maleic acid and fumaric acid. However, at that time the structure of the products obtained was unknown, because it was not until 1887 that Piutti determined the chemical structure of aspartic acid (Piutti, 1887, 1888).

The earliest attempts in Professor Fox's laboratory to synthesize polyamino acids by heating were made in 1954 (Fox and Middlebrook, 1954; Fox, 1956). In 1955, Professor Fox reinvestigated Dessaignes's synthesis of aspartic acid by heating ammonium malate (Fox et al., 1955). It was found that the hydrolyzate of the resulting material gave a small amount of alanine and β-alanine, together with the major product aspartic acid. Alanine and β-alanine probably was derived from aspartic acid by thermal decarboxylation.

Later, similar thermal products were prepared by heating ammonium salts of malic acid, maleic acid and maleamic acid (Harada, 1959a). All these products yielded aspartic acid upon acid hydrolysis. The structure of the thermal product was determined mainly as anhydropolyaspartic acid by infra-red absorption spectra. The thermal polymers of aspartic acid could be formed by nonamino acid C_4 dicarboxylic acids, as shown in the following:

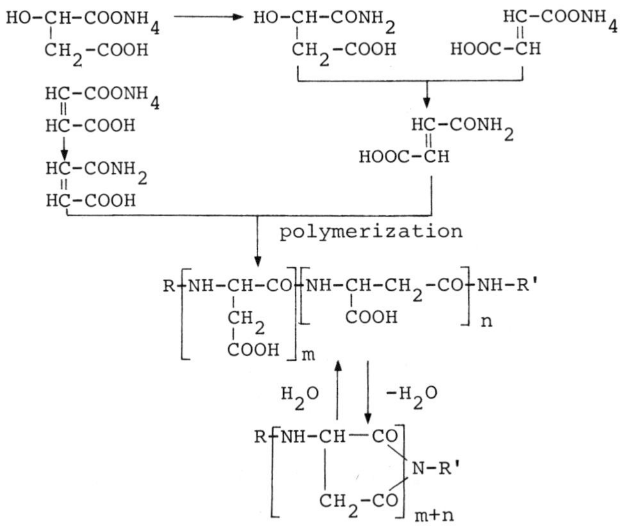

According to this polycondensation mechanism, the polymer should have no N-terminal amino acid. However, some N-terminal aspartic acid was found by the DNP method. This indicates that the last step of the mechanism could be reversible in the presence of water at elevated temperatures. Also, hydrolysis and formation of peptide bonds as well as transpeptidation may take place simultaneously, and these processes could result in the formation of N-terminal aspartic acid in the polymer. This fact is generally interesting in understanding the formation and structures of thermal polyamino acids.

Polyaspartic acid which was obtained by heating of aspartic acid was also studied (Vegotsky et al., 1958). The structure of the polymer was confirmed as anhydropolyaspartic acid by infra-red absorption spectra at 1780, 1710 cm^{-1}, which are characteristic of five-membered imide structures. The structure of the product is different from the diketopiperazine of aspartic acid.

APPROACHES TO THE THERMAL SYNTHESES OF PROTEINOIDS

Professor Fox's whole idea on the prebiological origin and development of bio-organic compounds and their systems starts from the abundant availability of thermal energy on the primitive Earth, as discussed in Professor Fox's earlier reviews (Fox, 1956; Fox,

1957; Fox et al., 1957; Fox et al., 1959; Fox, 1960; Fox et al., 1962). He thought that primitive protein-like polyamino acid could be formed from free amino acids under suitable thermal conditions. The formation of polyglycine and polyaspartic acid was already reported; however, no other homopolyamino acids nor heteropolyamino acids had been prepared by heating. One reason for the difficulty of peptide bond formation is the thermodynamic barrier. The difference of free energy for peptide bond formation in water is 2-4 Kcal/mol (Huffman, 1942).

$$H_2N\text{-}\underset{R}{\underset{|}{CH}}\text{-}COOH + H_2N\text{-}\underset{R}{\underset{|}{CH}}\text{-}COOH \rightleftharpoons H_2N\text{-}\underset{R}{\underset{|}{CH}}\text{-}CO\text{-}NH\text{-}\underset{R}{\underset{|}{CH}}\text{-}COOH + H_2O$$

$$\Delta F = 2\text{-}4 \text{ kcal/mol}$$

However, if the reaction is carried out under nonaqueous conditions with the removal of the resulting water from the reaction system, the formation of the peptide bond could occur. Therefore, the thermal condensation of amino acids would best occur under nonaqueous solution or in melting conditions. A number of conclusions regarding the formation of the "Proteinoid" which we have reached by reasoning and experiment in 1957-1959 (Harada, 1961, 1971) is summarized below:

1. Generally, chemical reactions are conducted in a homogeneous experiment such as in solution or under gaseous or melting conditions. However, amino acids are zwitter ions. These ionic properties contribute to their stability and high decomposition point of amino acids. Generally amino acids do not melt by heating, and their ionic property is a difficult barrier to overcome for their thermal copolycondensation of amino acids.

2. However, there are few exceptions. Heating of glutamic acid largely yields its lactam (pyrrolidon-2-carboxylic acid, or pyroglutamic acid). Pyroglutamic acid has no zwitter ion structure and it melts easily by heating. Similarly, a basic amino acid lysine is also converted by heating to a seven-membered lactam that also has no zwitter ion property and it melts easily.

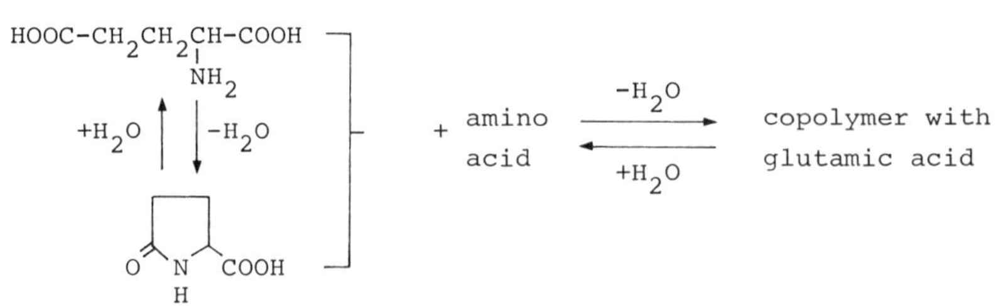

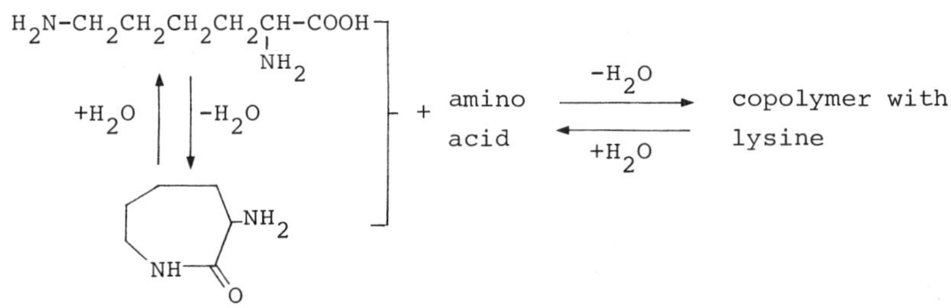

3. Therefore, it was thought possible to copolymerize an amino acid with pyroglutamic acid or lysine lactam by heating, if the amino acid is soluble in melted lactum. The molten homogeneous state is a favored condition for the thermal polycondensation of amino acids. Such was found to be the case.

4. The lactams of glutamic acid and lysine are both solvents and reactants in thermal polycondensations. The carboxyl or amino groups of these lactams either undergo condensation reactions with other amino acids, or the amino groups of other amino acids will aminolytically cleave the lactam and thus form new amide bonds. The lactams can also be hydrolyzed in the presence of water at high temperatures, and the glutamic acid or lysine thus produced can then react with other amino acid in a molten homogeneous solution.

5. Pyroglutamic acid and lysine lactam can promote the thermal polycondensation of amino acids.

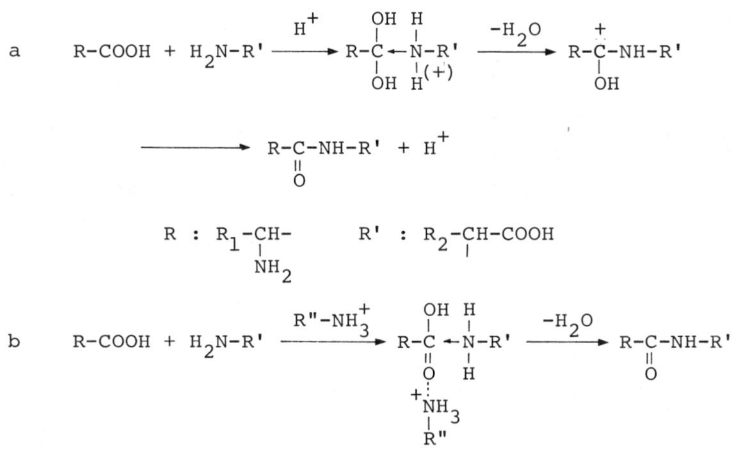

6. Diketopiperazines are known to form by heating amino acids, as solids or in solution. This indicates that amino acids are converted by dipeptides rather easily by heat. However, there are two ways dipeptide can react:

a) diketopiperazine formation by cyclization,
b) reaction to prolong peptide chain to tri-, tetra- and to polypeptide.

Case (a) is a unimolecular reaction involving the head-to-tail condensation of the dipeptide.

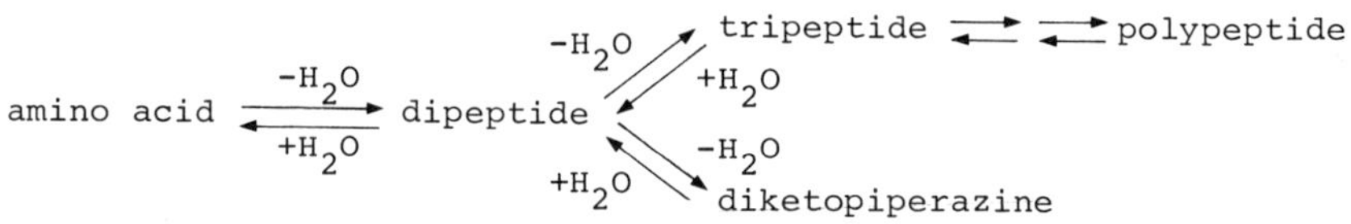

Therefore, if an inert solvent is used for the polycondensation, the proportion of diketopiperazine increases. Case (b) is a bimolecular reaction. These two reactions (a) and (b) are competitive. These limitations for the polycondensation of free amino acids therefore suggest melting of the reactant without any solvent. Pyroglutamic acid and lysine lactam adequately provide the liquid phase.

7. One aspect of note is that the thermal polycondensation is a reversible reaction and is thus distinguished from polymerization reactions initiated by radicals, or from polyamino acid formation using Leuchs anhydrides of amino acids. In the thermal polycondensation of amino acids, the resulting peptides could be hydrolyzed or be transpeptidated with other peptides or amino acids during the reaction. As a whole, the molecular weight of the thermal polyamino acid in the reaction mixture increases proportioned to the loss of water.

EARLY STUDIES ON THE THERMAL POLYCONDENSATION OF AMINO ACIDS

Guided by the reasoning summarized in the preceding paragraphs, a large number of condensation experiments have been carried out. Some of the very early experiments and their results are described below:

After conversion of glutamic acids into pyroglutamic acid, glycine was added. The mixture was heated for few hours at 170°C and a copolymer of glutamic acid and glycine [copoly(glu, gly)] was obtained (Harada and Fox, 1958).

The copoly(glu, gly) show pinkish biuret tests, and the infrared absorption spectrum reveals peaks at 1660 (amide I) and 1550 cm^{-1} (amide II) which are characteristic of polypeptides. The polymer was hydrolyzed to regenerate glutamic acid and glycine. Generally, amino acids cannot be homopolycondensed by heating with the exception of aspartic acid and glycine. However, when the amino acids are heated individually with pyroglutamic acid, all of the products showed clear biuret tests. These results indicated the formation of the peptide bond by heating.

On the other hand, when free lysine was heated at 180°C, it was converted to its seven-membered lactam, and prolonged heating resulted in the formation of thermal polylysine (Harada, 1959b). The thermal polylysine showed the characteristic infra-red absorption bands of polypeptides. It was found that the amount of ε-linked lysine residues was more prominent than the α-linked residue. The thermal polycondensation of neutral and acidic amino acids with lysine yielded several copolymers containing lysine.

Thermal polycondensation of aspartic acid and glutamic acid yielded copolymers containing these two amino acids (Harada and Fox, 1960; Fox and Harada, 1960b). The amino acid compositions were aspartic acid rich, and the structure was largely polyimide (anhydroaspartyl residue). The ratios of aspartic acid to glutamic acid in the amino acid composition and in the N-terminal composition were quite different, the proportion of glutamic acid in the N-terminal being much larger than that in the total amino acid composition. This indicates that the reactivity of N-terminal glutamyl residues is much less than that of N-terminal aspartyl residues during the thermal polycondensation. Similarly, thermal copoly(glu, asp) was prepared from L-glutamine and DL-malic acid by heating. The structure of the polymer was similar to the copoly(glu, asp) prepared from glutamic acid and aspartic acid by heating. In this reaction, L-glutamine which contains two amino nitrogens acts as a nitrogen donor to DL-malic acid which is converted to aspartic acid as described in the section on PRELIMINARY STUDY.

Similarly, several thermal copolymers containing glutamic acid, aspartic acid and other amino acids were prepared (Harada and Fox, 1965).

Copolymers of lysine with one other amino acid were prepared (Harada, 1959b), copoly(lys, gly), copoly(lys, asp), and copoly(lys, glu) being examples. Infra-red absorption spectra of polylysine, copoly(lys, neutral amino acid), copoly(lys, glu) indicated absorption bands typical of polypeptides; however, copoly(lys, asp) showed additional bands at 1720 and 1780 cm^{-1}, which are characteristic of the five-membered cyclic imide structure (anhydroaspartyl residue). The latter copolymer was highly branched and insoluble in water.

By using sufficient proportions of glutamic acid or lysine in the presence of other amino acids, it became possible to prepare copolymers of glutamic acid and lysine containing eighteen amino acids common to protein by heating the appropriate amino acid mixtures under molten conditions. The polymers thus obtained were similar to protein. Therefore they have been termed "Proteinoids"

(Fox and Harada, 1958; Fox and Harada, 1960a) (Figs. 1 and 2). Various comparisons of proteinoid with natural proteins have been made in detail (Fox and Dose, 1977). Proteinoids show several properties characteristic of proteins, such as several color reactions, salting in and salting out effects. They were hydrolyzed to amino acids and these were analyzed by bioassay and by amino acid analyzer. The infra-red absorption spectra showed characteristic bands for polypeptides in addition to the bands due to anhydro-aspartyl residues. The proteinoid was consumed by L. arabinosus and produced lactic acid. The anhydroaspartyl residue in the proteinoid could be hydrolyzed by weak alkali to form α- and β-aspartyl residues, and it also yielded asparginyl and isoasparginyl residues by ammonolysis. Therefore, the contents of the anhydroaspartyl residue in polyamino acids could be analyzed by the determination of

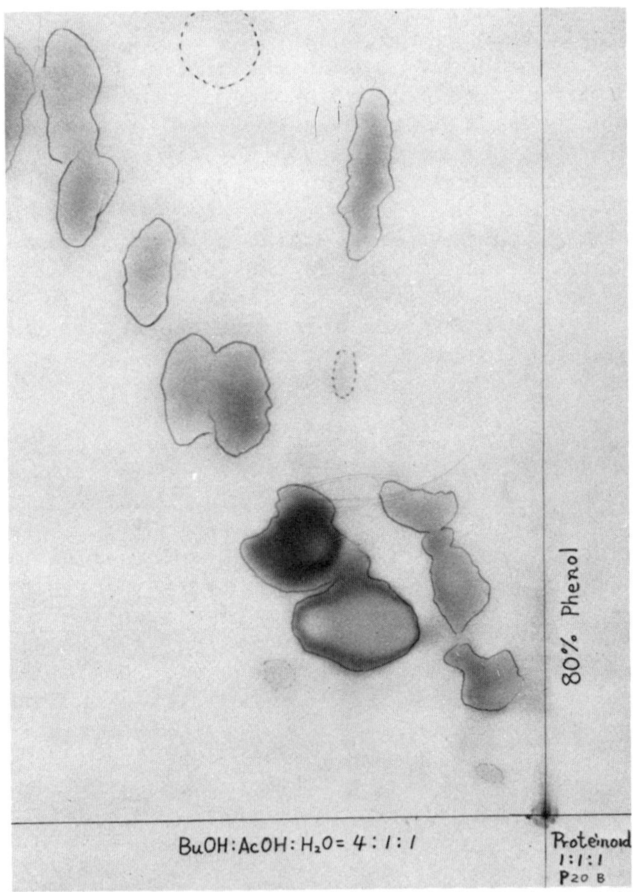

Fig. 1. Two dimensional paper chromatogram of the hydrolyzate of acidic proteinoid.

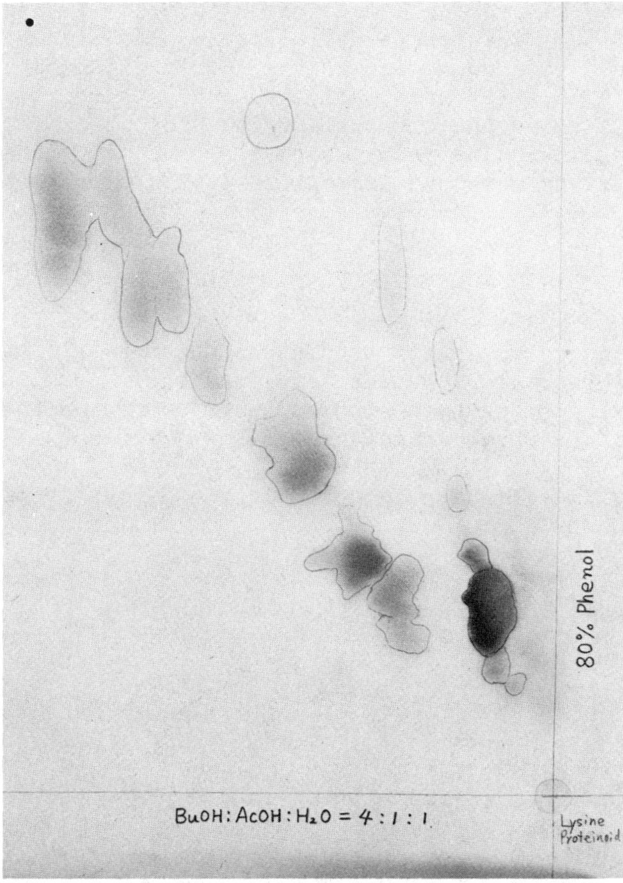

Fig. 2. Two dimensional paper chromatogram of the hydrolyzate of lysine proteinoid.

the amide group by the Kjeldahl method after ammonolysis (Fox and Harada, 1963; Harada and Fox, 1975). Two dimensional paper chromatography of the hydrolyzate of acidic and basic proteinoid are shown in Figs. 1 and 2. The amino acid composition of proteinoid is quite constant before and after purification by microsphere formation (Fox and Harada, 1963; Harada and Fox, 1975). One interesting aspect in the structure of the proteinoid is that the total amino acid composition is quite different from N-terminal amino acids composition of proteinoid.

Professor Fox's experimental approach to the chemical evolution advanced to a new multimolecular system composed of proteinoid (Fox et al., 1959). When acidic proteinoid was dissolved in a diluted salt solution by heat, the solution became turbid when cooled. The solution showed a numerous number of uniform proteinoid droplets under microscope. These were called proteinoid microspheres and were considered as a model for the protocell produced under prebiotic conditions. Several copolymers with aspartic acid produce microspheres. Microsphere shows several properties in common with some bacteria, including gram positive and negative staining, budding, double-layer membrane and others (Fox and Dose, 1977) (see Fig. 3).

FINAL COMMENTS

Continuing to the early studies on the formation and characterization of thermal polyamino acids and their systems described above, many scientists have studied under Professor Fox for the development of the proteinoid theory (Fox and Dose, 1977).

In 1906, Emil Fischer lectured in the German Chemical Society as follows:

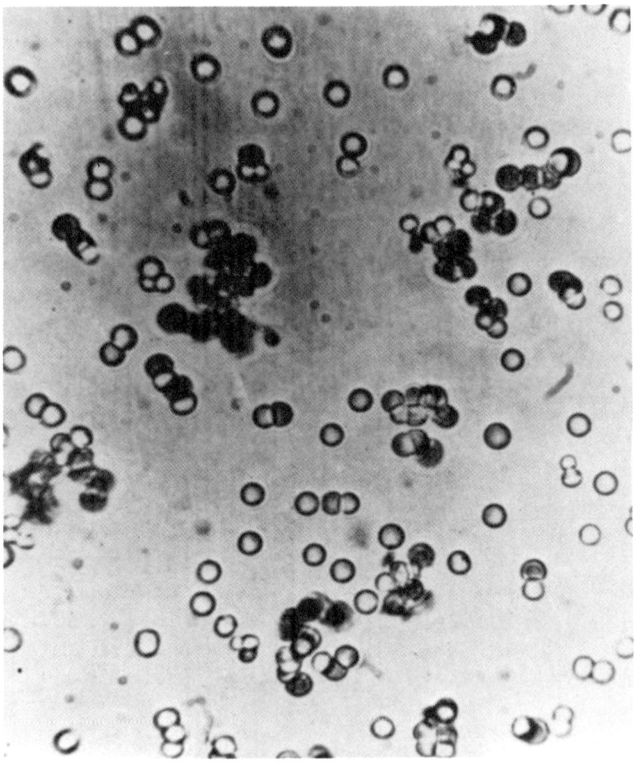

Fig. 3. Microphotograph of proteinoid microsphere. One of the earliest and the best picture taken in Professor Metz's laboratory in 1958-1959.

> Wenn es heute durch einen glücklichen Zufall, mit Hülfe einer brutalen Reaktion, z.B. durch Zusammenschmelzen der Aminosäuren in Gegenwart eines wasserentzichenden Mittels, geligen sollte, ein echtes Protein darzustellen, und wenn es weiter möglich wäre, was noch unwahrscheinlich ist, das künstliche Produkt mit einem naturlichen Körper zu identifizieren, so würde demit für die Chemie der Eiweisstoffe wieng und für die Biologie so gut wie garnichts erreicht sein.

E. Fischer thought that the step-by-step synthesis of peptide is the best way for the synthesis of protein. Certainly such an organo-chemical approach successfully clarified various protein structures synthetically. However, just half a century after his lecture, we had succeeded in the synthesis of protein-like polyamino acids by heat in molten conditions, which, according to Fischer, was an organo-chemically meaningless method for an artificial protein formation. However, there were no organic reagents and solvents on the primordial Earth, and we had to find a new way to form peptide bonds under prebiotic conditions. Our reasoning for the formation of the peptide bond described in the section on APPROACHES TO THE THERMAL SYNTHESIS OF PROTEINOIDS was successful, and we believe that the thermal polycondensation is the best way to form the first protein-like polyamino acids (proteinoid). Half a century after E. Fischer's lecture, the organo-chemically meaningless method in synthesizing protein-like substances now has obtained a great meaning in prebiological chemistry. Prebiological chemistry has evolved to become biological chemistry after emergence of primitive life.

REFERENCES

Abderhalden, E., and Komm, E., 1924, Z. Physiol. Chem., 139:147.
Balbiano, L., and Trasciatti, D., 1900, Ber., 33:2323.
Dessaignes, V., 1850a, Compt. Rend., 30:324.
Dessaignes, V., 1850b, Compt. Rend., 31:432.
Fischer, E., 1906, Ber., 39:530.
Fox, S. W., 1956, Amer. Sci., 44:347.
Fox, S. W., 1957, J. Chem. Ed., 34:472.
Fox, S. W., 1960, Science, 132:200.
Fox, S. W., and Middlebrook, M., 1954, Federation Proc., 13:211.
Fox, S. W., Johnson, J. E., and Middlebrook, M., 1955, J. Amer. Chem. Soc., 77:1048.
Fox, S. W., Johnson, J. E., and Vegotsky, A., 1956, Science, 124:923.
Fox, S. W., Vegotsky, A., Harada, K., and Hoagland, P. D., 1957, Ann. N.Y. Acad. Sci., 69:328.
Fox, S. W., and Harada, K., 1958, Science, 128:1214.
Fox, S. W., Harada, K., and Kendrick, J., 1959, Science, 129:1221.

Fox, S. W., Harada, K., and Vegotsky, A., 1959, Experientia, 15:81.
Fox, S. W., and Harada, K., 1960a, J. Amer. Chem. Soc., 82:3745.
Fox, S. W., and Harada, K., 1960b, Arch. Biochem. Biophys., 86:281.
Fox, S. W., Harada, K., and Rohlfing, D. L., 1962, in: "Polyamino Acids, Polypeptides, and Proteins," M. A. Stahmann, ed., p. 47, Univ. Wisconsin Press, Madison.
Fox, S. W., and Harada, K., 1963, Federation Proc., 22:479.
Fox, S. W., Harada, K., Woods, K. R., and Windsor, C. R., 1963, Arch. Biochem. Biophys., 102:439.
Fox, S. W., and Dose, K., 1977, "Molecular Evolution and the Origin of Life," 2nd edn., Marcel Dekker, New York; and literatures listed therein.
Grimaux, E., 1882, Bull. Soc. Chim., 38:64.
Harada, K., and Fox, S. W., 1958, J. Amer. Chem. Soc., 80:2694.
Harada, K., 1959a, J. Org. Chem., 24:1662.
Harada, K., 1959b, Bull. Chem. Soc. Jpn., 32:1008.
Harada, K., and Fox, S. W., 1960, Arch. Biochem. Biophys., 86:274.
Harada, K., 1961, Tanpakushitsu Kakusan Koso (Protein, Nucleric Acid and Enzyme), 6:65.
Harada, K., and Fox, S. W., 1965, Arch. Biochem. Biophys., 109:49.
Harada, K., 1971, "Kagaku Shinka" (Chemical Evolution-Chemical Basis of the Origin of Life), Kyoritsu Pub. Co., Tokyo.
Harada, K., and Fox, S. W., 1975, BioSystems, 7:213.
Huffman, H. M., 1942, J. Phys. Cheml, 46:885.
Kovács, J., Könyves, J., and Pusztai, A., 1953, Experientia, 9:459.
Kovács, J., and Könyves, J., 1954, Naturwiss., 41:333.
Kovács, J., Könyves, J., and Casszar, J., 1954, Naturwiss., 41:575.
Kovács, J., Kovács, H. N., Könyves, J., Casszar, J., Vajda, T., and Mix, H., 1961, J. Org. Chem., 26:1084.
Maillard, L. C., 1911, Compt. Rend., 153:1078.
Maillard, L. C., 1914, Ann. Chim., 1:519.
Meggy, A. B., 1953, J. Chem. Soc., 851.
Okamoto, S., and Nakahara, A., 1964, Nogeikagaku kai shi, 38:28.
Piutti, A., 1887, Gazz. Chim. Ital., 17:519.
Piutti, A., 1888, Gazz. Chim. Ital., 18:457.
Polyakova, A. L., and Verechagin, M. F., 1949, Doklady Akad. Nauk U.S.S.R., 64:607.
Schaal, E., 1871, Ann., 151:26.
Schiff, H., 1897, Ber., 30:2449.
Schiff, H., 1899, Ann., 307:231.
Shibata, K., 1925, Acta Phytochim., 2:193.
Vegotsky, A., Harada, K., and Fox, S. W., 1958, J. Amer. Chem. Soc., 77:1048.
Watanabe, H., and Kosai, Y., 1963, Nippon Kagaku Zasshi, 84:744.

THE DEVELOPMENT OF THE PROTEINOID MODEL FOR THE ORIGIN OF LIFE

Duane L. Rohlfing

Department of Biology
University of South Carolina
Columbia, S.C. 29208

INTRODUCTION

It is a pleasure to contribute an article commemorating seven full decades of Sidney W. Fox and two-and-a-half decades of the proteinoid theory. This article in part reviews some of Fox's major research accomplishments, which have evolved to be known as the proteinoid model (or theory) for the origin of life. However, the range of his research and of his thinking are, I believe, better viewed after some focus on features of the unique arena in which the work is being done, and on some preconceptions and untested speculations that abound in that arena. Also, some ideas about the significances and the acceptance of the proteinoid model are presented.

The article is written from a perspective molded from experience as a former graduate student of Fox's, as an eye-witness to the first proteinoids and microspheres, as a researcher in the area for many years, as a coeditor of an earlier commemorative volume (Rohlfing and Oparin, 1972), and as a sometimes critic, both vocal and experimental. The viewpoint presented is undoubtedly biased, but it may be worthwhile to note that I am no longer active in the area; what results here thus may be more detached and more candid than if future professional achievements and fundings were a consideration, conscious or subliminal.

THE ORIGINS-OF-LIFE ARENA

Much of Fox's research has been devoted toward understanding how life originated. The "Origin-of-Life" arena has several unusual features, some perhaps unique, that have had and continue to have a bearing on thinking, on experimentation (or lack thereof), and on the acceptance of ideas and experimental results. A brief consideration of some of these is helpful in setting the stage for examining Fox's research.

First, the arena is an emotional one. It can and does conflict with theology in the minds of those who want it to, and in the past this has been a physically dangerous conflict. It also is an eye-catching arena, creating "good press" and oversimplification and overstatement (just as, for example, cancer research creates good press, oversimplification, and overstatement).

Second, it is an arena that deals with the past, with what was, not what is, with what happened, not what is happening. Accordingly, unequivocal proof is especially lacking; experiments and experimental models can at best only permit one to say "it could have happened this way." An offshoot is that concepts that are probably wrong but are not provably wrong tend to linger on. In most research areas, the contemporary is the issue and the contemporary provides an extant check of experiments designed to explore such issues; "truth" is more easily identified and accepted.

Third, "life" is a term that defies definition to everyone's satisfaction. It is borrowed from the layperson; it works well when contrasting rocks and elephants, but not so well in borderline situations. If an emperimental goal is to synthesize "life," that will be (is?) an accomplishment only to those who agree on the definition. (Perhaps the definition will emerge from the experiments.) Such a goal is a bit artificial, since the definition is also artificial. However, significant research can be done without having to use the term, within a framework of understanding that the term represents one of many levels of complexity, somewhere between chemical elements and super-societies.

The origins arena is multidisciplinary (as, of course, are others), where there is a role for physicists, geologists, chemists (perhaps especially organic chemists and biochemists), biologists, paleontologists, cosmologists, and undoubtedly a few other specialists. It is not easy to be good, especially experimentally, in all of these; most researchers stick to their area of experimental expertise--and probably tend to overemphasize in their own mind the importance of that special area. However, in a multidisciplinary area, a fragmented viewpoint often is not a very useful one; an overall perspective of the whole discipline and the connections and relationships between subdisciplines is important not only for balance but also for planning and interpreting experiments (cf. "evolutionary continuity," below). Fox has shown such an overall comprehension, in his writings and also to a great degree--perhaps more so than any other person in the area--in his research and that of his many associates.

Finally, much of experimental or observational science was learned reductionistically, taking the complex, tearing it up, and analyzing the simpler component parts; very relevant examples are Biology and Biochemistry. In contrast, experimental research in the origins area is of necessity mostly constructionistic, starting with the simple and seeing what can result that is more complex. It is an approach foreign to the training of many scientists. The experiments attempt to reiterate the route, called evolution, taken by nature. Various flow-diagrams result that serve as overall, undetailed models for evolution, supported in part by experiments, and with some easily recognized levels of complexity inserted as reference points. (Within the conceptual framework of such diagrams, the level "first life," not easily recognized, is not a prerequisite component for meaningful use of the models.)

PRECONCEPTIONS

Because unequivocal proof of things past is especially elusive, the origins-of-life arena was (is) probably more vulnerable than other areas to preconceptions and untested speculations. (Why bother to test if it can't really be proven?) An atmosphere of negativism easily results, sometimes propagated by critics who also were (are) in positions of authority and whose viewpoints--right or wrong--are given undue weight.

Aristotle, for example, was an authority whose views on "spontaneous generation" were used by theologians to support the Divine Creation concept. Darwin, an authority on evolution, stated in 1863 that " . . . it is mere rubbish thinking at present of the origin of life; one might as well think of the origin of matter." (Some eight years later he was somewhat more optimistic, but via a statement much less well known until recent years.) The authority whose expressed viewpoint probably did the most to stifle research in the origins area was Pasteur, who interpreted his sterility experiments correctly in one context but incorrectly in the one that is pertinent to the present topic. His 1864 statement, "Never will the doctrine of spontaneous generation recover from the mortal blow of this simple experiment," essentially quenched thinking and experiments about the origin of life for about 60 years. Pasteur also asked, "can matter organize itself?" His inferred answer was "no," although some 14 years later he conceded in a little noted statement that "I do not think that [spontaneous generation] is an impossibility."

Preconceptions of lesser magnitude have contributed to an inhibiting effect (see, e.g., Fox, 1973, for more detail). These include that DNA had to emerge before protein, that "life" violates the second law of thermodynamics (illustrating ignorance exceeding my own of nonequilibrium thermodynamics), that evolution had to be very slow (cf. Hsu, this volume), that heating amino acids yields only "gunk," and even that "life" is too hopelessly complex to understand its origin.

Presumptions also result from using the contemporary as a guide to the processes and compounds necessary for "first life." Certainly, it is better than no guide at all, and eventually we will want to understand how all biological processes and compounds originated. A very positive inference results from the unity of contemporary life, with many features (processes and compounds) common to all examined life; the commonality strongly suggests that "first life" utilized similar features (and, taken to an unwarranted extreme, that all living things evolved from a single living organism).

On the other hand, the contemporary may be a faulty guide for the minimum essentials for first life. Not all contemporary processes or compounds were necessary for "first life" (since not all are necessary for all contemporary life), and to insist that the features of first life include those of the simplest contemporary organisms is probably too stringent. An even greater potential pitfall is to exclude as being relevant to first life those features that are not utilized by the contemporary. The contemporary is not a good guide for such processes or compounds, and minds conditioned by the contemporary can be closed to alternatives. (A rough analogy is using a hummingbird as a guide to what an archeopteryx was like.)

CHANCE, SELF-ORDERING, AND INFORMATION

Years of loose usage have reinforced the concept that evolution is largely driven by chance, or by random processes. A consequence is the inference that meaningful (reproducible) experiments are not possible and that the whole subject of evolution is untestable. (For a long time, no one tested.) Experimentalists take (or, if skeptical, at least test) the viewpoint that evolutionary events are not random, not due to chance, and that their experiments--reflecting the experiments of nature--will be reproducible. The elements of self-ordering and predictability (if only in retrospect) thus enter.

Chance and randomness are more or less synonymous, and a random system can be defined as one in which there is an equal chance (probability) of all possible options being exercised (or, given a statistical number of opportunities, all options being exercised equally). In a non-random process, there is unequal chance of all possible options being exercised with equal probability (or, on a statistical basis, all options not occurring equally). When not all theoretically possible options are exercised equally, then one has, to greater or lesser extent, a self-ordering, informational system, one which, if tested again under a given set of experimental conditions, is reproducible. Such is the basis for the experimental approach to the study of the origins of life.

Which option(s) a self-ordering, informational system takes depends very much on the surroundings (conditions) in which the system occurs. For example, hydrogen gas (the system) remains hydrogen gas at one temperature (the surroundings, or conditions), but is converted to helium at higher temperatures and on to carbon, etc., at yet higher temperatures. Likewise ethanol in sulfuric acid is converted mostly to diethyl ether under some conditions but mostly to ethylene under other conditions. DNA acts as a biologically informational molecule when surrounded by a host of other molecules at 37°C, but not at 370°C nor at 37°C when surrounded by chloroform. In each case, the information content is present, but the expression of that content is mediated or constrained by the conditions. The "chance" element is the surroundings. In a similar vein, chance does have a role in evolution, but that chance is largely associated with the unpredictability of the environment (surroundings) in which a system is evolving, rather than with the system that is evolving.

[In one sense, what may be chance is that the laws--or, rather, forces--of chemistry and physics are the way they are; since they are the way they are (without exception, or they wouldn't be called "laws"), the events shown by systems on which the forces work are predictable, although the way those forces are expressed is subject to environmental influence.]

RESEARCH ON THE ORIGIN OF LIFE--GENERAL COMMENTS

In view of the numerous preconceptions and inhibiting impact of authorities, it is perhaps remarkable that serious experiments in the origins area were even attempted. Pioneering spirit and professional courage were required. Oparin deserves great credit for leading the way, intellectually with his writings about a materialistic, evolutionary increase in complexity, and experimentally with his studies of coacervation phenomena as the basis of a protocell model. Herrera should be viewed as a poineer, although his thoughts and experiments, preceding Oparin's, were not and are not widely known. Urey helped to create a conducive environment; Fox in fact has noted the importance of the thoughts and statements of Urey and Oparin, "without which experiments on thermal origins of biochemical phenomena might not be recorded" (Fox, 1956). Miller deserves recognition for experiments not only done in the face of considerable opposition (Urey, his major professor, advised him against the project as being too risky for a graduate student), but also that served as a springboard for many subsequent experiments (in large part because they were published in a widely read journal).

Investigators in the area have some general guidelines to follow. An obvious one is the belief that evolution occurs, that it occurred before there was life as well as after, and that it largely proceeds from the simpler to the more complex. Another concerns a general knowledge of what was available on the primitive earth (substances and energy forces). A general experimental procedure is to start with the simple, subject it to relevant forces, and determine what results that is more complex. (Especially sought are substances of current biochemical importance; there is a tendency to ignore what results that is less complex or that is not important biochemically). But there is an often ignored, twofold element of "realism" that looms dictatorially in the evalution of the merits of an experiment. One is that of the geophysical relevance of the experimental conditions; the experiments (should) attempt to mimic what we think the conditions were. The other is that of evolutionary continuity. The processes used in one experiment should be feasibly connectable to or compatible with those of subsequent experiments, and the products of one process should have utility in subsequent experiments; otherwise, one has the experimental modeling of an evolutionary dead end. Any comprehensive, unified experimental model must consider both relevance and continuity. Fox has considered both aspects to the extent that his research has resulted in a relevant experimental continuum from simple atmospheric gases to functional protocell models.

THE PROTEINOID MODEL

Pre-Proteinoid Research

Some of Fox's earlier research is treated by Cherkin in this volume. Noteworthy is his experience in protein chemistry, with papers on protein sequencing and terminal analyses. Early interests in evolution and self-ordering at the molecular level are also evident. Perhaps the most significant finding showed that amino acids (tested as derivatives) express individuality in their interaction with the enzyme papain (Fox et al., 1953); there is an "informational" input, or component, from the (different) amino acids as well as from the enzyme.

In much of his origins research, Fox has used thermal energy to overcome thermodynamic barriers. The prebiotic availability of heat is, of course, well documented (Fox and Dose, 1977). The quality of heat, however, is especially noteworthy, providing a range sufficient for the most demanding as well as for the more subtle processes, from over 1000°C (from the interior of the Earth) to temperatures generated by sunlight (including "room temperature"). Heat also is local and intermittent; rapid transitions in temperature are common (e.g., via contact with rain), providing facile means for protecting substances that would be destroyed on long exposure to high temperatures.

Prior to the research on proteinoid, Fox used thermal energy to convert biochemically important compounds to other metabolites (Fox et al., 1955). For example, ammonium salts of malic or fumaric acids yielded aspartic acid and α- & β-alanine; urea with malic acid yielded carbamylaspartic acid (ureidosuccinate), an intermediate in the biosynthesis of pyrimidines. Subsequently, uracil was shown to result from malate and urea (Fox and Harada, 1961). The results were suggested as representing a kind of thermal recapitulation of biochemical anabolic pathways. The relationship of some amino acids as precursors (thermal and biochemical) of purines and pyrimidines fortified the inference that amino acids (proteins) preceded nucleotide bases (nucleic acid) evolutionarily.

Although chronologically out of place, the synthesis (Harada and Fox, 1964) at 1000°C of several amino acids from methane, ammonia, and water is important in extending the range of Fox's research back to the simplest molecular starting components. Twelve proteinous and three nonproteinous amino acids were identified; the reactions took place in columns of sand or silica gel. Subsequently (Fox and Windsor, 1970), formaldehyde, water, and ammonia, heated at 180°C, gave products that on hydrolysis yielded several amino acids. Both syntheses were in <u>open systems</u>, as aspect that is a key one--discussed later--that shows Fox's cognizance of the importance of the geophysical relevance aspect of the discipline.

<u>Proteinoid</u>

A major misconception to be overcome was the "fact" that heating amino acids leads only to "gunk," tars, etc. [There were reports around the turn of the century of the thermal synthesis of polyglycine and polyanhydroaspartic acid; the former was reinvestigated by Meggy (1953), the latter by Kovacs (1953).] Indeed, most amino acids do not homopolymerize thermally, but most will copolymerize readily with glycine, with aspartic acid, with lysine (which also homopolymerizes), and with glutamic acid (which does not homopolymerize). The latter three form molten anhydro derivatives at suitable temperatures (imide, lactam, and lactam, respectively), which serve as solvents for the other amino acids. (Glycine does not form a melt; syntheses with glycine plus other amino acids are solid phase syntheses, macroscopically.) Katchalski (1951) mentioned thermal energy as a possible means for polymerizing amino acids, a suggestion noted by Fox.

In 1954, Fox and Middlebrook (1954) reported the thermal formation of polyaspartic acid and also of copolymers of leucine or valine with aspartic acid. Harada, who did much of the early work on proteinoids in Fox's lab, showed that glutamic acid is suitable for copolymerization with other amino acids (Harada and Fox, 1958). Copolymers of amino acids with lysine were reported in 1959 (Harada, 1959), and with glycine in 1972 (Saunders and Rohlfing, 1972). The first report of a thermal polyamino acid containing most of the amino acids common to protein was in 1958 (Fox and Harada, 1958; Fox and Harada, 1960).

A typical synthesis of proteinoid, simple enough to be repeated in hundreds of high school science projects, involves heat for a few hours at 165-175°C (or at 190-200°C for lysine-rich proteinoids). Alternatively, heating as low as 65°C for months, rather than hours, suffices (Saunders and Rohlfing, 1976; sunlight generates surface temperatures of over 85°C, a fact that connotes widespread and common production of prebiotic polyamino acids).

The thermal polyamino acids show many properties in common with protein, one reason for using the term "proteinoid." The molecular weights are in the thousands, or, in lysine-rich proteinoid, closer to 100,000; accordingly, the often-abused prefix "poly-" is accurately applied. The presence of the peptide bond is shown by a variety of techniques, including biuret reaction, infrared analysis, attack by proteolytic enzymes, utility as nutrients for microorganisms, and, after repeated purification, 100% recovery of amino acids on acid hydrolysis. When trifunctional amino acid residues are present, some of the peptide linkages are non-α and some are α.

Each of the 18 proteinous amino acids are co-incorporated (glutamine and asparagine typically are not tested); the compositions of the dialyzed products can be varied and controlled, to include those more neutral in composition as well as those rich in trifunctional residues or glycine. The polymerization process is also applicable to nonproteinous amino acids, including many reactant sets that simulated prebiotic syntheses or extra-terrestrial analyses suggest were abundant prebiotically.

Catalytic activity on a number of different substrates is one property illustrating the utility of proteinoid and one that relates to the origins of enzymes. Extensive lists of protein-like properties are published elsewhere (e.g., Fox and Dose, 1977), including the incorporation of non amino acid groups and morphogenicity (treated later), but one of particular note is that of nonrandomness. Evidences for nonrandomness are numerous, the most definitive being sequence studies (Nakashima et al., 1977). The results reconfirm Fox's earlier finding (Fox et al., 1953) that the carboxyl and amino groups of the various amino acids are not of equal reactivity and that each amino acid has its own "individuality" which is expressed in the polymerization process. There is thus "information" in the individual amino acids, as expected, that is expressed in the polymerization process, as expected, making the process and the products nonrandom, as expected. (Incidentally, the way that such information is expressed is reproducibly influenced by the experimental conditions, as expected; e.g., Rohlfing and McAlhaney, 1976).

Microspheres

The continuum of increasing complexity reaches the morphological level when proteinoids provide the structural materials for microscopic-sized units called microspheres. (The idea and initial set of instructions for forming the first proteinoid microspheres are preserved in a note from Fox to a technician, Jean Kendrick; the note is quoted in Cherkin's paper in this volume.) Microspheres form readily and in great number from acidic proteinoid by changing the temperature (e.g., heating to dissolve amorphous proteinoid, then cooling to form spheres, or simply adding water--mimicking rain--to molten, unpurified proteinoid). Lysine-rich proteinoids form microspheres when the ionic strength of the solvent is sufficient; oceanic salinities serve nicely. Glycine-rich proteinoids also form spheres. The spheres contain no lipid, unless intentionally added; the proteinoid obviously has lipoid, hydrophobic character.

Of course, sphere formation is not unique to proteinoid, but rather represents one application of general forces involving phase separations and compacting the most volume into the smallest surface area. What _is_ special about proteinoid microspheres is the scenario of evolutionary continuity represented by their formation from proteinoid--and their functional utility.

A large number of properties and phenomena of envisionable evolutionary importance are shown by microspheres (Fox and Dose, 1977). These include: stable, selectively permeable boundary with osmotic and conductance properties; catalytic activities; motility; compartmentalization; communication between spheres via transfer of material through morphological junctions; proliferation via fission and budding; and growth by accretion. The spheres serve as models for dynamic, propagative heterotrophic protocells.

Continuations

Proteinoids and/or microspheres have been used as components of systems that mediate other phenomena of envisioned prebiological significance (Fox and Dose, 1977). Lysine-rich proteinoids promote the room-temperature formation of aminoacyl adenylates from amino acids and ATP; the latter provides not only the adenylate moiety but also the energy. Aminoacyl adenylates serve as the starting materials and the energy source (as in contemporary protein syntheses) for polyamino acids, formed especially in the presence of lysine-rich proteinoid; oligoadenylic acid is also formed in the same experiment.

A mixture of lysine-rich proteinoid and homopolynucleotides results in particulate associations that have been studied as models for primitive ribosomes. Such complexes promote the formation of polyamino acids from the adenylates; the kind of polyamino acid formed in greatest abundance can be, under empirical conditions, related in a codonic or an anticodonic way to the kind of homopolynucleotide complexed with proteinoid. The self-ordering in these experiments was especially influenced by the conditions, but the experimental approach to understanding the origins of coding relationships is suggested from the results.

Principles Illustrated by the Proteinoid Model

Fox's experiments illustrate the modulation toward milder and milder energy sources. Higher energetic forces, such as U.V., electric discharge, ionizing radiations, or, in Fox's case, temperatures of ca. 1000°C, are used to convert simple molecules to amino acids and other micromolecular substances. Much milder energy, e.g., 150-200°C or as low as 65°C for prolonged periods, suffices to polymerize free amino acids. Milder energy yet, 100°C to room temperature, or a change in ionic strength of aqueous solutions, is adequate for forming microspheres. Proteinoids and microspheres function at room temperature, using mild chemical energy (or light energy to form ATP), as in extant biochemical processes. Milder energy is also seen to be important in the self-ordering of processes. When the difference in the energy levels of reacting groups is small, expression of such differences can occur only when the energy force for the process does not swamp out such differences in reactivity. In the extant, living system, the basis for expression of the differences is found largely in very weak interactions (hydrogen bonds, ionic bonds, hydrophobic interactions), as in association of enzyme and substrate, and requires mild chemical energy. The swamping out of such "individuality" by too-strong energy forces is apparent to anyone who has fried an egg.

Also seen in Fox's experiments is the role of water. Some processes require hypohydrous conditions (proteinoid formation), but the products of such processes require solvent water for most of their functions (e.g., catalytic activity, microsphere formation and function.) Relics of the hypohydrous environment are yet identifiable in contemporary processes occurring in microscopically hydrophobic environments (e.g., membrane phenomena, particulate processes like protein biosynthesis).

The proteinoid model also illustrates experimentally the phenomenon of emergent properties. At each transition from one level of complexity to another, properties result that were not shown by the starting materials and which would not be predictable without prior knowledge. Proteinoid, like protein, has properties largely different from those of free amino acids, and microspheres exhibit a set of properties largely different from those of proteinoid. (Such has been traced by E. Mayr to the subatomic level). Knowledge of the properties at one level does not provide enough information to predict those of the next level. However, an experimental, constructionistic model readily reveals the new properties. Experience has taught us to expect new properties, the contemporary provides clues to the kinds of properties to expect and look for, but the experiments are necessary to verify or refute the expectations. Enough experience has been gained to confidently predict that even more complex, and more subtle, phenomena will emerge as constructionistic experiments continue.

The continuity of the transitions from one level of complexity to another (which transcends in importance the phenomena of any one level) permits Fox to experimentally document one further aspect. Although much time was available for evolution, much time was not needed for some of it. The laboratory transitions from simple molecular precursor substances to functional microspheres require only hours, not centuries nor millennia (cf. the article by Hsu in this volume). Opportunities for natural experiments are multiplied accordingly. Most of the period of time on Earth has been devoted to evolutionary experiments after there was life.

THE VALUE OF THE PROTEINOID MODEL

Levins (1968) states that a model is "The basic unit of theoretical investigation [and] . . . is a reconstruction of nature for the purposes of study." [It has been stated of Fox's work that "These aren't model experiments . . . ; they're an experimental reconstruction."] The proteinoid model has provided and is yet providing an expanding framework for further "purposes of study," yielding ideas for further experiments, the results of which add detail to and expand the model; the one complements the other, as in any functional model.

A model can have various fates. One is to be expanded and refined, a dynamic situation. Another is to remain as it is; a stagnant model, however, is of little use. Another is to become obsolete or discarded (one important function, incidentally, for a model). The proteinoid model, I believe, is an example of the former category.

A model is judged by its utility, by how productive it is. But any model has constraints, and in the origins area, the value of a model has the overriding constraints of geophysical relevance (the conditions aren't wrong) and of utility in subsequent processes (evolutionary continuity). Most models fall short in one or the other or both constraints.

Some models "look good" in theory, but simply don't work when tested by experiment. (Most of these are unpublished experiments, although some do get published.) One example is the claim for polyamino acids from HCN and intermediates, bypassing the free amino acid stage. Although the hydrolizates of the products contain amino acids, the insistence that the products are polyamino acids is not substantiated by experiment and is refuted by experiment. Another example is the concept of introducing side chains into polyglycine, to form heteropolyamino acids. Flow diagrams show in theory how the various amino acid residues could be introduced, but experiments yield only a few kinds of residues and in very low proportion relative to glycine.

Some experimental models look good initially and appear to work, but do so in the wrong way, within the constraints stressed above. The action of electric discharge on simple gases in a closed system is an interesting way to form amino acid precursors and other compounds, but not on the prebiotic Earth. The conditions of closed system and lightning (electric discharge) can be found separately in and above the surface of the Earth, respectively, but not together. Until it is shown by experiment that similar results are obtained in an open system (which is not likely because of the probable rapid dissipation of critical volatile intermediates), the model in my opinion has only historical significance in creating experimental interest in the origin of life area. (The blending of geophysical relevance and experimental expediency can be difficult.)

An example of a model lacking evolutionary continuity is the coacervate model for protocells. Biopolymers such as gum arabic, histones, and bioenzymes were used as structural and functional materials for coacervate droplets, and thus the model, while being both structrual and functional, does not explain experimentally the prebiotic source of materials.

The proteinoid model looks good (or bad, depending on one's preconceptions and prejudices--see ACCEPTANCE OF THE PROTEINOID MODEL) and it works, possibly in the right way. The processes are geophysically plausible and the products of the processes (amino acids, proteinoids, microspheres) show the kind of utility in subsequent processes that makes the overall model a functional, dynamic one.

The continuing usefulness of a model may well depend on how well it withstands reasonable, testable criticisms. Indeed, some of the conditions used by Fox have been questioned by experiment (cf. Rohlfing, 1974). These include: the use of only proteinous sets of amino acids in the formation of proteinoid instead of sets indicated by simulation experiments and extra-terrestrial analyses; the use of nitrogen gas (to simulate a nonoxidizing atmosphere) instead of probable atmospheric components; the omission of probable geophysical "contaminants" in the syntheses of proteinoid; the use of intially dry reactant amino acids; and the use of temperatures available only where heat from the interior of the Earth reaches the surface. The experimental answers to these criticisms show that the processes yet work--sometimes better--regardless of the conditions tested. (A minor exception is noted by McAlhaney and Rohlfing, 1976). The model for the processes is thus strengthened and expanded by the testing of the criticisms. However, the nature of the products of the processes was in some but not all cases affected by the conditions tested. The experiments suggest that the model for the products (i.e., the compounds) may need revising--or may be expanded, depending on one's viewpoint.

One major inference results from such experiments; if prebiotic protein was formed from sets of reactant amino acids that were readily available, then the experiments indicate that such protein contained high proportions of glycine and some content of some non-proteinous amino acid residues. However, an interesting argument partially documented by experiment (Rohlfing, 1979) can be made against this premise, based on the concept that the important aspect may not be so much what was readily available, but rather what was available in adequate amount and useful in connected processes. Glycine-rich sets of amino acids probably were more abundant than sets containing only proteinous amino acids and enriched with either acidic amino acids or lysine. However, the latter sets yield proteinoid in greater yield than do some of the former, and in a subsequent step proteinoid from the latter sets forms microspheres in greater yield than does proteinoid from the former. An enrichment factor of about 30 was noted for the two processes. While tested only to a limited extent, proteinoid and microspheres from the latter sets of amino acids may be found to have more utility than those from glycine-rich sets (in part because of the lack of limited amount of key functional groups in the glycine-rich sets). Accordingly a double or triple sieving mechanism may have operated to enrich and preserve that which was less abundant but more useful. (Double sieving is seen in the charging of tRNA molecules, with the purpose of enhancing the fidelity of the protein synthesizing machinery.)

ACCEPTANCE OF THE PROTEINOID MODEL

If a model or theory evolves slowly, it typically gains general acceptance even more slowly, for a variety of reasons. One of these is basically defensive, i.e., "He can't be right because I've spent my whole professional career trying to prove a different viewpoint." Ego is a strong driving force! A related factor can be expressed as "Don't confuse me with facts, my mind is made up." For example, questions or criticisms of Fox's work have been voiced in print repeatedly, Fox has answered them repeatedly, the answers have been ignored repeatedly, the criticisms revoiced repeatedly, and reanswered repeatedly. Some often-voiced examples include; heat was not frequent prebiotically, heat is too hot (destructive as well as constructive), the wrong atmosphere was used, etc. These are, as indicated, answered elsewhere, repeatedly, including to some extent earlier in this paper. A third factor stems from entrenched familiarity with the contemporary. "It can't be right because it isn't exactly the way it works today." Examples include the presence in proteinoid of some non-α-peptide linkages, absence of phospholipids in microsphere boundaries, and protein being informational before nucleic acid.

A fourth, likely major, factor concerns specialized definitions. Many terms, while useable in a general (lay-level) sense, have also specialized biological or biochemical meaning. Examples include "information," "random" and "nonrandom" (with all-or-nothing connotations), "conjugation," "budding," "second generation," "menbrane," even "molecular evolution" and "proteinoid." (The latter term and "protein, thermal" regularly occur in Chemical Abstracts.) When encountered by the specialist, such terms trigger a specialized image or meaning; recognition and acceptance in the general sense requires rethinking or even reconditioning. Communication accordingly suffers.

More and more the proteinoid model is being accepted as useful (cf. Dyson, 1982); general acceptance probably will be slow, but it is informative to look at it in a comparative sense. Daily (1938) has stated that "The best model is the one that works best. The perfect model, working infinitely well, is not for men now living." The proteinoid model shows by experiment geologically plausible processes leading from simple gases to amino acids to polyamino acids to functional microspheres to integrated phenomena of multicomponent systems. The most outstanding feature of Fox's model is its range of evolutionary continuity, which in turn is a reflection of the range of his thinking and knowledge and capabilities. Most investigators have confined their research to a narrow area. I firmly believe that the proteinoid model currently is the best model available; there simply are no other comprehensive experimental models to challenge it.

That is not to say that a better experimental model may not emerge (although the prospects for this seemingly decrease as competing theoretical ideas continue over the years to lack comprehensive experimental support.) Nor is that to say that it is the right model. Its real value will no doubt be determined in retrospect, with more facts available concerning its internal productiveness. Aging and unbiased evaluation from a historical perspective will be needed. (My evaluation is, of course, biased.)

It may become obsolete, it may be proven wrong and discarded, or it may be viewed as the best. Regardless, I feel confident that the proteinoid model is and will continue to be a springboard for future experimental investigations of the origins of an integrated variety of life's processes and of life itself. The origins arena, and science, is benefitting because S. W. Fox is immersed in it. Rather than just commemorating the seven decades of Fox and two-and-a-half decades of proteinoid, it is more fitting to hail in with exciting expectations the eighth decade of Fox and the fourth of proteinoid.

REFERENCES

Daily, R. A., 1938, "Architecture of the Earth," p. 1, Appleton-Century, New York.

Dyson, F., 1982, J. Mol. Evol., 18:344.

Fox, S. W., 1956, Amer. Sci., 44:347.

Fox, S. W., 1973, Naturwiss., 60:359.

Fox, S. W., and Dose, K., 1977, "Molecular Evolution and the Origin of Life," 2nd edn., Marcel Dekker, New York.

Fox, S. W., and Harada, K., 1958, Science, 128:1214.

Fox, S. W., and Harada, K., 1960, J. Am. Chem. Soc., 82:3745.

Fox, S. W., and Harada, K., 1961, Science, 133:1923.

Fox, S. W., and Middlebrook, M., 1954, Federation Proc., 13:211.

Fox, S. E., and Windsor, C. R., 1970, Science, 170:984.

Fox, S. W., Johnson, J. E., and Middlebrook, M., 1955, J. Am. Chem. Soc., 77:1048.

Fox, S. W., Winitz, M., and Pettinga, C. W., 1953, J. Am. Chem. Soc., 76:5539.

Harada, K., 1959, Bull. Chem. Soc. Japan, 32:1008.

Harada, K., and Fox, S. W., 1958, J. Am. Chem. Soc., 80:2694.

Harada, K., and Fox, S. W., 1964, Nature, 201:335.

Katchalski, E., 1951, Adv. Protein Chem., 6:123.

Kovacs, J., Konyves, J., and Pusztai, A., 1953, Experientia, 9:459.

Levins, R., 1968, "Evolution in Changing Environments," Chpt. 1, Princeton Univ. Press, Princeton, New Jersey.

McAlhaney, W. W., and Rohlfing, D. L., 1976, BioSystems, 8:45.

Meggy, A. B., 1953, J. Chem. Soc.: 851.

Nakashima, T., Jungck, J. R., Fox, S. W., Lederer, E., and Das, B. C., 1977, Int. J. Quantum Chem., QBS4:65.

Rohlfing, D. L., 1974, in "The Origin of Life and Evolutionary Biochemistry," K. Dose, S. W. Fox, G. A. Deborin, and T. E. Pavlovskaya, eds., p. 397, Plenum, New York.

Rohlfing, D. L., 1979, Polymer Preprints, 20:4.

Rohlfing, D. L., and McAlhaney, W. W., 1976, BioSystems, 8:139.

Rohlfing, D. L., and Oparin, A. I., 1972, "Molecular Evolution: Prebiological and Biological," Plenum, New York.

Saunders, M. A., and Rohlfing, D. L., 1972, Science, 176:172.

Saunders, M. A., and Rohlfing, D. L., 1976, Science, 193:68.

PREBIOLOGICAL EVOLUTION: THE CONSTRUCTIONIST APPROACH TO THE ORIGIN OF LIFE

Richard S. Young

American Society for Cell Bioligy
Bethesda, Maryland 20814

It is difficult to estimate the number of scientific laboratories today whose primary efforts are directed toward a translation of the history of the origin and early evolution of life. If one uses the membership of the International Society for the Study of the Origin of Life as a guide, we can estimate that more than 350 scientists representing 90-100 laboratories around the world are engaged in origin of life research. The field has been pursued on an experimental basis for around 30 years, in only a few laboratories for the first 10 years and then a period of rapid growth in the early 1960s with the influx of funding dollars in the United States from the National Aeronautics and Space Administration.

The earliest laboratory efforts focused on the early chemical events leading to the formation of simple organic molecules of biological interest (e.g., amino acids and purine and pyrimidine bases). The goal was to understand how these molecules could have been synthesized by non-biological processes under the conditions presumed to have prevailed on the primitive Earth. This work attracted the attention of the Astronomers and Planetologists who began to model primitive environments in an attempt to be more precise about the nature of the early Earth. This early work was essentially constructionist in concept. Starting from simple elements in a model environment, the objective was to develop more and more complex molecules aiming ultimately toward the same highly complex molecules from which contemporary life forms are constructed. This work became known as Chemical Evolution--the chemical events leading to the origin of life.

By the early 1960s the Palebiologists and Organic Geochemists were attracted to this field. Study of precambrian rocks in the

1950s had shown that the advent of life on the Earth was much earlier than previously conceived, dating back to three or more billion years ago. Their approach was, of course, quite different from the constructionists--trying to work their way back through history from the recent to the ancient, always seeking older and older material, searching the fossil record for evidence of the "first" organisms and the early molecules that they were composed of. During this time the Biochemists, Geneticists, Microbiologists and others were beginning to "dissect" microorganisms and trace amino acid sequences in their proteins, trying to establish evolutionary relationships between contemporary life forms and morphological and molecular remnants in the ancient rocks. This "reductionist" approach attempts to establish a continuum from early to modern life, by analysis of modern life. More recently, attempts to understand the genetic code and its origin have been made. To many scientists, the universality of the DNA molecule and its preeminence as the ultimate regulator of life processes argues that life actually began with the advent of DNA. Others argue that modern DNA may bear little relationship to the "original" information containing molecule, and that in any case, DNA in a dilute "sea" of organic and inorganic molecules would find little to direct. What is needed in the minds of some is a special environment where conditions are to some degree controlled, and where a select family of molecules can be "directed" to perform the rituals of life--replication, metabolism, and mutation. To create such an environment, a selectively permeable barrier is required, at least similar to contemporary membranes.

There is continuing controversy over the question of the nature of the first living unit--whether it was essentially the first DNA molecule, or the natural construction of a special environment (e.g., cell) in which the first information containing molecule could function. There is also continuing debate over whether the best approach to the origin of life question is the constructionist or reductionist approach. It seems appropriate that both should proceed. Clearly the past 25 or more years in Chemical Evolution has been fruitful in indicating that nature would inevitably have produced the necessary building blocks for life as we understand it. The problem remains to determine how and when those products of chemical evolution became organized in such a way as to permit them to function as a living system. Did it happen "correctly" only once, or many times in many places? If the origin of life is an inevitable consequence of the application of the laws of Physics and Chemistry, does it follow that there is one pathway to success--that only one kind of life will evolve? If so, the constructionist approach may actually lead to <u>the</u> solution of the problem.

In this context it is interesting to note that the earliest thinkers in the origin of life context, A. I. Oparin and J. B. S. Haldane, both thought in terms of understanding the origin of life, by synthesis and assembly of component parts. Oparin devoted much of his laboratory program for many years to the study of coacervate droplets as primitive cell models.

It is of interest to note that these two men who independently wrote in the 1920s about the origin of life, never met until 1963 in Wakula Springs, Florida, at a meeting arranged by Sidney W. Fox on the Origins of Prebiological Ssytems. Haldane (1965) presented a paper entitled "Data Needed for a Blueprint of the First Organism," which I find interesting in the context of this discussion. A series of quotes will make my point.

> . . . , if we have not committed planetary suicide, some of us, or the next generation, will try to make a living organism.
>
> Thus, it does not seem necessary to include DNA in our specification for a minimal organism. I suggest that life, by which I mean indefinite replication of patterns of large molecules, can be based on RNA without DNA.
>
> If the minimal organism involves not only the code for its one or more proteins, but also twenty types of soluble RNA, one for each amino acid, and the equivalent of ribosomaly RNA, our descendants may be able to make one, but we must give up the idea that such an organism could have been produced in the past, except by a similar pre-existing organism or by an agent, natural or supernatural, at least as intelligent as ourselves, and with a good deal more knowledge.
>
> I suggest that the possibility is still open but, in favorable circumstances, a single RNA chain could suffice to specify a peptide. The study of existing organisms and viruses may be of little help to us here.
>
> I suggest that the first organisms may have been something like a tobacco mosaic virus, but including the enzymes needed for its own replication. More verifiably, I suggest that the first synthetic organisms may be so constituted. For natural, but not for laboratory life, a semipermeable membrane is needed.
>
> But I would suggest that the DNA was an adaptation to increase complexity in the system rather than part of the primitive system.

Very few laboratories have seriously devoted themselves to trying to model the transition from chemical to biological evoluation. One of the first to do so was that of Sidney W. Fox. In the middle 1950s he had developed a method of thermal polymerization of dry amino acids to form "proteinoids" which on further heating with water form microspheres. These microspheres have very interesting properties and have over the years served as a very interesting model for the primitive cell. Whether or not microspheres truly represent the evolutionary pathway taken in the earliest history of life, they represent a rational approach to the understanding of the crucial events in the origin of life story--the transition of the inanimate to the animate. I would argue that it is only in this fashion that the total story can ever be unraveled. No amount of dissection of contemporary cells will ever unveil this phase of evolution, nor is it likely to be found in the fossil record.

I have always been puzzled by the fact that of the 100 or so laboratories working in the origin of life field only one has devoted itself in some way to the synthesis of a living system. A few have worked with model systems and none have stated in a grant proposal that their goal was to synthesize life. Perhaps this is a commentary on present-day grant review and the political pressures brought to bear. In any case, if the understanding of the origin of life is a serious subject for scientific study (and surely it is) then the synthesis of a living system, by competent scientific persons and methodology is a necessary subset of the subject.

In my view, the synthesis of a living system is the only hope for development of the complete origin of life story.

REFERENCE

Haldane, J. B. S., 1965, in "The Origins of Prebiological Systems and their Molecular Matrices," S. W. Fox, ed., pp. 11-18, Academic Press, New York and London.

THE ORIGIN AND EVOLUTION OF SIDNEY WALTER FOX, EXPERIMENTALIST

Arthur Cherkin

Veterans Administration Medical Center
Sepulveda, California 91343

> As for me, reason and experiment take the place of the authorities. (Paracelsus, 1493-1541)

> The proteinoid theory is based primarily on experiments in which mixtures of amino acids yield, under geologically relevant conditions, heteropolyamino acids that are highly ordered internally and limited in heterogeneity. (Fox, 1974)

My purpose in underscoring "experiment(s)" in the above quotations is to emphasize the theme of this chapter. Experiment is of course a *sine qua non* for most research aimed at answering questions in science. When the question is "How did life begin?", however, the situation becomes quite different. Man has puzzled about this question for millenia and has sought answers through religion and philosophy, theory and speculation. Such roads of thought have led to much heat but little light. Witness, for example, the ongoing arguments over creationism (Numbers, 1982). The experimental road paved by Sidney Walter Fox during the past four decades has provided a far more intellectually satisfying solution of the puzzle of our primordial origins. The construction of that road has encountered many formidable thought obstacles, rooted in religious, philosophical and theoretical preconceptions. Fox has at times been disappointed but not discouraged by what amounted to anti-experimental opponents. For example, in a letter of 7 February 1965, he commented: "I don't believe that they can indefinitely prefer undisciplined conjecture to vigorous experimentation bonded with careful interpretation, in this particular area."

More recently, Fox commented to me:

> I think theorists are driven to define and explain knowledge. It remains that experimentalists are driven to define and dispel ignorance. . . . I think the motto of an experimentalist is 'What you don't know won't hurt you as much as what you do know that isn't so.'

This volume adds to the argument for persistent creative experimentation in dispelling ignorance.

ORIGINS OF EXPERIMENTAL RESEARCH INTEREST

The origin of an experimental bent of mind, the inclination to do experiments with one's own hands and to reason from the results, is presumably rooted in one's genes and brought into expression by one's environment. In the Fox family, the genetic component awaited the birth of Sidney Walter Fox in Los Angeles on March 24, 1912, for the environmental trigger to be effective, since we find no experimental scientists among his forebears. However, his three sons--chemist J. Lawrence Fox, physicist Ronald F. Fox, and biologist Thomas O. Fox--provide some evidence for this genetic notion by their accomplishments as research scientists in their fields. (Fox maintains that the most important genes of his sons came through their maternal line.)

The motivation of an individual to become a research scientist has interested Fox. He considers that neither the desire to contribute to society nor to acquire wealth provides the impetus to do research; rather, it is intellectual curiosity which drives the scientist (Thomas, 1963, p. 92). More recently, I heard a Caltech friend of Fox's, the late Max Delbrück, carry this thought further in a discussion with students. He excluded not only the motivations to aid society or to earn a living, but also the drives to satisfy curiosity or to gain fame. Delbrück considered that a scientist does research because of some irresistible inner compulsion, a compulsion which is hard to define. He also cautioned students that those who responded to this driving force would often be considered to be a bit strange by their fellowmen who did not share it. Fox agrees. (With characteristic foresight, Delbrück also advised that future research funding would diminish, so that students should think in terms of important but simple, inexpensive experiments. The major experiments carried out by Fox fit these specifications.)

Fox identifies his initial commitment to a career in chemistry as occurring in 1928 when Paul H. Waldraff, his chemistry teacher at Los Angeles High School, invited him to carry out experiments in Waldraff's small laboratory. As Fox put it (Thomas, 1963, p. 94): "This work served to confirm all my thoughts--I knew I wanted to follow chemistry." (As a coincidence, I was making the same decision at about the same time while running experiments in the chemistry laboratory of Monrovia High School, 25 miles east of Los Angeles High).

With this brief reflection on origins, we may proceed to outline the evolution of Sidney Fox as an experimentalist. The point of view of this chapter is that of a close personal friend for more than a half-century, illuminated by a large and cherished correspondence and discussions (and tempered in respect for Fox's distaste for brouhaha). Broader biographical material is available in Thomas (1963), in Oparin and Rohlfing (1972), and in a personal perspective prepared by Fox (1974) in response to an editor's solicitation. The important experiments carried out by Fox and his colleagues are described in detail and references are cited in other chapters in this volume.

EVOLUTION OF FOX AS AN EXPERIMENTAL SCIENTIST

Los Angeles, 1929-1933

My first meeting with Fox was in the middle of an experiment he was running. Very clear in my memory, it occurred in the UCLA organic chemistry laboratory of the late Professor G. Ross Robertson, one morning in the Fall of 1930. Fox was wearing a long black laboratory apron and holding a round-bottom flask containing some organic liquid. We met, chatted for a few minutes, then Fox said: "Look, I have an experiment to run now. You have a facile mind. Let's get together for lunch." One lunch led to another and to a firm friendship ever since. We used to talk about life, chemistry, experiments, writing a better formulary for household chemical products, and everything else. I distinctly recall the creative sandwiches which Fox's mother invented. Anything which could be contained within two slices of bread found its way into the Fox lunch box. This included spaghetti, canned peaches, stewed prunes, and other gustatory surprises which I admired and enjoyed sharing. The relationship of nutrition to brain function is an active area of psychobiology today but I abjure the temptation to suggest that creative sandwiches might nourish creative research.

It cannot be denied, however, that the creative brain of Fox does express itself in many ways besides his experiments, e.g., in his strong preference for creative jazz rather than classical symphonies to satisfy his musical talents, in his social outlook, and especially in his sparkling sense of humor. The latter sometimes finds expression in the opening sentences of his speeches of acceptance of awards. For example, when he received the Florida Academy of Sciences' Honors Medal as the state's Outstanding Scientist of the Year, in 1968, he opened with: "I express my sincere thanks for this medal and for the citation. This citation is much more welcome than the one I received earlier in the year from Metropolitan-Dade County. That citation was for 'Blocking the Intersection.'"

Chemistry students at UCLA in the early 1930s were fortunate. We were in the first class at the brand new Westwood campus, opened in 1929. Our chemistry faculty--including Chairman William Conger Morgan and Professors Blacet, Crowell, Dunn, Ramsey, Robertson, Stone, and Young--were all strongly motivated to build a strong department which could compete with those at Caltech and UC Berkeley. The severe economic depression which had started in 1929 made employment opportunities scarce in the 1930s and gave students a sober stimulus to work hard. Our classes were small and some of our classmates, e.g., Glenn Seaborg and the late Saul Winstein, were exceptionally brilliant contributors to class discussions. Perhaps of greatest relevance to Fox, UCLA had no graduate courses in those early days, so that we undergraduates were encouraged to substitute our own experiments for the standard curriculum experiments. We were introduced to the miracles of Beilstein and Chemical Abstracts in our sophomore year. Max Shaw Dunn, our Professor of Biochemistry, was notable for his stimulation of research by the juniors and seniors in his classes. Thus, Fox was able to develop a new synthesis of DL-aspartic acid and to publish this as his first paper, while still an undergraduate student (Dunn and Fox, 1933). This encouragement of independent research carried out by undergraduates had a profound effect upon Fox and his classmates who were interested to engage in a research career. I believe that this early research experience was a factor in the decision by Professor Max Bergmann, a year after Fox received his B.A. in chemistry, to appoint him to his staff at the Rockefeller Institute in New York, when Bergmann had many excellent applicants who already had the Ph.D. degree in chemistry and were eager to join his laboratory.

Before Fox moved to New York and Bergmann's laboratory in 1934, we had spent several happy months together as partners in our two-man chemical company called SYNTHETICS. Fox and I had been among the fortunte few to obtain positions as research chemists, in those dark days of the depression. Fox joined a large petroleum company and I joined a large cosmetics manufacturer. At night and during weekends, we worked to get SYNTHETICS going, to provide custom organic synthesis. After a few months, we devoted full time to the enterprise. While I acquired equipment and installed the laboratory benches, plumbing and the like, Fox was running experiments appropriate to our limited apparatus. I can still see him, again in his long, black, laboratory apron, crouched outdoors in the California sunshine where he was studying photochemical reactions of chlorinated aliphatic hydrocarbons. Fox and I agree, in retrospect, that if we had persisted with our SYNTHETICS enterprise, we would have developed it into a major producer of fine chemicals. But Fox wisely decided to move to New York, an area richer in basic research opportunities (for family reasons, I needed to remain in Los Angeles).

New York, 1934-1935

The association of Fox with Max Bergmann was particularly fortunate. Bergmann had worked with Emil Fischer, a giant in synthetic biochemistry. I remember how impressed we had been in Max Dunn's class when we had learned about Fischer's synthesis of a polypeptide by the stepwise addition of one amino acid residue at a time, starting with a single amino acid. In the 1930s, the biosynthesis of proteins in living systems was a murky mystery. It seemed totally improbable that 300 amino acid molecules could simultaneously snap together to form 299 peptide bonds and one protein molecule. Max Dunn and others speculated that each "protein" existed as a family of closely related molecular species rather than as a unique molecule. The elegant DNA-RNA-protein mechanism was of course unimagined in those days, just as the question of how the first protein molecules were formed in evolution was an untouched area of research. But an important first step in the growth of Fox's attention to this key problem of evolution now occurred. As he states (Fox, 1974): "Bergmann's interest in various aspects of synthesizing proteins left me with the feeling that molecular evolution at the level of proteins deserved exploration, especially through studies of amino acid sequence." This feeling, when Fox was 22 years old, is what focused his lifetime attention upon protein synthesis.

Pasadena, 1935-1940

After a year with Bergmann, Fox returned to California as a research assistant at the California Institute of Technology. There he was encouraged by Thomas Hunt Morgan to work for a Ph.D. under Professor Hugh Huffman, studying thermodynamic properties of amino acids and peptides. Morgan was chairman of Fox's doctoral committee, which included Huffman, biochemist Henry Borsook and biologists Tyler, Went, and Wiersma. Fox (1974) recalls that at Caltech: "Morgan aroused my interest in evolution by making comments that I interpreted as meaning that biochemistry could make as much contribution to broad concepts of evolution as genetics had." Morgan also commented that "All the important problems of biology are problems of proteins." Two important interests were now firmly joined: protein synthesis and evolution. Experimentation to link these two interests had yet to evolve.

Ames, Iowa, 1943-1955

With his fresh Ph.D. degree from Caltech and his attractive, vivacious wife Raia, Fox first spent three years doing research in industry and at the University of Michigan, then moved to Iowa State College as an assistant professor, where he taught protein chemistry. It was at this time that he recognized "that study of the organic chemical aspects of proteins had lagged far behind the physical" (Fox, 1974). He therefore spent about ten years in an experimental attack upon the primary structure of proteins, based upon sequence studies of the peptides resulting from partial hydrolysis of proteins. By the end of this period, Fox's evolution as an experimental scientist had provided him with an exceptionally strong foundation for his line of experiments to this day. He had published major papers on the thermodynamics of amino acids and on the enzymatic synthesis of peptide bonds. He had demonstrated that an amino acid not found in nature (p-fluorophenylalanine) could be incorporated into a protein chain (Atkinson, Melvin, and Fox, 1951). With remarkable foresight, he pioneered methods for identifying the terminal amino acids of a protein molecule (Rosmus and Deyl, 1972) and the sequence of amino acids in a protein molecule. The stage was now set for exploring the key question of how the first protein molecules came into being, in nature.

The research scientist who is propelled by an inner impulse into the laboratory to do experiments must choose from among many opportunities the specific research strategy to be followed. Fox made a key decision in choosing a constructionist strategy to study the evolution of molecules and cells, in preference to the much more popular reductionist strategy. He has referred (Fox and Zworykin, 1976) to reductionism as:

> . . . the style of studying systems that is totally or largely analytical. This kind of science tends to draw its inferences from analysis of the contemporary cell. The catalogue of information thus obtained is a glorious one. To carry out such work, however, one must first break the cell apart, thereby destroying the crucial, integrated organization of the units. This is almost exactly not the way to find out how cells first originated. What was necessary instead was to guide, under geologically appropriate conditions, the assembly of components of a primitive cell. This was, then, a process of simple-to-complex. As such, it proceeded in the same direction as the first steps in evolution. The simple-to-complex direction is opposite to that of reductionism, which characteristically proceeds from complex to simple. The true relationships are further obscured by the fact that the primitive organism had to undergo eons of evolutionary change.

As he put it (Fox, 1974), starting in 1944:

> We then began to apply methods of determining amino acid sequence and composition to problems of Darwinian evolution at the level of proteins. In the results of these stuies, I thought I saw a clue to how a first prebiotic protein could have arisen on the Earth, and

> that an evolutionary continuality could result from a prebiotic protein. Our main experiments thus emerged from studies of molecular evolution in organisms; I am sure I would not have made a frontal attack on the problem of the origin of life.
>
> The essential, and perhaps only, assumption in the beginning of our research was that amino acids could order themselves during the formation of peptides and higher polymers. This was not pure assumption, however. In experiments that we had been doing with the peptide anilide synthesis model introduced by Bergmann and Fraenkel-Conrat, we had observed that identity of products was often a function of identities of the reactant amino acid residues. . . . Experiments in thermal polymerization of amino acids have seemed to some to be a purely empirical 'shot-in-the-dark.' In fact, however, they were based on reasoning from a careful unwillingness to accept several assumptions prevalent at the time. A most significant consequence of the studies was the [experimental] finding that thermally intractable amino acids could co-polymerize in useful ways. The finding of enzymelike activities in proteinoids was generally not so surprising as the order induced by reactant amino acids in a mixture.

The "shot-in-the-dark" comment above is a mild statement of a set of strong biases among some scientists against the thermal polymerization experiments originated by Fox. I believe one bias arose because of an intuitive expectation that heating a dry mixture of amino acids should always lead to decomposition, with formation of a dark brown tarry mess. I suppose that I would have shared this bias if it were not for an experiment I had run at UCLA in 1932 or so. Fox and I had each been impressed with the scarcity and high cost of L-proline, at a time when Max Dunn was trying to make available to biochemists pure amino acids and peptides at reasonable costs. Proline synthesis by chemical reduction of L-pyroglutamic acid (5-oxoproline) had been attempted without success, so I decided to try electrolytic reduction (also without success, incidentally). The relevant point here, however, is that a well-known synthesis of L-pyroglutamic acid at that time involved cyclizing L-glutamic acid, with elimination of one molecule of water. I still remember my surprise that simply heating dry L-glutamic acid in a porcelain evaporating dish, set in a sand bath, could result in a white product which was readily recrystallized from hot aqueous solution to provide L-pyroglutamic acid of high purity in a yield of over 90%. Fox was familiar with this experiment and therefore understood that a dry amino acid could be strongly heated without necessarily decomposing it.

The thermal synthesis of L-pyroglutamic acid did not lead to co-polymerization of L-glutamic acid under the conditions used. Under the conditions of temperature and time worked out later by Fox, however, polymerization of amino acid mixtures (Fox and Middlebrook, 1954) or of L-glutamic acid alone (Harada and Fox, 1958) to linear polypeptides did occur. The expected dark brown tar did not inevitably appear. More important, this discovery challenged another bias; the anticipated random linkage sequence of amino acids was not found. Instead, what was revealed was a clear tendency for certain peptide linkages to occur in proportions far higher than random. Fox quickly grasped the fundamental significance of this experimental demonstration of nonrandom ordering of amino acids as they assembled themselves into thermal polymers. He has referred to this as one example of a "chemical kismet," in this case the destiny of a given amino acid to favor formation of peptide bonds with certain amino acids over others.

The polymers of amino acids produced by heating under conditions which are plausible for prebiotic times had an array of molecular properties so similar to those of present-day proteins that Fox justifiably applied the term "proteinoids" to them. These properties, described more fully in other chapters in this volume, include: macromolecular structure; molecular weights in the thousands; compositions which include most of the twenty amino acids found in protein hydrolyzates; nonrandom distribution and arrangement of amino acids in the macromolecule; solubility, precipitability and ionic behavior; color reactions (biuret, ninhydrin, Millon's reagent); infrared absorption spectra; hydrolysis (acidic, alkaline); nutritive capacity; and enzyme-like reactions. The latter property is of great interest because its potential evolutionary significance far exceeds that of the artificial enzymes (Breslow, 1982) which are not chemically related to modern proteins.

In a letter of 3 January 1955, Fox wrote:

> We have definite evidence, incidentally, that our polymers are not all, if any, diketopiperazine. Leucine-phenylalanine, for instance, responds to chymotrypsin and also has terminal residues before proteolysis. It's slow but sure work, like recapitulating the gradual acceptance of the Peptide Theory itself; in fact, we are now looking at the Peptide Theory in a new light.

The "slow but sure" comment has proved to be prophetic. The stage was now set to consider experiments aimed at evolution at the cellular level of organization.

Tallahassee, Florida, 1955-1964

After twelve years at Iowa State, Fox was appointed Professor of Chemistry at Florida State University. There he devised the simple experiment which made possible a quantum leap into cellular evolution. On August 14, 1958, while attending a symposium in Ireland, Fox mailed to his laboratory assistant, Mrs. Jean Kendrick, the following notes:

> There is one more experiment I have been thinking of in preparation for Oct., and which I would like you to place on high priority, as follows:
>
> Make proteinoid (2:2:1) in usual way (2g. or so enough), lift hot melt from oil bath with clamp and immediately add to tube 20 ml. of boiling H_2O (Protect eyes!). When this settles down, swirl, pour off supernatant while warm. When cool, examine this under microscope.
>
> If there is anything cell-like or globular, repeat with proteinoid-nucleicoid when and if latter is obtained and characterized.

In a letter of December 3, 1958, Fox wrote with obvious delight:

> At Berkeley we had a conference attempting to define life and otherwise wasting time. Latest news here is that by pouring boiling water over our synthetic proteinoid we get spheres that look like cell outlines. Each smudge of protein gives 10^6-10^8 of the little darlin's, they are of uniform size (2-4μ) and they can be centrifuged, washed, and resuspended without loss of shape. This is something we can really feel paternal about, particularly when they are activated by Brownian motion.

As Fox (1974) describes this step forward:

> In perhaps the only prediction I have made in this program I proposed . . . that a primitive cell would arise by contact of thermal polyamino acids with hot water. We did the model experiment in 1958, it was published in 1959. The most astonishing consequence of that experiment is the roster of properties that the resultant proteinoid microsphere is found to have in common with contemporary cells, including a tendency to reproduce by any of several physical mechanisms. At this level, the findings were largely empirical. (Fox now refers to the reproduction as "heterotrophic protoreproduction.")

The roster of properties included: size, morphology and mode of association, all associated with those of coccoid bacteria; stability in aqueous suspension; staining characteristics, including gram-positive or gram-negative forms; ultrastructural appearance by electron microscopy; structured boundary; selective passage of molecules through the microsphere boundary; catalytic activities; growth by accretion; budding and fission; binding of polynucleotides; and electrical characteristics. Fox continued:

> When one recognizes the accumulated evidence that such protocells could have evolved to contemporary cells, the picture for macromolecular and cellular origins is a complete outline. An eventual evolutionary continuality was hazily visualized throughout and from the beginning of the laboratory research, although the clarity increased at each stage. Until about the age of forty [in 1952], I had favored a highly logical and rational science, and distrusted intuitive approaches (which I now recognize as rational but at a subconscious level). At that time my feeling about intuition changed, but I may merely have reached a stage at which I had acquired an adequate matrix of [experimental] information for intuitive insights. I had enjoyed especially finding unique papers in the older and foreign literature, e.g. Schiff's 1897 paper that described thermal homopolymerization of aspartic acid. At that time, also, I felt I had to choose between increased involvement in national and international biochemical affairs, or deeper immersion in the laboratory. I chose the latter.

This is hardly the place to dive into the deeps of free will versus determinism. But I believe that Fox's "choice" of laboratory experimentation over biochemical statesmanship was another example of kismet, in this case the subconscious drive to do experimental research which Max Delbrück had discussed with his students.

In any event, the key microsphere experiment is elegant in its simplicity. All it requires is to dissolve proteinoid in hot water and allow the solution to cool. Back at UCLA, in our undergraduate research with Max Dunn, we considered ourselves fortunate when this common recrystallization technique resulted in the successful formation of crystals. Viewing these under the microscope provided a major thrill, after days spent in synthesizing an amino acid or peptide. Only too often, no crystals would separate (unless we had a sample of the pure compound to provide seed crystals) despite prayerful scratching of the interior wall of the flask and days of refrigeration of the solution. Evaporation of the solvent would typically leave a thick gummy or oily residue which might or might not later crystallize. The same result could reasonably be expected of the proteinoid solution, that is, crystal formation at best or an oily residue at worst. The result, however, was far different from such plausible expectations. What occurred was the formation of myriad tiny spherules, 0.5-7.0 μm in diameter. The remarkable morphologic, structural, chemical, and other properties of these spherules, named "proteinoid microspheres" by Fox, are described in other chapters of this book.

Miami, Florida, 1964-Present

The detailed exploration of the above properties led Fox to a long series of experiments which continue to this day at his Institute for Molecular and Cellular Evolution at the University of Miami. One by one, properties hitherto regarded as unique to contemporary proteins and living organisms have been discovered in these laboratory-created proteinoids and microspheres, by careful experiments. For example, the demonstration that the oxidation of glucose could be catalyzed by proteinoid (Fox and Krampitz, 1964), provided early evidence of this important biochemical property of proteinoid, later confirmed with diverse substrates and reactions (Fox et al., 1982).

In view of the long history of foresightful Fox experiments with unexpected important outcomes, interesting future experiments can be expected. A precedent may be found in the problem of identifying the N-terminal amino acids in proteins, which Fox had addressed experimentally in the early 1940s and later reviewed (Fox, 1945). A more recent review (Rosmus and Deyl, 1972) states that: "Apart from the experimental details . . . there is nothing to be added to the ingenious foresight of S. W. Fox expressed nearly thirty years ago." I foresee a particularly exciting future for Fox's current experiments on the origins of behavior in macromolecules and protocells (Fox, 1980; Fox et al., 1982). More than a decade ago, Fox had demonstrated informational transfer between microspheres, as conveyed by endoparticles passing through the junction between adjacent microspheres; he considered this to be a model of primordial communication (Hsu, Brooke, and Fox, 1972). At that time, Fox and I were exchanging ideas about communication between nerve cells and wondering about questions such as "What was the primordial origin of learning and memory?" A start toward answering such questions is provided by experiments with artificial cells made from proteinoid; these cells, when suspended in appropriate solution, have the ability to generate electrical pulses, measurable by the intracellular microelectrode technique (Ishima, Przybylski, and Fox, 1981). Recently, the case for a molecular biology of learning has been strongly advanced by Kandel and Schwartz (1982). They have made brilliant use of the marine mollusk Aplysia californica as an experimental model to provide insights into the molecular basis of learning, insights which are relevant to learning in mammals. The time appears ripe for bridging the wide evolutionary gap between primordial learning and neuronal learning. As Fox (1980) has put it: ". . . we are now in a position to attempt to trace each kind of behavior from its ultimate molecular and evolutionary roots."

It is of interest that the topics to be considered in Mainz at the Seventh International Conference on the Origins of Life, in July, 1983, include not only such areas as Cosmochemistry and Prebiotic Systems but also Evolution of Intelligence. Science has come a long way!

Finally, the Mainz Conference includes another topic of interest to Sidney Fox: the Evolution of Social Systems. Even at this complex level, Fox's experimental nature is clearly evident in his letter of 2 January, 1981, to a journal editor, in which he wrote:

> As I understand dialectics, I believe the objective of experimental science is to be nondialectical, to make choices by controlled experiment such that argument becomes superfluous. Dialectical materialists may insist that the goal is unattainable, like the absolute truth, but I believe that it should be striven for, and that the emphasis should be experimental rather than dialectical.

AFTERWORD

The felicitous opportunity to write this chapter has led me to reflect upon the situation of young research scientists today as compared to the 1930s, when Sidney Fox started his research career. Several trends concern me: the notion that bigger is better; the excessive bureaucratization, "Federalization" and industrialization of our universities; the efforts devoted to developing curriculum vitae that will impress: the preoccupation with preparing competitive grant applications that will sell, with the consequent drift away from daring research creativity toward safe and popular strategies; the erosion of scholarly interactions in a collegial atmosphere; the reduced opportunities for tenure appointments. These concerns recall to mind the advice which Max Delbrück gave to students interested in a research career, namely, that they think in terms of important but inexpensive experiments. The research career of Sidney Fox provides a model of such experiments and of a happy, productive career based on them.

REFERENCES

Atkinson, D. E., Melvin, S., and Fox, S. W., 1951, Arch. Biochem. Biophys., 31:205.
Breslow, R., 1982, Science, 218:532.
Dunn, M. S., and Fox, S. W., 1933, J. Biol. Chem., 101:493.
Fox, S. W., 1945, Advances Prot. Chem., 2:155.
Fox, S. W., 1974, Molec. Cell. Biochem., 3:129.
Fox, S. W., 1980, Comp. Biochem. Physiol., 67B:423.

Fox, S. W., and Krampitz, G., 1964, Nature, 203:1362.
Fox, S. W., and Middlebrook, M., 1954, Federation Proc., 13:211.
Fox, S. W., Nakashima, I., Przybylski, A., and Syren, R. M., 1982, Int. J. Quant. Chem.,QBS9:195.
Fox, S. W., and Zworykin, V., 1976, in "Concepts, Critiques and Comments: Wide and Varied," B. Dibher and M. Rubien, eds., pp. 91-96, A. Colish, New York.
Harada, K., and Fox, S. W., 1958, J. Am. Chem. Soc., 80:2694.
Hsu, L. L., Brooke, S., and Fox, S. W., 1971, Curr. Mod. Biol., 4:12.
Ishima, Y., Przybylski, A. T., and Fox, S. W., 1981, BioSystems, 13:243.
Kandel, E. R., and Schwartz, J. H., 1982, Science, 218:433.
Numbers, R. L., 1982, Science, 218:538.
Oparin, A. I., and Rohlfing, D. L., 1972, in "Molecular Evolution, Prebiological and Biological," D. L. Rohlfing and A. I. Oparin, eds., pp. 1-5, Plenum Press, New York and London.
Rosmus, J., and Deyl, Z., 1972, J. Chromatography, 70:221.
Thomas, S., 1963, "Men of Space," pp. 90-111, Chilton Company, Philadelphia.

ONE-CARBON COMPOUNDS IN THE PREBIOTIC SYNTHESES OF BIOMOLECULES

A.S.U. Choughuley

Bio-Organic Division
Bhabha Atomic Research Centre
Trombay, Bombay 400 085, India

INTRODUCTION

The main one-carbon compounds that may have participated in the primitive Earth's chemical syntheses are methane, carbon monoxide, carbon dioxide, formaldehyde, formic acid, formamide, hydrogen cyanide, cyanamide and urea. Except urea, all these compounds are also known to be present in the atmosphere of other planets or in interstellar dust clouds. Some other one-carbon species detected in interstellar medium include methylidyne (CH), methylidyne ion (CH^+), cyanogen radical (CN), hydrogen isocyanide (HNC), formyl radical (HCO), formyl ion (HCO^+), methyl alcohol (CH_3OH), isocyanic acid (HNCO), carbon monosulfide (CS), carbonyl sulfide (OCS) and thioformaldehyde (H_2CS). In addition, methyne (CH_2), CO_2^+ and CO^+ have been detected in comets (Oro, 1972; Schwartz, 1981). These highly reactive species are not found in the atmosphere, geosphere, or hydrosphere of the earth, and as such may not have played any significant role in the prebiotic chemistry of our planet--except to form relatively more stable compounds, if they did exist for very short periods.

Simulated Primitive Earth experiments have been carried out using methane, carbon monoxide or carbon dioxide as the carbon source and ammonia or nitrogen as the source of nitrogen. [In some cases ethane, acetylene and its derivatives have been used, but they would not be particularly discussed in this article.] As long as the overall composition of the experiment is reducing, the formation of amino acids and other biologically important nitrogen compounds does take place under the influence of a suitable energy source. In the presence of water in these experiments, the two most important intermediates formed are formaldehyde and hydrogen cyanide (Miller, 1953; Fox, 1963, 1965; Ponnamperuma and Gabel, 1968; Lemmon, 1970; Oro, 1963; Stephen-Sherwood and Oro, 1973; Schwartz, 1981). However, when a mixture of methane and ammonia is subjected to electric discharge, aminonitriles are formed directly (Chadha et al., 1975 and references cited therein).

What kind of atmosphere did the primitive earth have? Was it as reducing as suggested by Oparin (1924); Haldane (1928) and Urey (1952)? The success of Miller's classic experiments and scores of others that followed his work, have had a strong bias in favour of the reducing atmosphere. However, Rubey (1951, 1955), Abelson (1966) and Holland (1962, 1978) have forcefully suggested that methane may not have lasted for more than 10^8 years, and CO and CO_2 must have made their appearance as a significant component of the primitive Earth's atmosphere fairly early. Similarly there is a strong case made against the presence of ammonia as a major component for more than 10^6 years. Abelson (1956) showed that various mixtures of CO_2, CO, N_2 and H_2 gave amino acids under experimental conditions similar to those of Miller's. Later Ring et al. (1972) demonstrated that a mixture of methane, nitrogen and water and traces of ammonia when subjected to electric discharge gave at least ten amino acids, well characterized by amino acid analyzer and GC-MS analysis.

The formation of formaldehyde and formic acid by the reduction of carbon dioxide in aqueous solutions by electron irradiation was shown by Garrison et al. (1951), and Miller (1953) demonstrated that formaldehyde and hydrogen cyanide were the primary products in his experiments with methane-ammonia-water system. Later in his detailed paper (Miller, 1955) he also described the formation of formic acid and other aliphatic acids. Thus, formaldehyde and formic acid are the products in both types of experiments, carried out using oxidizing as well as reducing atmospheres. Various energy sources have been used for both types of experiments, and it has been found that qualitatively the products formed are comparable if not identical--whatever be the source of energy used.

Several books and review articles have been written on chemical evolution and the origin of life. The list is long, but the articles referred to above, while discussing the formation of formaldehyde and hydrogen cyanide in reducing as well as partially oxidizing conditions, can give a comprehensive overview of the subject. In this article I will briefly deal with the experiments carried out in different types of atmospheres and the role of hydrogen cyanide. I will discuss the role of formaldehyde and formic acid somewhat in detail for two reasons: one being that much of the work would be discussed together perhaps for the first time and the second being that it would give me an opportunity to summarize the work done in our laboratory during the last ten years.

EXPERIMENTS USING DIFFERENT COMBINATIONS OF METHANE, CARBON MONOXIDE, CARBON DIOXIDE, AMMONIA, NITROGEN, HYDROGEN AND WATER

Amino Acids

Ever since Miller (1953) reported the formation of glycine, alanine, β-alanine, aspartic acid and glutamic acid in his electric discharge through methane-ammonia-water system, several authors have used methane as a reactant in their experiments. Abelson (1956) subjected 20 different mixtures of methane, carbon monoxide, carbon dioxide, ammonia, water, hydrogen, nitrogen and oxygen to electric discharges and observed that the amino acids, alanine, β-alanine, glycine and sarcosine, were formed when the experimental conditions were non-oxidizing. Similar experiments using different combinations of the above gases and electric discharge as the energy source were also carried out by Heyns et al. (1957), Pavlovskaia and Pasynskii (1959), Oro (1963), Grossenbacher and Knight (1965), Ponnamperuma and Flores (1966) and Ring et al. (1972). Dose and Rajewsky (1957) used X-rays and γ-rays, while Palm and Calvin (1962) used electrons from a linear accelerator. Heat as an energy source was employed by Harada and Fox (1964) and Oro (1965), and shortwave ultraviolet light was used by Groth and Wessenhoff (1957), Terenin (1959), and Dodonova and Sidorova (1961). Bar-Nun et al. (1970) used shock waves for similar experiments. Although in the earlier experiments the products, especially amino, acids, were characterized on the basis of paper chromatography only, later workers, for example, Ponnamperuma and Flores (1966) and Ring et al. (1972), used GC and GC-MS for the characterization of amino acids, along with ion-exchange chromatography and paper chromatography. The amino acids characterized in these experiments include glycine, alanine, valine, leucine, isoleucine, serine, threonine, proline, aspartic acid, glutamic acid, lysine, ornithine and tyrosine. Choughuley and Lemmon (1966) subjected a mixture containing CH_4-NH_3-H_2O-H_2S to electron irradiation and observed the formation of cysteic acid and cysteine, along with other products. Later Khare and Sagan (1971) and Sagan and Khare (1971) reported the formation of cysteine and other amino acids by the action of heat and ultraviolet light on a primitive gaseous mixture in which ethane and hydrogen sulfide were additional reactants.

In a slightly different set of experiments, water was excluded. Ponnamperuma and Woeller (1967), Woeller and Ponnamperuma (1969) and Chadha et al. (1971) showed that when methane-ammonia mixtures were subjected to electric spark or semicorona discharge, a number of products were formed. The gaseous products were hydrogen, nitrogen, hydrogen cyanide and acetylene. The volatile liquid was made up of alkylamines and alkyl nitriles. The aminonitriles characterized by GC included aminoacetonitrile, α-aminopropionitrile, methylaminoacetonitrile and β-aminopropionitrile. On hydrolysis, the corresponding amino acids were formed, which were characterized by GC and GC-MS techniques of the appropriate derivatives.

Hayatsu et al. (1971) described the formation of amino acids in Fischer-Tropsch type synthesis, heating CO, H_2 or D_2 and NH_3 or ND_3 as reactants at 200-700°C in presence of Ni, Alumina or clay.

Shimoyama et al. (1978) and Ponnamperuma and Shimoyama (1982) passed electric discharge through a mixture of methane and nitrogen in the presence of sodium-montmorillonite suspended in water. The amino acids formed were glycine, alanine, α-amino-n-butyric acid and sarcosine. The effect of clay seems to be to produce larger molecules and a variety of amino acids in a shorter period; the amino acids may then polymerize to peptides.

Nucleic Acid Bases

Adenine was the first base to be detected in a simulated prebiotic experiment by the action of heat on ammonium cyanide (Oro, 1960; Oro and Kimball, 1961). However, Ponnamperuma et al. (1965) detected it first in a system containing methane, ammonia and water, subjected to the bombardment with 5 MeV electrons. No other biologically important purine or pyrimidine has been detected so far in any combination of primitive Earth gases except the formation of a pyrimidyl type of heterocyclic compound in the spark discharge experiment, with methane ammonia mixture, by Chadha et al. (1971, 1975).

Adenine, guanine and cytosine are, however, formed in low yield along with other N-heterocyclic compounds, when CO, H_2 or D_2 and NH_3 or ND_3 are heated in the presence of nickel-iron catalysts (Hayatsu et al., 1968; Yang and Oro, 1971). Hayatsu et al. (1971 and references cited therein) have also reported the formation of urea, guanidines, uracil, 1,3-dimethyl uracil, thymine and pteridine derivatives in these Fischer-Tropsch types of syntheses.

Porphyrins

Hodgson and Ponnamperuma (1968) and Simonescu et al. (1978) have detected porphyrins in experiments where electric discharge was passed through methane-ammonia water mixture. Other relevant syntheses will be discussed later while discussing the role of formaldehyde.

Sugars and Fatty Acids

Although formation of sugars in the experiments in which formaldehyde is a major product is conceivable, no sugars seem to have been detected in the experiments in which reducing or oxidizing gaseous mixtures were subjected to suitable energy sources.

The formation of fatty acids such as formic acid and acetic acid was observed by Miller (1955). The higher fatty acids are known to be formed in industrial Fischer-Tropsch syntheses using CO and H_2 and cobalt or iron catalysts (Storch et al., 1951). It is possible that many products of biological interest may have originated on the primitive Earth by Fischer-Tropsch types of syntheses near volcanic regions (Studier et al., 1968; Anders et al., 1974).

EXPERIMENTS WITH HCN, ITS OLIGOMERS, CYANATE, CYANAMIDE AND UREA

Formation of These Compounds

Hydrogen cyanide is a primary product in the Miller type of experiment (Miller, 1953; Wolman et al., 1972). It is also a major product in simulated Jovian atmosphere experiments (Chadha et al., 1975 and references cited therein). It is formed when compounds of carbon, hydrogen and nitrogen are heated (Harada and Fox, 1965) or exposed to high energy sources (Palm and Calvin, 1962; Abelson, 1966). It can also be formed by the dehydration of formamide.

Various oligomers of HCN are formed by base catalysis of HCN solutions, and these must have also played a key role in the syntheses of amino acids and nucleic acid bases (Harada, 1967; Moser et al., 1968; Ferris et al., 1974, 1978).

The formation of cyanate and carbamyl phosphate has been shown by Yamagata and Mohri (1982) by passing electric discharge through a mixture of carbon dioxide, nitrogen and hydrogen.

Cyanamide and dicyandiamide formation was shown by Steinman et al. (1964) and Schimpl et al. (1965) upon UV irradiation of dilute HCN solutions and also by electron irradiation of a mixture of methane, ammonia and water.

Urea as such is a hydrolytic product of cyanamide. Lohrmann (1972) observed that urea and guanidine are formed during the photolysis of ammonium cyanide.

Hydrogen Cyanide and Its Oligomers in Amino Acid Syntheses

Oro and Kamat (1961) reported the formation of glycine, alanine and aspartic acid by heating an aqueous solution containing HCN and NH_3. This work was confirmed by Lowe et al. (1963). They identified aspartic acid, threonine, serine, glycine, glutamic acid, alanine, isoleucine, leucine, β-alanine, β-diaminopropionic acid and α-aminobutyric acid among the products formed. Sanchez et al. (1966) showed that HCN, NH_3 and cyanoacetylene react together to give asparagine and aspartic acid. Later, several authors looked into different aspects of the chemistry of HCN and its oligomers. Some of the pertinent references relating to the formation of amino acids by the reaction of HCN with appropriate species or by hydrolysis of the oligomer or polymer to yield amino acids are, Matthews and Moser (1967), Moser et al. (1968), Sanchez et al. (1967), Labadie (1968), Sweeney (1972), Ferris et al. (1973), 1974, 1981), and Draganic et al. (1978a, 1978b). In their recent work, Ferris et al. (1981) have shown that diaminomaleonitrile is formed rapidly and irreversibly from HCN before the oligomers are observed, hence diaminomaleonitrile is the molecule that condenses to HCN oligomers. The heteropolypeptide structure, suggested by Matthews and coworkers (Matthews, 1979, and references cited therein) from the hydrolytic products of the polymer, is not consistent with the observations made by Labadie et al. (1968), and Ferris et al. (1973, 1981).

Purines and Pyrimidines

1. Purines. The first synthesis of a purine under possible prebiotic conditions was achieved by Oro (1960), who showed that adenine formation did take place on heating a concentrated soluion of ammonium cyanide. This synthesis was later confirmed by Lowe et al. (1963). The intermediates in this reaction are believed to be diaminomaleonitrile and 4-amino-imidazole-5-carbonitrile. The latter then reacts with HCN, cyanogen or cyanate to give adenine, guanine and other purines (Oro and Kimball, 1961; Sanchez et al., 1966, 1967, 1968; Ferris and Orgel, 1966; Ferris et al., 1969, 1978). Recently Ferris et al. (1977, 1978) have shown that hydrolysis of HCN oligomers under mild conditions gives rise to the formation of adenine, orotic acid 4-amino-imidazole-5-carboxamide and a number of amino acids and other products.

2. Pyrimidines. In earlier work direct formation of pyrimidines from HCN oligomers was not reported. The earliest synthesis of a pyrimidine was reported by Fox and Harada (1961). On heating urea with malic acid they obtained uracil. Later Oro (1965) reported that uracil could be formed by the reaction of potassium cyanate with propiolamide, acrylonitrile or β-amino-propionamide. Sanchez et al. (1966) and Ferris et al. (1968, 1974) observed that heating potassium cyanate or urea with cyanoacetylene or cyanoacetaldehyde yielded cytosine. Cytosine on deamination gave uracil. Recent work of Ferris et al. (1977, 1978) has shown that during hydrolytic decomposition of HCN oligomers at pH 8 to 9, in addition to amino acids, purines and pyrimidines are also formed. Orotic acid is formed first which decarboxylates to give uracil. The other pyrimidine formed is 5-hydroxyuracil. Recently Chittenden and Schwartz (1976) have shown that urea reacts with β-alanine to give dihydrouracil, which is converted to uracil by photodehydrogenation when the compound adsorbed on a clay mineral montmorillonite is irradiated with UV light. If an acetate salt is added to the system, thymine is formed in addition to uracil (Schwartz and Chittenden, 1977). Subbaraman et al. (1980) and Choughuley (1980)

have reacted urea with acetylenedicarboxylic acid in the presence of ammonium phosphate to give uracil in about 6% yield. In the absence of ammonium phosphate no uracil formation was detected. Very recently Voet and Schwartz (1982) have reported the formation of uracil by acid hydrolysis of HCN oligomers.

Cyanamide, Dicyandiamide and Urea in Condensation Reactions

The formation of these compounds under possible prebiotic conditions has been discussed in a previous section. Ponnanperuma and Peterson (1965) and Steinman et al. (1965) have shown that both cyanamide and dicyanadimide (the latter being the dimer) can be used to synthesize simple peptides such as glycylleucine, leucylglycine and alanylalanine from the corresponding amino acids. A related compound dicyanamide has also been used to convert up to 30% of glycine into diglycine and 15% and 7% respectively, to tri- and tetraglycines (Steinman et al., 1966). These compounds can be considered as carbodiimide analogs. In addition to acting as dehydration condensation agents in peptide synthesis, they also have been used to produce glucose-6-phosphate from glucose and H_3PO_4, adenosine 5'-phosphate from adenosine and H_3PO_4 and pyrophosphate from H_3PO_4 (Steinman et al., 1965). Lohrman and Orgel (1971) have used urea-inorganic phosphate mixture as a phosphorylating agent.

Ibanez et al. (1971) have used cyanamide to obtain dimer and trimers of thymidilic acid in low yield. Stephen-Sherwood et al. (1974) obtained up to the tetramer on heating deoxythymidine-5'-triphosphate at 90°C in the presence of cyanamide, 4-aminoimidazole carboxamide and deoxythymidine monophosphate.

FORMALDEHYDE, FORMIC AND FORMAMIDE IN PREBIOTIC SYNTHESES

As discussed in the introduction, formaldehyde and formic acid both are formed in most simulated primitive Earth experiments, and formamide is a simple derivative of formic acid, which is also formed by the reaction of CO and NH_3 under pressure. All these molecules have also been detected in interstellar clouds (Snyder et al., 1969; Zuckerman et al., 1971; Sandoval and Young, 1973). However, the role of formaldehyde in chemical evolution was not generally accepted till its discovery in the interstellar medium. Abelson (1966) argued that polymerization of HCHO to paraformaldehyde would remove it from circulation and thus it would not play a crucial role in the synthesis of compounds of biological interest. Gabel and Ponnamperuma (1967) suggested that paraformaldehyde could get broken down to formaldehyde in hydrothermal springs. Oro et al. (1959) used paraformaldehyde with hydroxylamine for the synthesis of amino acids. Choughuley et al. (unpublished results) used paraformaldehyde with equal ease as formaldehyde solution for the synthesis of methyleneaminoacetonitrile and hydroxymethyl derivatives of amino acids and uracil.

Formaldehyde in Sugar Syntheses

The condensation of formaldehyde to give sugars has been known since 1861 (Butlerow, 1861; Loew, 1886, 1887, 1889; Euler and Euler, 1906; Orthner and Gerisch, 1933). The mechanism of this reaction has been discussed by Breslow (1959) and Wanzlick (1962). All these reactions were carried out in concentrated solutions and therefore their relevance to the prebiotic formation of sugars could not be taken for granted. Relevant experiments keeping primordial conditions in view were started by Oro and Cox (1962), who demonstrated the formation of 2-deoxyribose by heating an aqueous solution of formaldehyde and acetaldehyde in the presence of divalent metal oxides. Ponnamperuma (1965) showed that UV irradiation of formaldehyde gave ribose, 2-deoxyribose and other sugars. Later Gabel and Ponnamperuma (1967) reported that very dilute solutions when refluxed over kaolinite yielded trioses, tetroses, pentoses and hexoses.

Formaldehyde and Formic Acid in Combination with Ammonia or Hydroxylamine for the Synthesis of Amino Acids

Bahadur (1954) was one of the first authors to have used formaldehyde, potassium nitrate, ferric chloride and some other inorganic salts in aqueous solution for the synthesis of amino acids. He reported the formation of aspartic acid, serine and lysine when such a solution was irradiated with visible light. Later Abelson (1956) described the synthesis of glycine by UV irradiation of ammonium formate, ammonium hydroxide, sodium cyanide and ferrous sulfate. Reid (1959) used short wave-length UV for the irradiation of a reaction mixture consisting of formaldehyde, ammonia, carbon dioxide and inorganic salts. He obtained glycine and alanine. Pavlovskaia and Pasynskii (1959) subjected a solution of formaldehyde, ammonium nitrate and ammonium chloride to irradiation with a high-pressure mercury lamp, and reported the formation of serine, glycine, glutamic acid, alanine and valine. Around the same time Oro et al. (1959) showed that an aqueous solution of formaldehyde and hydroxylamine at moderate temperature and in slightly acidic solution could yield glycine and smaller amounts of alanine, β-alanine, serine, threonine and aspartic acid. Subsequently Oro and Kamat (1961) refluxed a mixture of formaldehyde and hydrazine and obtained glycine, valine and lysine. In a majority of these experiments, the conditions were not very much in keeping with the accepted primordial environments. The concentrations were very high, and also very highly oxidized compounds were used as reactants. The claim for the identification of some of the amino acids may also be exaggerated as only paper chromatography was used for the identification of amino acids (often one may mistake iminodiacetic acid for aspartic acid, and the situation may be similar with regard to the other amino acids).

Fox and Windsor (1970) heated formaldehyde and ammonia at temperatures of up to 200°C and the reaction mixture, which did not give a positive biuret reaction for peptides, on hydrolysis gave glycine and other amino acids. On the basis of amino acid analyzer and TLC data, these amino acids were characterized as aspartic acid, serine, glutamic acid, valine, and alanine. Harada and Suzuki (1977) subjected a solution of ammonium bicarbonate or ammonium formate to contact glow discharge electrolysis and identified glycine, alanine, aspartic acid, glutamic acid, serine and threonine among the products. Egami (1981) has recently reported the formation of glycine, alanine, serine and aspartic acid as the reaction products from formaldehyde and hydroxylamine, and serine and aspartic acid from glycine and formaldehyde in his primeval sea-type experiments.

Reactions of Formaldehyde and Formic Acid with Ammonium Cyanide and Other Amino Acids

Formation of methyleneaminoacetonitrile. Formaldehyde reacts with HCN and NH_3 to give aminoacetonitrile (AAN). If excess HCHO is present methyleneaminoacetonitrile is formed (Adams and Langley, 1956). Similarly other aldehydes on reaction with ammonium cyanide give corresponding aminonitriles, which on hydrolysis give corresponding amino acids. This reaction is known as the Strecker synthesis of amino acids, and Miller (1955) showed that amino acids in his experiments were formed predominantly by this reaction. Later Ponnamperuma and Woeller (1967) and Chadha et al. (1971) obtained aminonitriles by passing electric discharges through anhydrous methane-ammonia mixtures. This has been discussed in the preceding sections.

Choughuley et al. (1972, 1975) showed that aminoacetonitrile and methyleneaminoacetonitrile (MAAN) could be easily formed even in very dilute solutions containing HCHO, HCN and NH_3. It was also shown that aminoacetonitrile and HCHO react in dilute solution to give MAAN. This molecule is far more stable than the corresponding AAN, and it may have been formed under prebiotic conditions.

Hydrolytic products of MANN. Methyleneaminoacetonitrile and AAN both, on hydrolysis with 6N HCl, are known to give glycine quantitatively. However, Choughuley et al. (1972) observed that under mild hydrolytic conditions (pH 5.5-10) a number of products were formed. These included glycine, iminodiacetic acid, serine, hydroxymethylserine, sarcosine, alanine, glycineamide, diglycine and triglycine. The optimum yields are obtained at pH 8.5, and alanine formation is noticed at higher temperatures (140°C) or when phosphoric acid, its salts or formic acid is included in the system.

Later it was shown that MAAN reacts with other amino acids to give peptides in addition to the products listed above (Choughuley et al., 1975). The possible mechanism for the formation of these amino acids has also been discussed by these authors (Fig. 1).

Reaction of formic acid with serine and other hydroxyamino acids. Subbaraman et al. (1972) observed that formaldehyde on reaction with glycine in dilute and mildly alkaline conditions gave serine, hydroxymethylserine, sarcosine, imino-di-acetic acid and alanine. When other aldehydes, viz acetaldehyde and benzaldehyde, were used, the corresponding hydroxyamino acids threonine and β-phenylserine were obtained. It was observed that when these hydroxyamino acids were heated in a solution with formic acid, corresponding reduced amino acids were formed. Thus serine gave alanine, threonine gave α-aminobutyric acid and β-phenylserine gave phenylalanine (Subbaraman et al., 1979). It is not unlikely that this may also have been a route for the synthesis of amino acids under prebiotic conditions (Fig. 2).

Reactions of Formaldehyde and Formic Acid with Uracil

The prebiotic synthesis of uracil by the reaction of urea with various other molecules has been discussed earlier. The reaction of formaldehyde with glycine to give serine, and reduction of serine to yield alanine was extended to uracil (Choughuley et al., 1977), as shown in Fig. 3.

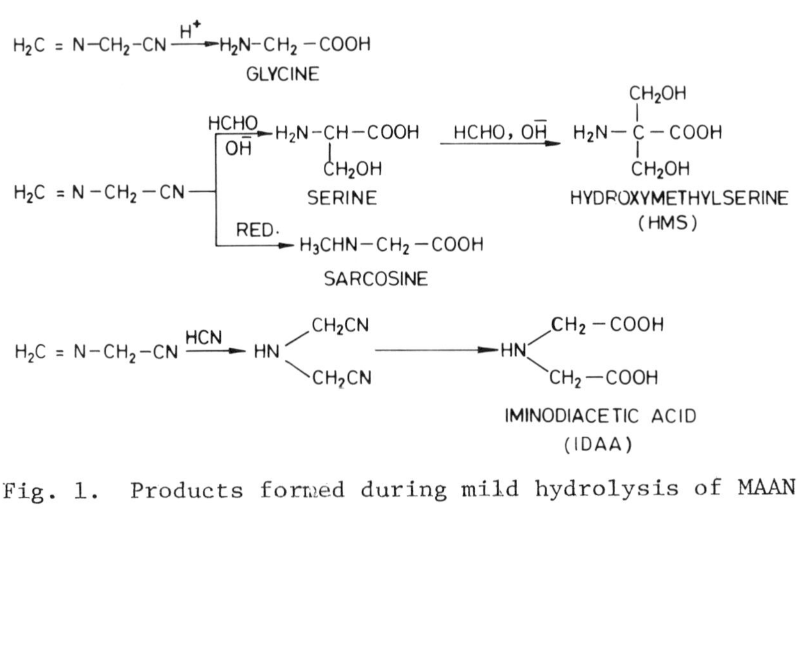

Fig. 1. Products formed during mild hydrolysis of MAAN.

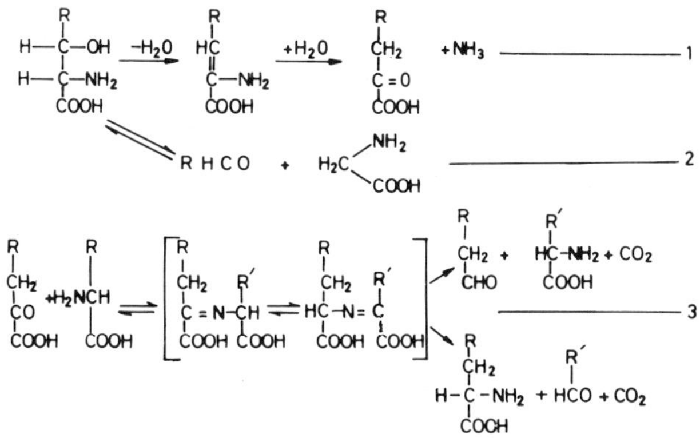

Fig. 2. Possible mechanism for the formation of reduced amino acids from hydroxy amino acids.

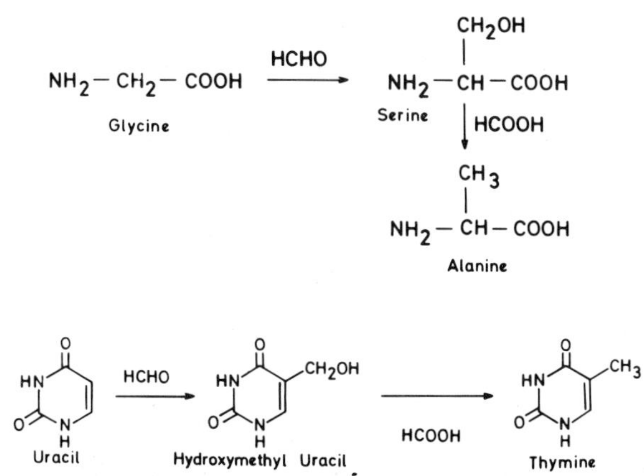

Fig. 3. Reaction of formaldehyde with glycine and uracil to give corresponding hydroxymethyl derivatives and their reduction with formic acid to give alanine and thymine respectively.

It was observed that uracil on reaction with formaldehyde and formic acid in dilute aqueous solution at 100-140°C gave 5-hydroxymethyl-uracil (5-HMU), methylenebiuracil (MBU) and thymine. Cline et al. (1959) have reported the formation of 5-HMU on reaction of uracil with HCHO. When 5-HMU was synthetically prepared and reduced with formic acid, the same profile of uracil derivatives was obtained as mentioned above. The mechanism for the formation of thymine is believed to be the general hydride transfer mechanism (see Figs. 4 and 5).

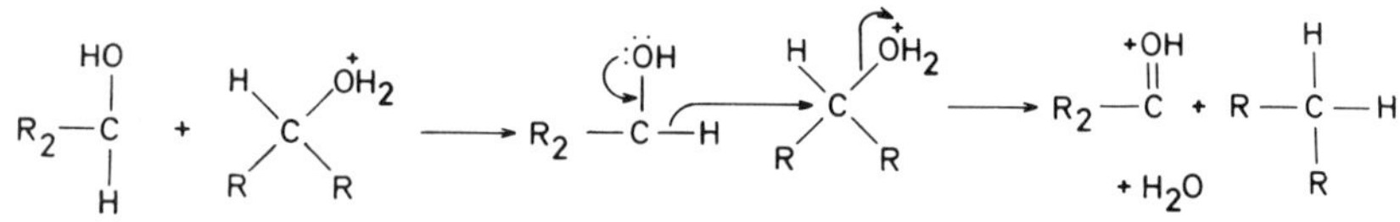

Fig. 4. General hydridge transfer mechanism.

The possible reaction steps for the formation of various products are shown in Fig. 5.

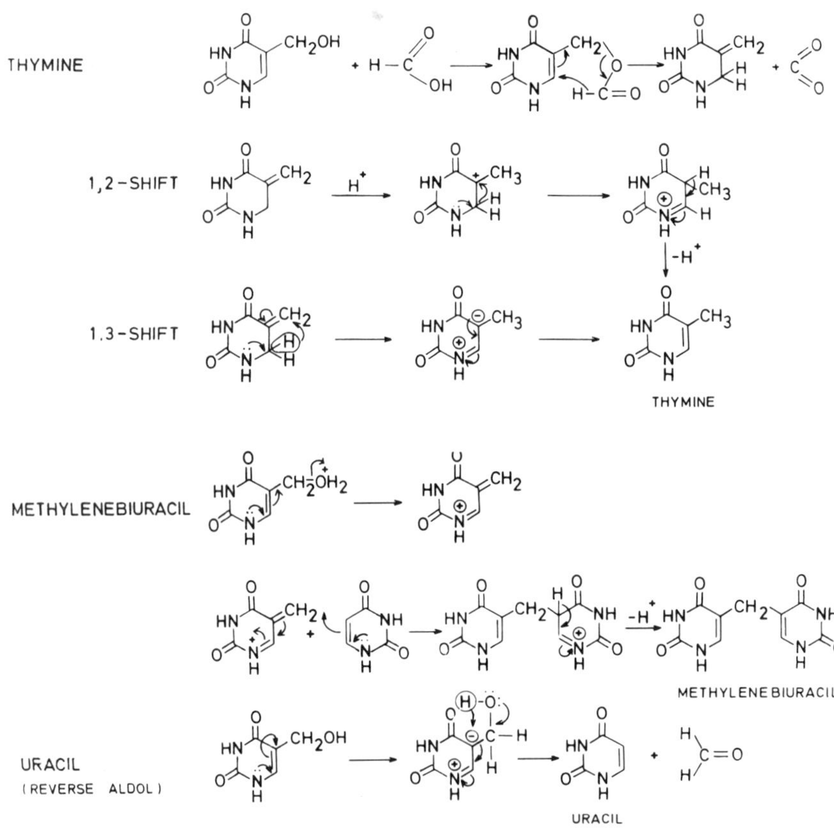

Fig. 5. Possible reaction mechanism for the formation of thymine, methylenebiuracil and uracil from hydroxymethyluracil.

It ought to be stated here that Stephen-Sherwood et al. (1971) also described a synthesis of thymine in 0.1% yield by heating dilute aqueous solution of uracil, formaldehyde, ammonia and hydrazine. The possible intermediate has been suggested as 5-HMU. We tried to compare the products formed by the reduction of 5-HMU with hydrazine with those obtained by the use of formic acid. In our hands, hydrazine gave a very complicated set of products and the yield of thymine was in traces only (unpublished results). The yield of thymine was up to 5% in some of the experiments carried out by us, MBU being the major one (62%) in most cases.

Reaction of Formaldehyde with Pyrrole

Szutka (1964) and Hodgson and Baker (1967) showed that pyrroles and aldehydes react to give corresponding porphyrin derivatives. In almost all the experiments solvents such as pyridine and metal ion catalysts have been used. The energy sources employed were γ-irradiation or heat. The yields reported are of the order of 0.1% and the procedures for separation of the products employed have been elaborate.

We became interested in porphyrins because of an observation in another study that pyrrole formation took place when serine was heated at temperatures of 250-300°C. It was speculated that under primitive Earth conditions pyrrole could have been formed by similar reactions near volcanic regions, and if passed into nearby lagoons containing aldehydes, the two by interaction could lead to the formation of porphyrin. Various conditions were used for such a study, and it was found that pyrroles and aldehydes best reacted in dilute aqueous solutions left at room temperature for a week or more. It was also observed that most of the porphyrins and other related compounds could be extracted with benzene or chloroform, and preparative TLC was the best method for the separation and isolation of the compounds (Choughuley, 1980). The porphyrin obtained by us and some relevant data are shown in Table 1.

Besides the data given in the table, the compounds all gave the characteristic emission spectra in the red region on excitation in the Soret region (Choughuley, 1980; Subbaraman et al., unpublished results).

Formaldehyde in the Oligomerization of HCN

Schwartz and Goverde (1982) have reported that cyanohydrins, which are readily obtained by the reaction of HCHO and other carbonyl compounds with HCN, accelerate the rate of oligomerization both in homogeneous solution as well as in the frozen state.

Table 1. Porphyrin Formation by Pyrrole-Aldehyde Reaction

Aldehyde used	Porphyrin formed	Molecular ion m/e	λ_{max} in Soret region	Yield %
Formaldehyde	Porphyrin	310	395	0.4
Benzaldehyde	Tetraphenyl porphyrin	614	414	0.5
Acetaldehyde	Tetramethyl porphyrin	366	413	0.03

Formamide Reactions

Adenine synthesis by heating ammonium cyanide (Oro, 1960) is well known. Another interesting synthesis is described by Morita et al. (1968). In this case, formamide and $POCl_3$ are heated in a sealed tube at 120°C for 15 hours, and the adenine formed is in 40-50% yield. The authors have excluded any conversion of formamide to HCN which could give adenine on polymerization as a possible route, by adding K ^{14}CN to the reaction mixture and showing that the radioisotope was not incorporated into the adenine molecule to any significant extent. Yamada and Okamoto (1972) have described the formation of purine in 71% yield on heating formamide alone at 160-200° in a sealed tube.

In another experiment Phillip (1977) has shown that spontaneous phosphorylation of nucleosides takes place in formamide ammonium phosphate mixtures.

DISCUSSION AND CONCLUSION

In the present review, rather than describing how amino acids, bases and other compounds could be formed by various methods, an effort has been made to place emphasis on the role of compounds like CO, CO_2, CH_4; HCHO, HCOOH, HCN and its oligomers and a few other simple compounds which may have relevance in primordial organic chemistry, in the formation of amino acids, bases and sugars. It is fairly clear now that in the primitive Earth's atmosphere the intermediate molecules such as formaldehyde, formic acid, hydrogen cyanide and urea could have been formed, and these in turn may have played an important role in the synthesis of larger molecules. Some of these may have also helped in polymerization and condensation reactions.

An attempt has been made to include as many different types of reactions as possible and to cite the necessary references. However, this review on the subject is by no means an exhaustive one. On the other hand, it complements other reviews in the literature.

ACKNOWLEDGMENTS

I thank Dr. M. S. Chadha, Head, Bio-Organic Division, B.A.R.C., Trombay Bombay, India, for his comments and suggestions while preparing this manuscript.

REFERENCES

Abelson, P. H., 1956, Carnegie Inst. Wash. Yearbook, 1955-1956, 55: 171.
Abelson, P. H., 1966, Proc. Nat. Acad. Sci., U.S., 55:1365.
Adams, R., and Langley, W. D., 1956, Organic Synthesis, Coll., Vol. 1, Gilman and Blatt, eds., 2nd edn., pp. 355-357, Wiley, New York.
Anders, E., Hayatsu, R., and Studier, M. H., 1974, Origins Life, 5:57.
Bahadur, K., 1954, Nature, 173:1141.
Bar-Nun, A., Bar-Nun, N., Bauer, S. H., and Sagan, C., 1970, Science, 168:470.
Breslow, R., 1959, Tetrahedron Lett., 21:22.
Butlerow, A., 1861, Ann., 120:296.
Chadha, M., Lawless, J., Flores, J., and Ponnamperuma, C., 1971, in: "Molecular Evolution," Vol. 1, C. Ponnamperuma and R. Buvet, eds., pp. 143-151, North-Holland, Amsterdam.
Chadha, M. S., Molton, P. M., and Ponnamperuma, C., 1975, Origins Life, 6:127.
Chittenden, G. J. F., and Schwartz, A. W., 1976, Nature, 263:350.
Choughuley, A. S. U., and Lemmon, R. M., 1965, Nature, 210:628.
Choughuley, A. S. U., Subbaraman, A. S., and Kazi, Z. A., 1972, Indian J. Biochem. Biophys., 9:144.
Choughuley, A. S. U., Subbaraman, A. S., and Kazi, Z. A., 1975, Origins Life, 6:527.
Choughuley, A. S. U., Subbaraman, A. S., Kazi, Z. A., and Chadha, M. S., 1977, BioSystems, 9:73.
Choughuley, A. S. U., 1980, in: "Origin and Evolution of Life and Intelligence in the Universe," M. S. Chadha, S. M. Chitre, and R. R. Daniel, eds., pp. 85-90, Indian Space Research Organization, Banglore.
Cline, R. E., Fink, R. M., and Fink, K., 1959, J. Amer. Chem. Soc., 81:2521.
Dodonova, N. Y., and Sidorava, A. I., 1961, Biophysics, 6:14.
Dose, K., and Rajewsky, B., 1957, Biochem. Biophys. Acta, 25:225.
Draganic, I., Draganic, Z., Shimoyama, A., and Ponnamperuma, C., 1978, Origins Life, 8:371.

Draganic, I., Draganic, Z., Shimoyama, A., and Ponnamperuma, C., 1978, in: "Origin of Life," H. Noda, ed., pp. 129-133, Japan Sci. Soc. Press, Tokyo.
Egami, F., 1981, Origins Life, 11:198.
Euler, M., and Euler, A., 1906, Ber., 39:50.
Ferris, J. P., Donner, D. B., and Lobo, A. P., 1973, J. Mol. Biol., 74:511.
Ferris, J. P., and Edelson, E. H., 1978, J. Org. Chem., 43:3989.
Ferris, J. P., Edelson, E. H., Auying, J. M., and Joshi, P. C., 1981, J. Mol. Evol., 17:69.
Ferris, J. P., and Joshi, P. C., 1978, Science, 201:361.
Ferris, J. P., Joshi, P. C., Edelson, E. H., and Lawless, J. G., 1978, J. Mol. Evol., 11:293.
Ferris, J. P., Joshi, P. C., and Lawless, J. G., 1977, BioSystems, 9:81.
Ferris, J. P., Kuder, J. E., and Catalono, A. W., 1969, Science, 166:765.
Ferris, J. P., and Orgel, L. E., 1966, J. Amer. Chem. Soc., 88:3829.
Ferris, J. P., Sanchez, R. A., and Orgel, L. E., 1968, J. Mol. Biol., 33:693.
Ferris, J. P., Wos, J. D., Nooner, D. W., and Oro, J., 1974, J. Mol. Evol., 3:225.
Ferris, J. P., Zamek, O. S., Altbutch, A. M., and Freiman, H., 1974, J. Mol. Evol., 3:301.
Fox, S. W., 1963, in: "Organic Geochemistry," I. A. Berger, ed., pp. 36-48, Pergamon Press, London.
Fox, S. W., ed., 1965, "The Origins of Prebiological Systems and of Their Molecular Matrices," Academic Press, New York.
Fox, S. W., and Harada, K., 1961, Arch. Biochem. Biophys., 86:3745.
Fox, S. W., and Windsor, C. R., 1970, Science, 166:984.
Gabel, N. W., and Ponnamperuma, C., 1967, Nature, 216:453.
Garrison, W. H., Morrison, D. C., Hamilton, J. G., Benson, A., and Calvin, M., 1951, Science, 114:416.
Grossenbacher, K. A., and Knight, C. A., 1965, in: "The Origins of Prebiological Systems and of Their Molecular Matrices," S. W. Fox, ed., pp. 173-186, Academic Press, New York.
Groth, W., and Von Wessenchoff, H., 1957, Naturwiss., 44:510.
Haldane, J. B. S., 1928, Rationalist Annual, 148:3.
Harada, K., 1967, Nature, 214:479.
Harada, K., and Fox, S. W., 1964, Nature, 210:335.
Harada, K., and Fox, S. W., 1965, in: "The Origins of Prebiological Systems and of Their Molecular Matrices," S. W. Fox, ed., pp. 187-201, Academic Press, New York.
Harada, K., and Suzuki, S., 1977, Naturwiss., 64:484.
Hayatsu, R., Studier, M. H., Oda, A., Fuse, K., and Anders, E., 1968, Geochem. Cosmochim. Acta, 32:175.
Hayatsu, R., Studier, M. H., and Anders, E., 1971, Geochim. Cosmochim. Acta, 35:939.
Heyns, K., Walter, W., and Meyer, E., 1957, Naturwiss., 44:385.
Hodgson, G. W., and Baker, B. L., 1967, Nature, 216:29.

Hodgson, G. W., and Ponnamperuma, C., 1968, Proc. Nat. Acad. Sci. U.S., 59:22.
Holland, H. D., 1962, in: "Petrologic Studies, A Volume to honour A. F. Buddington," A. E. J. Engel, H. L. James, and B. F. Leonard, eds., pp. 447-477, Geol. Soc. Amer., Boulder, Colorado.
Holland, H. D., 1978, "The Chemistry of the Atmosphere and Oceans," p. 351, Wiley - Interscience, New York.
Ibanez, J. D., Kimball, A. P., and Oro, J., 1971, J. Mol. Evol., 1:112.
Khare, B. M., and Sagan, C., 1971, Nature, 232:577.
Labadie, M., Jensen, R., and Neuzil, E., 1968, Biochim. Biophys, Acta, 165:525.
Lemmon, R. M., 1970, Chem. Rev., 70:95.
Loew, O., 1986, U. Prakt. Chem., 34:51.
Loew, O., 1887, Ber., 20:144.
Loew, O., 1889, Ber., 22:475.
Lohrmann, R., 1972, J. Mol. Evol., 1:263.
Lohrmann, R., and Orgel, L. E., 1971, Science, 171:490.
Lowe, C. U., Rees, M. W., and Markham, R., 1963, Nature, 199:219.
Matthews, C. N., and Moser, R. E., 1967, Nature, 215:1230.
Miller, S. L., 1953, Science, 117:528.
Miller, S. L., 1955, J. Amer. Chem. Soc., 77:2351.
Morita, K., Ochiai, M., and Marumoto, R., 1968, Chem. Ind., 1117.
Moser, R. E., Clagget, A. R., and Matthews, C. N., 1968a, Tetrahedron Lett., 13:1605.
Moser, R. E., Clagget, A. R., and Matthews, C. N., 1968b, Tetrahedron Lett., 13:1599.
Moser, R. E., and Matthews, C. N., 1968, Experientia, 24:658.
Oparin, A. I., 1924, Proiskhozhdenie Zhizni (The Origin of Life) Izd. Moskovskiy Rabochiy, Moscow.
Oro, J., 1960, Biochem. Biophys. Res. Commun., 2:407.
Oro, J., 1963, Ann. N.Y. Acad. Sci., 108:464.
Oro, J., 1965, in: "The Origins of Prebiologival Systems and of Their Molecular Matrices," S. W. Fox, ed., pp. 137-171, Academic Press, New York.
Oro, J., 1965, in: "Current Aspects of Exobiology," G. Mamikunian, and M. H. Briggs, eds., pp. 13-76, Permagon, New York.
Oro, J., 1972, Space Life Sci., 3:507.
Oro, J., and Cox, A. C., 1962, Federation Proc., 21:80.
Oro, J., and Kamat, S. S., 1961, Nature, 190:442.
Oro, J., Kimball, A., Fretz, R., and Master, F., 1959, Arch. Biochem. Biophys., 85:115.
Oro, J., and Kimball, A. P., 1961, Arch. Biochem. Biophys., 94:217.
Palm, C., and Calvin, M., 1962, J. Amer. Chem. Soc., 84:2115.
Pavlovskaia, T. E., and Pasynskii, A. G., 1959, in: "The Origin of Life on Earth," A. I. Oparin, ed., pp. 151-157, Pergamon Press, Oxford.
Philipp, M., and Seliger, H., 1977, Naturwiss., 64:273.

Ponnamperuma, C., 1965, in: "The Origins of Prebiological Systems and of Their Molecular Matrices," S. W. Fox, ed., pp. 221-241, Academic Press, New York.
Ponnamperuma, C., and Flores, J., 1966, Abs. No. C-33, Amer. Chem. Soc. 152nd Meeting, Sept. 11-16.
Ponnamperuma, C., and Gabel, N. W., 1968, Space Life Sciences, 1:64.
Ponnamperuma, C., Lemmon, R. M., Mariner, R., and Calvin, M., 1963, Proc. Nat. Acad. Sci. U.S., 49:737.
Ponnamperuma, C., and Peterson, E., 1965, Science, 147:1572.
Ponnamperuma, C., and Shimoyama, A., 1982, Origins Life, 12:9.
Ponnamperuma, C., and Woeller, F., 1967, Curr. Mod. Biol., 1:156.
Reid, C., 1959, in: "The Origin of Life on the Earth," A. I. Oparin, ed., pp. 619-625, Pergamon Press, New York.
Ring, D., Wolman, V., Friedmann, N., and Miller, S. L., 1972, Proc. Nat. Acad. Sci. U.S., 69:765.
Rubey, W. W., 1951, Geol. Soc. Amer. Bull., 62:1111.
Rubey, W. W., 1955, in: "Crust of the Earth," A. Polelervart, ed., pp. 631-650, Geol. Soc. Amer., New York.
Sagan, C., and Khare, B. M., 1971, Science, 173:417.
Sanchez, R. A., Ferris, J. P., and Orgel, L. E., 1966, Science, 153:72.
Sanchez, R. A., Ferris, J. P., and Orgel, L. E., 1966, Science, 154:784.
Sanchez, R. A., Ferris, J. P., and Orgel, L. E., 1967, J. Mol. Biol., 30:223.
Sanchez, R. A., Ferris, J. P., and Orgel, L. E., 1968, J. Mol. Biol., 38:121.
Sandoval, A. A., and Young, V., 1973, BioSystems, 5:161.
Schimpl, A., Lemmon, R. M., and Calvin, M., 1965, Science, 147:149.
Schwartz, A. W., 1981, in: "Marine Organic Chemistry," E. K. Duurama, and R. Dawson, eds., pp. 7-30, Elsevier, Amsterdam.
Schwartz, A. W., and Chittenden, G. J. F., 1977, BioSystems, 9:87.
Schwartz, A. W., and Goverde, M., 1982, J. Mol. Evol., 18:351.
Shimoyama, A., Blair, N., and Ponnamperuma, C., 1978, in: "Origin of Life," H. Noda, ed., pp. 95-99, Japan Sci. Soc. Press, Tokyo.
Simionescu, C. I., Simionescu, B. C., Mora, R., and Leanca, M., 1978, Origins Life, 9:103.
Snyder, L. E., Buhl, B., Zuckerman, B., and Palmer, P., 1969, Phys. Rev. Lett., 22:679.
Steinman, G. D., Kenyon, D. H., and Calvin, M., 1965, Nature, 206:707.
Steinman, G. D., Kenyon, D. H., and Calvin, M., 1966, Biochem. Biophys. Acta, 124:339.
Steinman, G. D., Lemmon, R. M., and Calvin, M., 1964, Proc. Nat. Acad. Sci., U.S., 52:27.
Steinman, G. D., Lemmon, R. M., and Calvin, M., 1965, Science, 147:1574.
Stephen-Sherwood, E., and Oró, J., 1973, Space Life Sciences, 4:5.

Stephen-Sherwood, E., Odom, D. G., and Oró, J., 1974, in: "Cosmochemical Evolition and the Origin of Life (1974)," J. Oró, S. L. Miller, C. Ponnamperuma, and R. S. Young, eds., Vol. II, pp. 301-308, Reidel, Dordrecht, Holland and Boston.

Stephen-Sherwood, E., Oro, J., and Kimball, A. P., 1971, Science, 173:446.

Storch, H. H., Golumbic, M., and Anderson, R. B., 1951, "The Fischer-Tropsch and Related Synthesis," Wiley, New York.

Studier, M. H., Hayatsu, R., and Anders, E., 1968, Geochem. Cosmochim. Acta, 32:151.

Subbaraman, A. S., Kazi, Z. A., and Choughuley, A. S. U., 1972, Indian J. Biochem. Biophys., 9:268.

Subbaraman, A. S., Kazi, Z. A., and Choughuley, A. S. U., 1979, Indian J. Biochem. Biophys., 16:253.

Subbaraman, A. S., Kazi, Z. A., Choughuley, A. S. U., and Chadha, M. S., 1980, Origins Life, 10:343.

Szutka, A., 1964, Nature, 202:1231.

Terenin, A. M., 1959, in: "The Origin of Life on the Earth," A. I. Oparin, ed., pp. 136-139, Pergamon Press, Oxford.

Urey, N. C., 1952, "The Planets, Their Origin & Development," Yale Univ. Press, New Haven.

Voet, A. B., and Schwartz, A. W., 1982, Origins Life, 12:45.

Wanzlick, H. W., 1962, Angew. Chem. Int. Ed., 1:75.

Woeller, F., and Ponnamperuma, C., 1969, Icarus, 10:386.

Wolman, Y., Haverland, W. J., and Miller, S. L., 1972, Proc. Nat. Acad. Sci. U.S., 69:809.

Yamada, H., and Okomato, T., 1972, Chem. Pharm. Bull., 20:623.

Yamagata, Y., and Mohri, T., 1982, Origins Life, 12:41.

Yang, C. C., and Oró, J., 1971, in: "Chemical Evolution and the Origin of Life," pp. 152-167, North-Holland, Amsterdam.

Zuckerman, B., Buhl, J. A., and Gottlieb, C. A., 1971, Astrophys. J., 163: L 41.

FORMATION OF BIOORGANIC COMPOUNDS IN AQUEOUS SOLUTION INDUCED BY CONTACT GLOW DISCHARGE ELECTROLYSIS

Kaoru Harada

Institute of Chemistry
University of Tsukuba
Sakura-mura, Ibaraki 305, Japan

INTRODUCTION

There are several types of electric discharge, which have been studied mainly with the reactions in gaseous phase and under relatively low pressure. Such discharges are certainly an interesting problem relevant to the field of chemical evolution. Past work has shown that electric discharges carried out under reducing atmospheres yielded several amino acids, both those common to protein and others (Miller, 1953, 1955; Fox and Dose, 1977). Electric discharge has been used in chemical evolution because it is considered to be a model of lightning in the atmosphere on the primitive Earth. Among other energy sources (ultraviolet rays, γ-rays, charged ionizing radiations, heat and shock waves, etc.) available on the primitive Earth, electric discharge is chemically effective and can easily be reproduced in the laboratory for model experiments on the prebiotic formation of bioorganic compounds.

However, most of the electric discharge studies, including the prebiotic formation of bioorganic compounds, have been carried out in the gaseous phase. Few studies on high voltage electric discharge over aqueous solutions using a Tesla coil have been reported (Steinman and Lillevik, 1964; Steinman, 1966; Ebisawa et al., 1979). In the present study, a type of electric discharge called "contact glow discharge electrolysis" (CGDE), which differs from high voltage AC current electric discharge, was employed (Hickling and Ingram, 1964a; Hickling, 1971).

In this review, the chemistry of organic compounds in aqueous solution by CGDE are described. We would like to clarify the reaction mechanism and find applications for this new type of chemical reaction, and also search for chemical evolutionary pathways for the formation of biologically important compounds. From these studies, we have found several new chemical reactions which cannot be carried out by more conventional organic reactions. We would like to relate here some typical examples of the CGDE reactions obtained in our laboratory after a brief introduction on some historical background.

HISTORICAL BACKGROUND

It was early shown that electric discharge against aqueous solutions containing metal ions resulted in the oxidation of the ions (Denaro and Hickling, 1958). Hickling confirmed several products from the electric discharge against liquid ammonia (Hickling and Newns, 1959; Hickling and Newns, 1961), aqueous ammonia (Mazzocchin et al., 1973) and diluted sulfuric acid (Hickling and Ingram, 1964b). Hydrazine was produced from liquid ammonia; hydrazine, hydroxylamine, hydrogen peroxide from aqueous ammonia; and hydrogen peroxide from dilute sulfuric acid. One of the interesting results in these reactions, is that the amount of the products obtained by electric discharge is larger than the amount obtained by electrolysis according to Faraday's law. For example, in electric discharge against liquid ammonia, 2.5 equivalents of hydrazine/F was produced in the beginning of the reaction. The formation of hydrogen peroxide by electric discharge in dilute sulfuric acid was 1.8-2.5 equivalents/F. When Fe^{++} in a dilute sulfuric acid solution was treated with electric discharge, the formation of Fe^{+++} was 7-8 equivalents/F. These facts indicate that the reactions induced by electric discharge differ from the electrolysis reaction. It has been suggested that electrolysis is a charge transfer type and CGDE is an energy transfer type reaction (Hickling, 1971).

Several studies on the effect of electric discharge on inorganic compounds have been reported, however, few studies on the effect of electric discharge or organic compounds in aqueous solution have been reported. Hickling described the formation of oxalic acid from formic acid, and the formation of succinic acid from acetic acid in his review (Hickling, 1971). The formation of oxamide from formamide was reported (Brown et al., 1962). In the latter reaction, oxamide was produced in an electric efficiency of 2.9-3.1 equivalents/F.

In CGDE, the ions (H_2O^+, Ar^{n+}, etc.) produced by electric discharge are accelerated in an electric field and the ions gain energy (∿100 eV). The charged particles strike the aqueous solution, and dissociate plural numbers of water molecules, and produce several hydroxyl and hydrogen radicals (primary reaction). The radicals react with the substrate molecules and abstract a hydrogen atom, and the resulting substrate radicals react with other molecules (radicals) to form various reaction products (secondary reaction) (Hickling and Ingram, 1964). In the CGDE reaction, hydroxyl radicals which are produced by the dissociation of water, could play an important role in the abstraction of hydrogen, hydroxylation and further oxidation of the substrates in the aqueous solution.

APPARATUS, ANALYTICAL METHODS

A typical apparatus for CGDE is shown in Fig. 1 (Harada, 1974). The reaction tube holds 20-30 ml of an aqueous solution containing 1-3 m mol of the substrate. The negative electrode is immersed in the aqueous solution, and the positive electrode is kept in contact with the surface of the solution. During CGDE, argon gas was introduced in the reaction mixture and the solution was stirred by a magnetic stirrer and the reaction temperature was measured by an immersed thermometer. The reaction temperature was controlled at about 10-20°C by using a dry ice-methanol bath. The applied voltage was 400-600 V resulting in 30-60 mA. CGDE took place continuously and smoothly on the surface of the aqueous solution.

The reaction products were withdrawn at appropriate times, and were analyzed by an amino acid analyzer (Yanagimoto model LC-55),

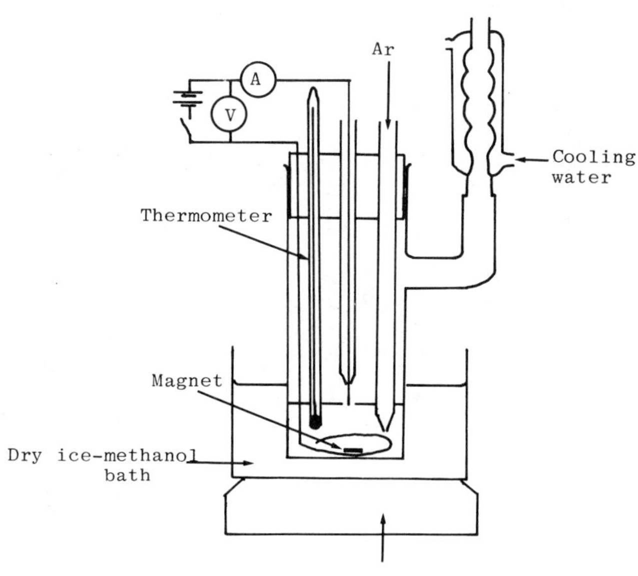

Fig. 1. Apparatus for CGDE

liquid chromatography (JASCO Familic-100), gas chromatography (Hitachi model 163), isotacho electrophoresis (Shimazu IP-2A), and thin layer chromatography. Some of the isolated samples were identified by IR, UV and NMR spectroscopy. In some cases, in order to identify the products, the samples were derivatized to compare with authentic samples which were prepared separately.

CHEMICAL REACTIONS INDUCED BY CGDE

Dehydrogenation

The primary product obtained by the CGDE against the aqueous solution are hydroxyl and hydrogen radicals. These hydroxyl radicals can abstract hydrogen from other compounds. Therefore, by the use of appropriate chemicals in the aqueous solution, a dehydrogenation reaction can take place by CGDE. For example, when aqueous solutions of dihydrouracil (DHU), dihydrothymine (DHT) and dihydro-orotic acid (DHOA) were treated with CGDE, pyrimidines, which are the constituents of nucleic acid, were obtained in the aqueous solution (Harada et al., 1978). The time course of the dehydrogenation reactions by CGDE is shown in Fig. 2. Uracil, thymine and orotic acid were separated by chromatography and were confirmed by UV and IR spectra. The yields of the pyrimidines reach maximum values after 30 min and the yields are 8-15% under the conditions used.

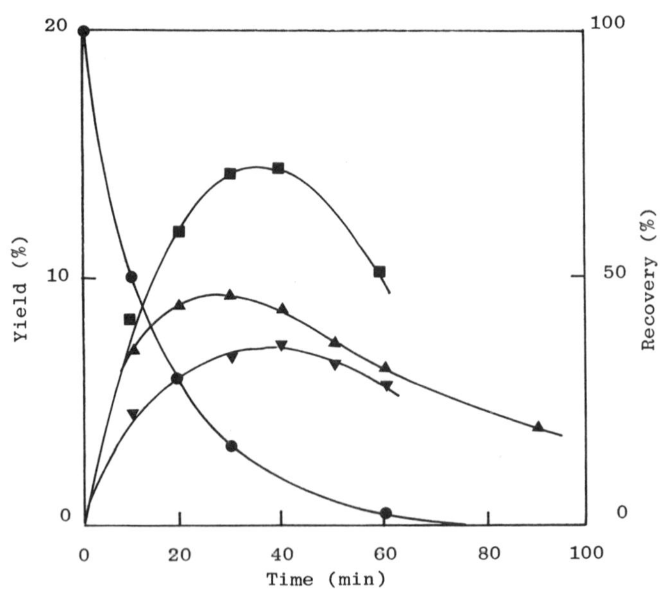

Fig. 2. Formation of pyrimidines by CGDE. Conditions: Reactant 5×10^{-4} mol/20ml H_2O, 1200-600V, 45mA, 20-25°C. ●: Dihydrouracil, ■: Thymine, ▲: Uracil, ▼: Orotic acid.

Oxidative Conversion of β- and γ-Amino Acids

When an aqueous solution of β-alanine was treated with CGDE, a considerable amount of glycine (10-25%) was formed. In addition, when CGDE was applied to an aqueous solution of γ-aminobutyric acid, the formation of β-alanine (3-9%) and glycine (7-17%) were observed (Suzuki et al., 1978). Apparently, in these reactions, a methylene group was eliminated from β-alanine and γ-aminobutyric acid.

$$H_2N\text{-}CH_2\text{-}CH_2\text{-}CH_2\text{-}COOH \longrightarrow$$

$$H_2N\text{-}CH_2\text{-}CH_2\text{-}COOH \longrightarrow$$

$$H_2N\text{-}CH_2\text{-}COOH$$

If such a methylene group elimination reaction were effective with glutamic acid, the formation of aspartic acid would be produced. Glutamic acid was actually converted into aspartic acid (8%) by CGDE. Therefore, the methylene group elimination reaction takes place generally through CGDE in aqueous solutions.

A detailed study of CGDE was carried out on aqueous solutions of β-alanine. The products analyzed by an amino acid analyzer are shown in Fig. 3. The small peak (B) after glycine (A) was identified as isoserine, and the time course of the CGDE of β-alanine is shown in Fig. 4.

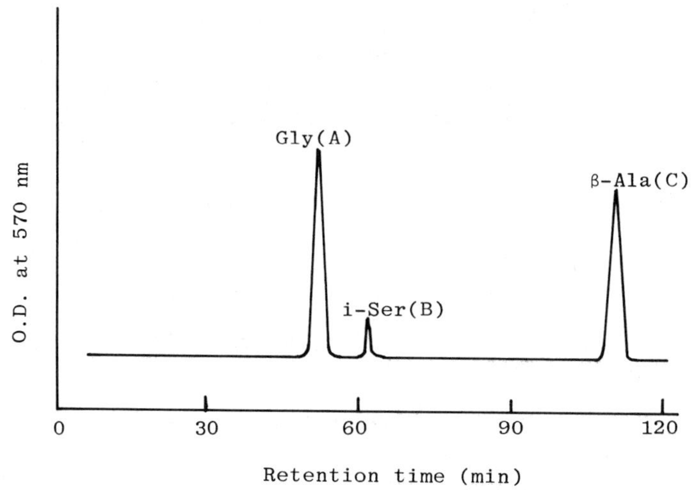

Fig. 3. Oxidative modification of β-alanine by CGDE. Conditions: β-Ala 0.001mol/20ml H_2O, 1500-600V, 50mA, 20°C, 40 min. A: Gly(13.4%), B: i-Ser(3.5%), C: β-Ala(58.9%).

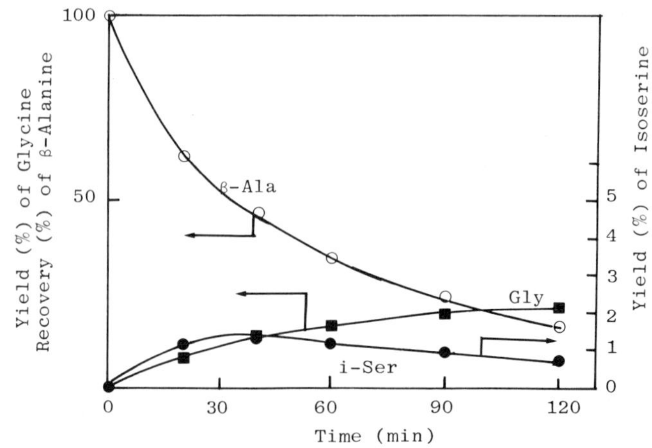

Fig. 4. Time course of oxidative modification of β-alanine by CGDE. ○: β-Ala, ●: i-Ser, ■: Gly.

These results suggest that β-alanine (C) was oxidized first to isoserine (B), and that isoserine (B) is unstable under the reaction conditions and was further oxidized to glycine (A). Therefore, the possible mechanism for oxidative conversion of β-alanine by CGDE is: β-alanine → isoserine → glycine (Harada and Terasawa, 1980a).

In order to confirm the oxidation mechanism of β-alanine by CGDE, an aqueous solution of isoserine was treated with CGDE. The time course of the oxidation is shown in Fig. 5. After 1 hr, about 80% of isoserine was consumed, and the yield of glycine reached about 50%. This result indicates that isoserine is easily converted into glycine under the conditions used in the electric discharge.

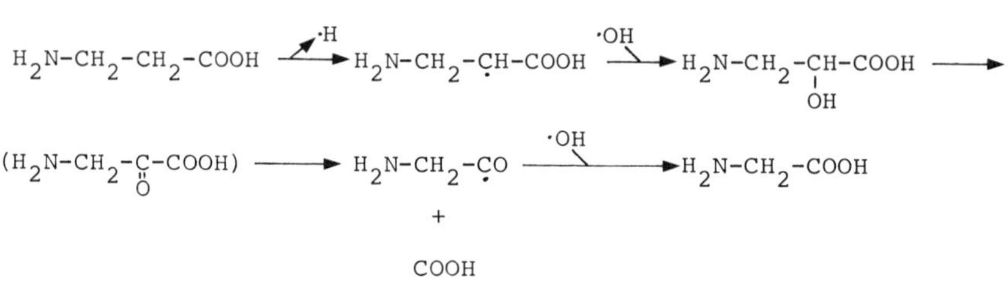

Therefore, the oxidative processes by CGDE is possibly as shown in the above scheme.

A detailed CGDE study of γ-aminobutyric acid was carried out. The amino acid analysis of the oxidative conversion of γ-aminobutyric acid is shown in Fig. 6. The amino acids found in the reaction mixture were γ-amino-α-hydroxybutyric acid, γ-amino-β-hydroxybutyric acid, β-alanine, isoserine and glycine in addition to γ-aminobutyric acid. These results suggest that the conversion

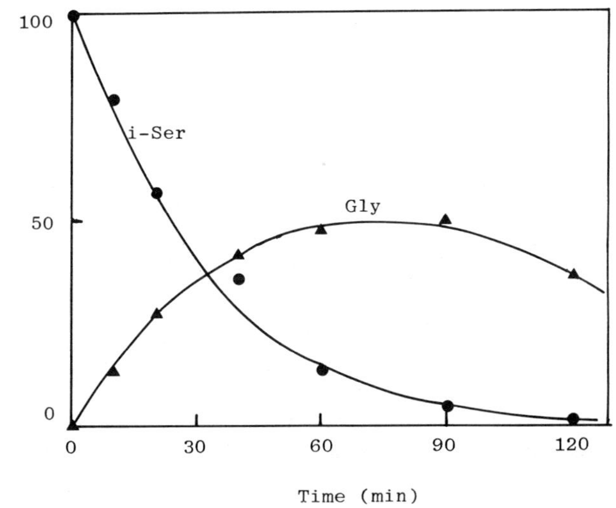

Fig. 5. Time course of oxidative modification of isoserine by CGDE. Conditions: i-Ser 0.001mol/20ml H_2O, 500-600V, 50-60mA, 10-20°C. ●: i-Ser, ▲: Gly.

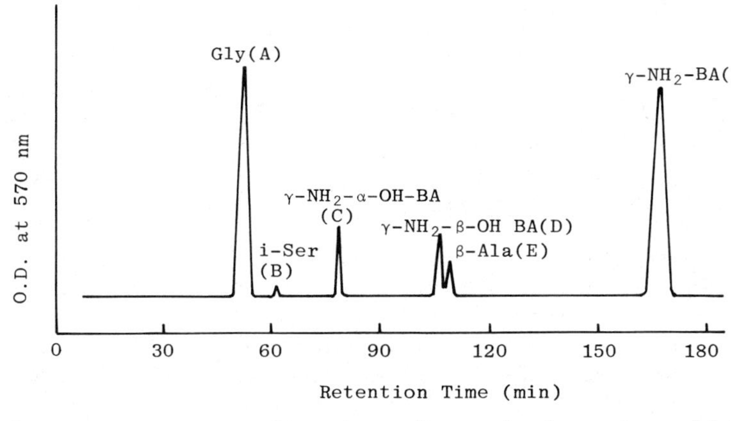

Fig. 6. Oxidative modification of γ-aminobutyric acid by CGDE. Conditions: γ-NH_2-BA 0.001mol/20ml H_2O, 600-1500V, 50mA, 20°C, 40min. A: Gly(8.5%), B: i-ser(1.5%), C: γ-NH_2-α-OH-BA(4.4%), D: γ-NH_2-β-OH-BA(3.6%), E: β-Ala(6.8%), F: γ-NH_2-BA(38%).

of γ-aminobutyric acid by CGDE proceeds as shown in:

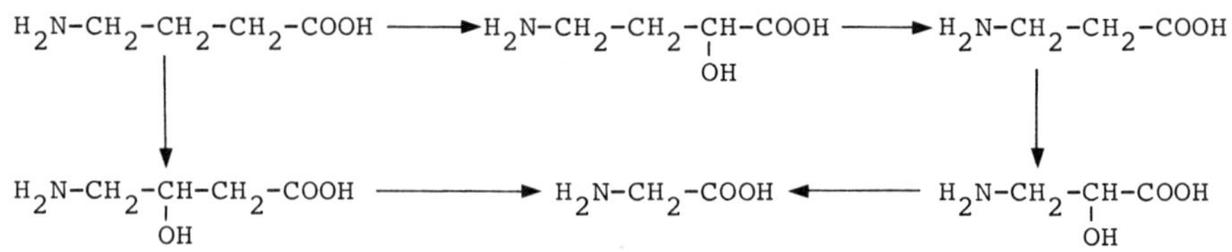

(Harada and Terasawa, 1980a; Harada et al., 1981).

These studies indicate that the apparent methylene group elimination reactions by CGDE can be explained by hydrogen abstraction of the substrate and a subsequent hydroxylation reaction. The hydrogen abstraction from the substrate took place on the carbon atoms situated far from the positively charged nitrogen atom. The oxidative conversion of β- and γ-amino acids shows that CGDE is a radical type reaction, and that the reaction proceeds in a relatively controlled way.

Several hydroxy amino acids were treated with CGDE and the time courses of the reactions were studied in order to confirm the reaction pathways. In all reactions the hydrogen abstraction took place on the carbon atom situated far from the charged nitrogen atom. The resulting substrate radicals react with hydroxyl radicals and the reaction proceeds repeatedly to form lower molecular weight oxygen containing compounds by cleavage of the carbon-carbon bond. And the substrates were oxidized to carboxylic acid and finally to carbon dioxide. The oxidation of γ-amino-α-hydroxybutyric acid, γ-amino-β-hydroxybutyric acid, β-amino-α-hydroxybutyric acid, serine, threonine and homoserine by CGDE were studied (Harada and Terasawa, 1980b; Harada et al., 1981).

Oxidative Conversion of Amines

The methylene group of the substrate was oxidized by CGDE as described above. Therefore, it should be possible to convert alkyl groups to oxygen containing groups by CGDE, whereas the alkyl groups are the most stable groups in organic compound. When CGDE was applied to an aqueous solution of ethylamine or propylamine, the alkyl groups were oxidized to form amino alcohols and amino acids (Harada et al., 1981). Fig. 7 shows the time course of the oxidation of ethylamine in acidic (pH 2.8) and basic (pH 12.1) conditions. The yield of glycine reaches about 10% after 4 hr in acidic conditions; however, the yield of glycine is only 1% in basic conditions. These results indicate that the amino group was protected from the radicals by forming a positively charged ammonium ion, and the hydrogen abstraction took place distal from the charged nitrogen atom. On the contrary, in the basic conditions, the amino group

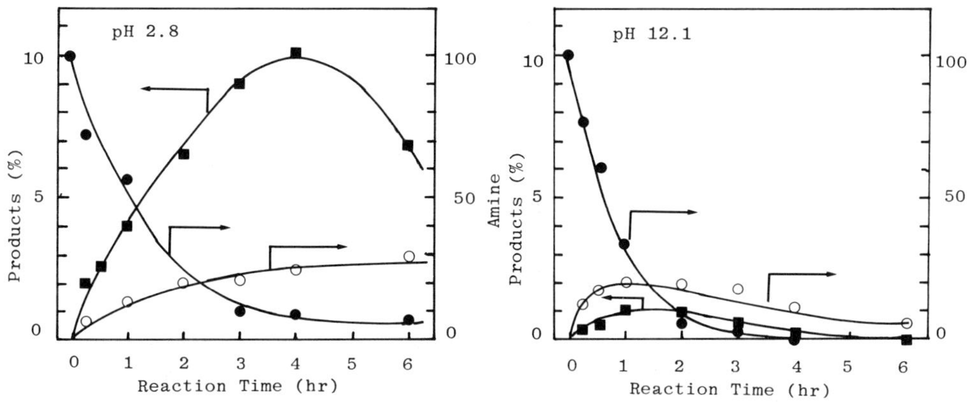

Fig. 7. Time course of oxidative modification of ethylamine by CGDE. Conditions: a, pH2.8; b, pH12.1, $EtNH_2$ 0.001mol/20 ml H_2O, 500-600V, 30mA, 10-20°C. ●: $EtNH_2$, ○: 1-NH_2-2-EtOH (including NH_3), ■: Gly.

was not charged, and was attacked by the radicals. The amino group was destroyed easily and the compound was converted to the lower molecular weight carboxylic acids. The confirmation of the intermediate amino alcohols support the following oxidation pathway as shown in:

$$\overset{+}{H_3N}\text{-}CH_2\text{-}CH_3 \xrightarrow{-H\cdot} \xrightarrow{+\cdot OH}$$

$$\overset{+}{H_3N}\text{-}CH_2CH_2\text{-}OH \xrightarrow{\cdot OH} \overset{+}{H_3N}\text{-}CH_2\text{-}COOH$$

These findings demonstrate that unusual oxidation reactions of alkyl groups, which are very difficult by conventional organic chemistry, are readily accomplished by CGDE.

Oxidative Conversion of Aminoalcohols

Hydroxy amino acids are formed as intermediates in the oxidation of β- and γ-amino acids (Harada and Nomoto, 1982). Amino alcohols were also found as intermediates in the oxidation of aliphatic amines. Therefore, amino alcohols should be converted to the corresponding amino acids through oxidation using CGDE. 2-Aminoethanol gave glycine in 43% yield in an acidic condition, however, under basic conditions the yield of glycine is only 1% (Fig. 8).

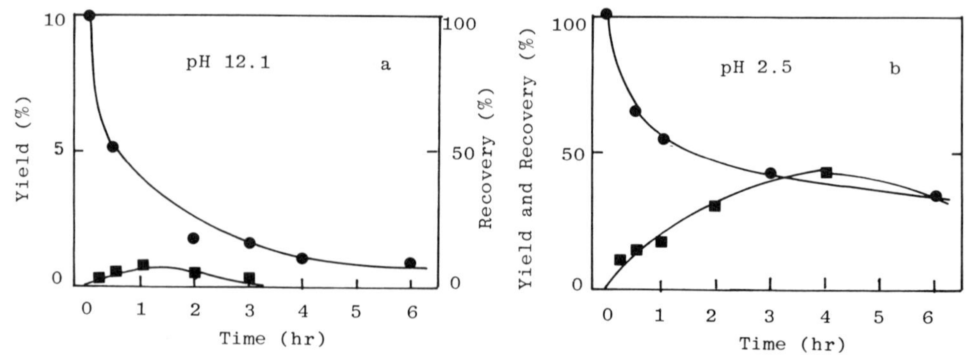

Fig. 8. Time course of oxidative modification of 1-amino-2-ethanol by CGDE. Conditions: a, pH12.1; b, pH2.5, 1-NH_2-2-EtOH 0.001 mol/20 ml H_2O, 500-600V, 30mA, 20°C. : 1-NH_2-2-ethanol (including NH_3), ■: Gly.

Direct Amination

The formation of hydrazine and hydroxylamine from liquid ammonia and aqueous ammonia by CGDE (Hickling and Newns, 1959), suggests the formation of amino radicals (•NH_2) in the reaction mixture. When aliphatic carboxylic acids in a concentrated ammoniacal solution were treated with CGDE, the carboxylic acids were directly aminated to form amino acids (Harada and Iwasaki, 1974; Harada et al., 1978). Aliphatic carboxylic acids such as acetic acid, propionic acid, succinic acid, glutamic acid, succinic acid anhydride and succinimide were aminated in an ammonia solution by CGDE and converted to various amino acids.

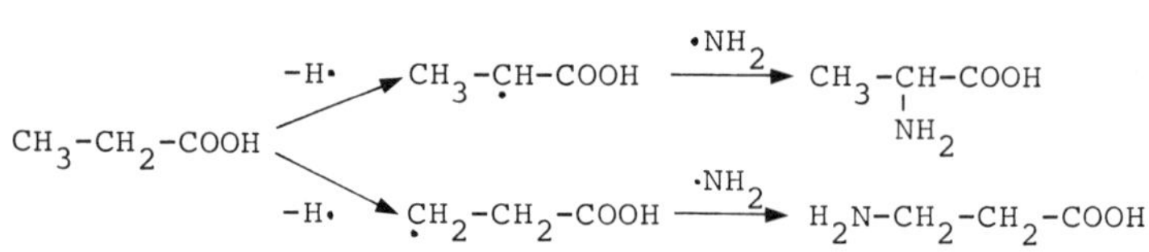

The above scheme and Fig. 9 show an example of the amination of propionic acid by CGDE. α,β-Unsaturated carboxylic acids can also be aminated easily to form amino acids including hydroxy amino acids (Fig. 10).

Direct Carboxylation

It has been shown that the application of CGDE to a formic acid solution, resulted in the formation of oxalic acid (Hickling, 1971). This implies the formation of the formic acid radical

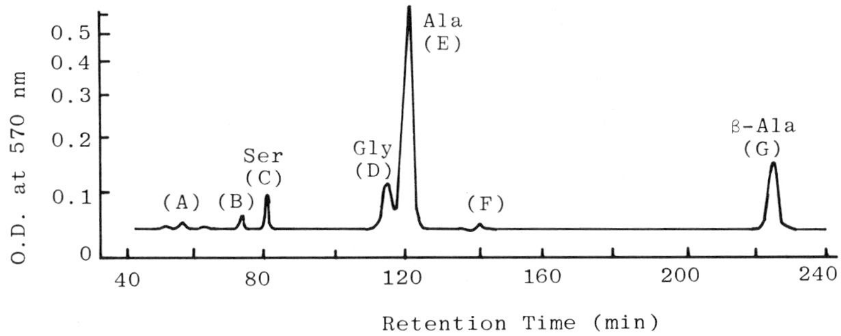

Fig. 9. Direct amination of propionic acid by CGDE. Conditions: Propionic acid 0.005Mol/25 ml H_2O, 450-600V, 75mA, 15°C, 3hr. A,B,F: Unknown, C: Ser, D: Gly, E: Ala(6.9%), G: β-Ala(5.3%).

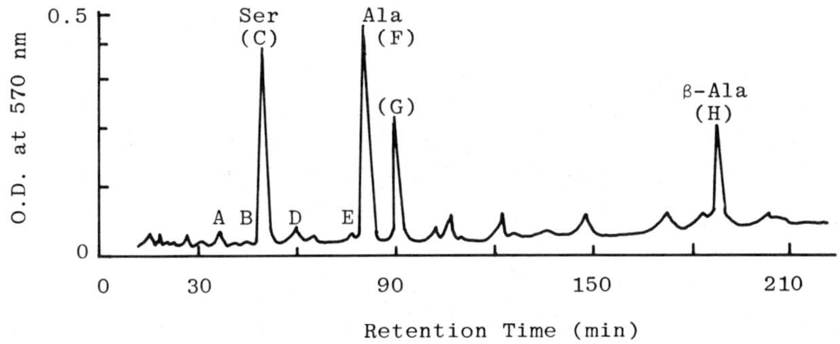

Fig. 10. Formation of amino acids from acrylic acid and ammonia by CGDE. Conditions: 0.005mol acrylic acid/conc NH_3 aq 15 ml, 400-600V, 50-60mA, 10-15°C, 1hr. A: Asp, B: Thr, C: Ser(3.1%), D: Glu, E: Gly, F: Ala(2.6%), G: α-NH_2-BA, H: β-Ala(1.8%).

(•COOH) in the reaction mixture. It was in fact shown that through the application of CGDE to aqueous mixtures of aliphatic amines and formic acid, the amines can be directly carboxylated to form various amino acids (Harada, 1974; Harada and Iwasaki, 1975; Harada et al., 1978). Several compounds containing amino groups such as methylamine, ethylamine, propylamine, pyrrolidine, piperidine, alanine, β-alanine and γ-aminobutyric acid were dissolved in a formic acid or a formamide solution and the mixtures were treated with CGDE.

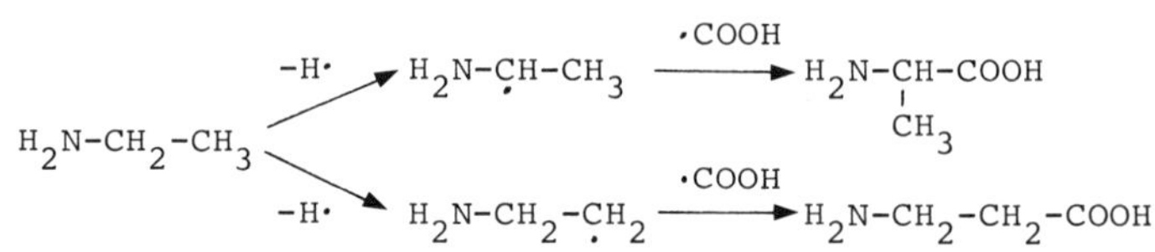

The above scheme and Fig. 11 show the amino acid analysis of the reaction product are alanine (2.6%), β-alanine (4.3%), and glycine (1.3%). The products obtained by carboxylation of propylamine are α-aminobutyric acid (0.9%), α-methyl-β-alanine (9.8%), γ-aminobutyric acid (8.1%) (Terasawa and Harada, 1980). Thus it was found that the direct carboxylation of alkyl groups takes place generally by CGDE. The yield of the carboxylation reaction can be improved by selecting the reaction conditions.

A sodium cyanide solution can be used for cyanization instead of carboxylation. The mixture of ethylamine with sodium cyanide was treated with CGDE, and the product was hydrolyzed and the resulting amino acids were analyzed (Harada et al., 1978). The major products are glycine (6.7%), alanine (3.9%), β-alanine (0.3%), serine (0.3%) and aspartic acid (0.1%). In this case also, the cleavage of α-β linkages of the resulting amino nitriles were observed, and the major product was glycine after hydrolysis.

The carboxylation of amines by using formic acid was carried out in acidic (pH 2.8) and basic (pH 12.5) conditions (Terasawa and Harada, 1980). A clear difference in the orientation of carboxylation was found between the two reactions. In acidic conditions, β- and γ-amino acids are the major products; and in basic conditions α-amino acid was found in the products; however, the yield was low. This orientation in the carboxylation reaction can be explained by a radical mechanism. The amines are protonated almost entirely at pH 2.8. The positively charged nitrogen atom attracts electrons strongly and the electron density of the α-carbon atom is low. The radicals having electrophilic properties would not attack the α-carbon atom, but the β- or γ-position to form substrate radicals by hydrogen abstraction. Therefore, the resulting β- and γ-substrate radicals react easily with formic acid radicals (•COOH) to achieve β- and γ-carboxylation of the aliphatic amine.

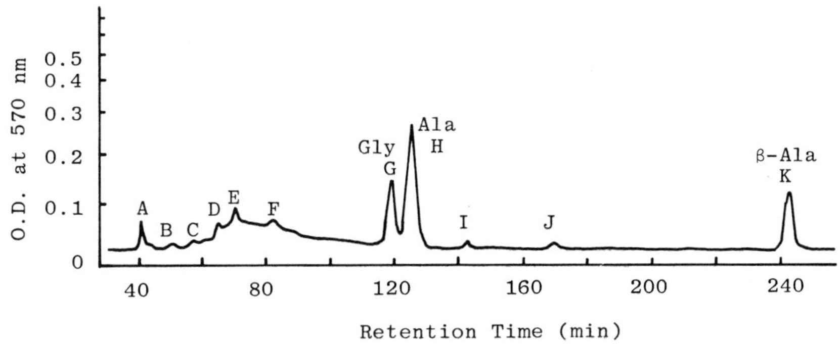

Fig. 11. Direct carboxylation of ethylamine by CGDE. Conditions: 0.0025mol $EtNH_2$ /20% HCOOH 20 ml, 450-500V, 50-60mA, 10°C, 2hr. E: Asp, F: Ser, G: Gly(1.3%), H: Ala(2.6%), K: β-Ala(4.3%).

On the other hand, in basic conditions, the amino nitrogen is not charged and the substrate does not show any clear orientation in carboxylation. All α, β, and γ-radicals were generated and carboxylated to form several amino acids. However, the amino group was not protonated and the free amino group was oxidized easily, and the oxidation conversion of the substrates took place rapidly.

Conversion of Elemental Carbon

Elemental carbon may be vaporized by CGDE, and the vapor (carbon radicals) may be converted into organic compound in an aqueous solution. The reaction apparatus is similar to that shown in Fig. 1. In this reaction, the platinum anode is replaced with a carbon electrode. CGDE took place between the sharpened carbon rod and aqueous ammonia. After the reaction was over, the ammoniacal solution was heated in a sealed tube, the reaction mixture evaporated and the remaining product hydrolyzed with 6N HCl. The amino acid analysis of the product shows clearly the formation of several amino acids (Harada and Suzuki, 1977a) (Fig. 12). The yield of the total amino acids based on the weight loss of the carbon electrode was calculated as 0.2%. The real mechanism of the conversion of elemental carbon to organic compounds in unknown. However, it is possible that the carbon radicals produced by CGDE react with ammonia to form nitrogen containing intermediates (e.g., HCN), and this is oligomerized or hydrolyzed to form amino acids. The resulting glycine was converted to its DNP-derivative and the compound was also characterized chromatographically.

Ammonium Carbonate, Ammonium Formate

An aqueous solution of ammonium carbonate was treated with CGDE, and the reaction mixture was concentrated and hydrolyzed. In both reactions, the formation of glycine was confirmed (Harada and

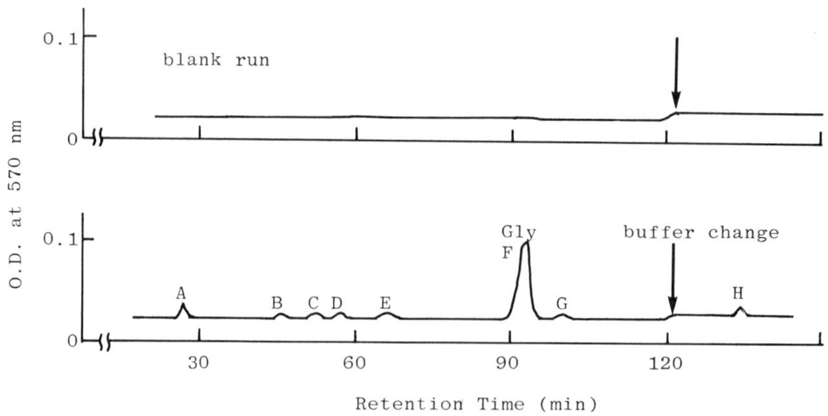

Fig. 12. Formation of amino acids from elemental carbon by CGDE. a: Control experiment, b: CGDE, 600-1200V, 60mA, 35min. A: Urea($6.1x10^{-7}$ mol), B: Asp($1x10^{-9}$ mol), C: Thr($1x10^{-9}$ mol), D: Ser($8x10^{-10}$ mol), E: Glu($5x10^{-10}$ mol), F: Gly ($4.5x10^{-8}$ mol), G: Ala($1x10^{-9}$ mol), H: Unknown, ↓: Buffer change.

Suzuki, 1977b). The yields of the amino acid were clearly above the control experiments

Oxidative Conversion of Alcohol, Aldehyde and Carboxylic Acid

It was found that the CGDE of amino acids yields oxygen containing compounds which do not contain nitrogen atoms. The time course of the oxidative reaction of several amino acids by CGDE were studied. The CGDE of alanine resulted in the formation of glycine, serine, aminomalonic acid together with carboxylic acids such as formic acid, oxalic acid, acetic acid, glycolic acid, pyruvic acid and lactic acid. In this way, the oxidations of oxygen containing compounds by CGDE were studied.

An aqueous solution of ethylalcohol was treated with CGDE. As shown in Fig. 13, the pH value decreased depending on the reaction time and reached a constant value (pH 2.3). On the other hand, the amount of the titratable acidic groups increased rapidly and reached a maximum value and then decreased. This phenomenon could be explained by the increase of carboxyl groups by CGDE, and the carboxylic acid could be oxidized further to carbon dioxide. The time course of this oxidative reaction of ethylalcohol was studied using an isotacophoretic analyzer (Fig. 14). It was found that the reaction mixture contains acetic acid, glycolic acid, glyoxal acid, oxalic acid, formic acid and lactic acid. A small amount of pyruvic acid, which could be derived from lactic acid, was also confirmed. The oxidation pathway of ethanol is shown in:

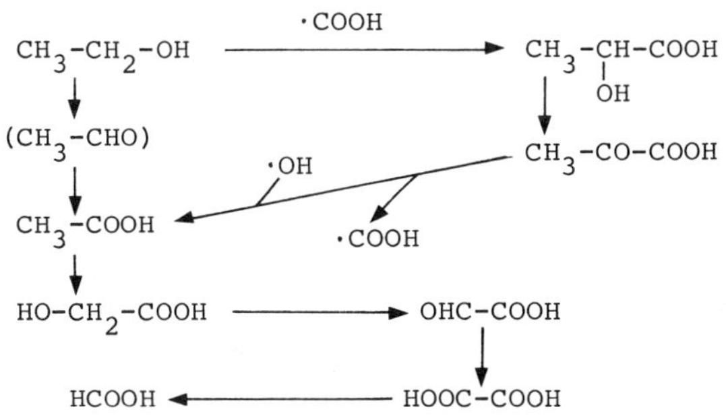

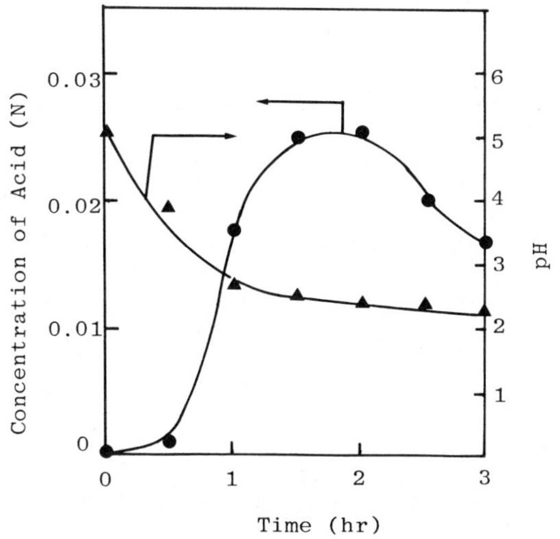

Fig. 13. Time course of acid formation from ethanol by CGDE. Conditions: 0.1M EtOH 25ml. ●: Acid concentration, ▲: pH of the reaction mixture.

The oxidation of methanol, formaldehyde, acetaldehyde, acetic acid, glycolic acid, glyoxylic acid and pyruvic acid were studied and the reaction pathways were clarified (Harada et al., 1982).

High Molecular Weight Compounds

The radiation chemistry of high molecular weight compounds have been studied extensively in connection with their physical and chemical properties. It was found that high molecular weight compounds were also affected by CGDE.

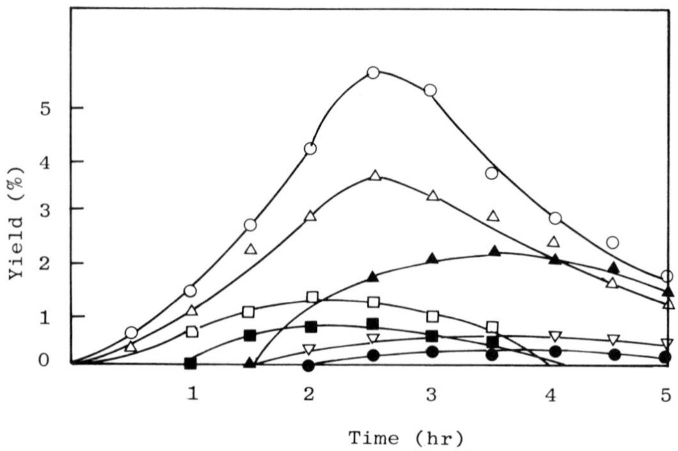

Fig. 14. Time course of oxidation of ethanol by CGDE. Conditions: 0.1M EtOH 20 ml. ○: Acetic acid, △: Formic acid, □: Glycolic acid, ■: Lactic acid, ▲: Oxalic acid, ▽: Pyruvic acid, ●: Glyoxylic acid.

Poly-DL-alanine (dp = 120) was dissolved in aqueous formic acid, and the solution was treated with CGDE. The reaction mixture was dialyzed and then hydrolyzed. Amino acid analysis shows the existence of serine and aspartic acid. Serine was formed by hydroxylation and aspartic acid was synthesized by carboxylation of the alanyl residue of polyalanine as shown in:

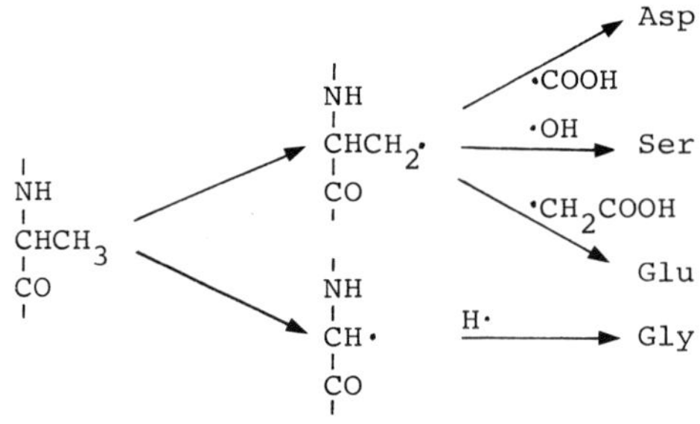

The IR spectrum of the dialyzed polymer shows an additional absorption based on the newly introduced carboxyl group (Nomoto et al., 1981). The carboxylation of the polyamino acid by CGDE is a new method for derivatization, and the chemistry is also interesting as a possible modification method for protoproteins on the primitive Earth.

CGDE has strong oxidative power, and the carbon-carbon bonds of any organic compounds could be cleaved easily. The aqueous solutions of polyacrylamide (Kokufuta et al., 1980) and polyethyleneglycol (Kokufuta et al., 1981) were treated with CGDE and the molecular weight distribution before and after CGDE were studied by gel permeation chromatography. The molecular weight distribution of the polymer shifts to low molecular weight in the chromatogram; however, the distribution is not as polydispersive as expected in the usual degradation of macromolecules. It was observed that the initial undegraded polymers remain for a long period in the reaction mixture. This phenomenon differs from the ozone oxidation of the macromolecules. The degradation products obtained by CGDE are also different from that obtained by ozone oxidation. In the oxidation of polyethyleneglycol, CGDE gave carboxylic acid and aldehyde. On the other hand, ozone oxidation gave formic acid ester, ethyleneglycol, and diethyleneglycol as the products.

CONCLUSIONS

The CGDE can be regarded as a model reaction induced by a lightning striking the hydrosphere on the primitive Earth. Several bioorganic compounds have been synthesized by CGDE as model experiments of the chemical evolution on the primitive Earth. On the other hand, CGDE is a new type of chemical reaction for organic compounds. CGDE is a kind of plasma chemistry and also a kind of radical chemistry in aqueous solutions. These reactions induced by CGDE, are strong but relatively controlled reactions compared with the reactions in gaseous electric discharge. And some of the CGDE reactions could be applied for the synthesis of organic compound.

The CGDE reaction starts with the homolytic dissociation of the water molecules to hydroxyl and hydrogen radicals (primary reaction). yielding substrate radicals. The resulting radicals react with other radicals and molecules to form various reaction products (secondary reaction). By the addition of various compounds in the reaction mix-tion). By the addition of various compounds in the reaction mixture in addition to the substrate, many kinds of reactions can be performed (amination, carboxylation, coupling reactions, etc.). However, the oxidation reaction of CGDE is the most characteristic among the various CGDE reactions. The oxidation reactions probably proceed by hydrogen abstraction from the substrate and the subsequent addition of the hydroxyl radical. And the processes proceed repeatedly to complete oxidation, that is, to carbon dioxide. The most characteristic points of the CGDE oxidation reactions are: (a) The oxidation reaction is so powerful as to oxidize an alkyl group to carboxylic acid. (b) The oxidation reaction does not require any conventional oxidizing agent. Therefore, the oxidation by CGDE is a clean reaction without using any oxidizing agent and the process stops when the electric current stops. The clean and powerful oxidation of CGDE could be used for water treatment. There are many unexplored areas left in the study of CGDE both of a fundamental but also of an applied nature, and we are continuing these studies on CGDE. Along with the study of CGDE, similar chemical reactions in aqueous solution induced by argon plasma and burning flames are now under investigation.

REFERENCES

Brown, E. H., Wilhide, W. D., and Elmore, K. L., 1962, J. Org. Chem., 27:3698.
Denaro, A. R., and Hickling, A., 1958, J. Electrochem. Soc., 105:265.
Ebisawa, H., Mitsugi, E., Shirai, T., and Yanagisawa, S., 1979, Nippon Kagaku Zasshi, 1304.
Fox, S. W., and Dose, K., 1977, "Molecular Evolution and the Origin of Life," 2nd edn., Marcel Dekker, New York (literatures listed therein).
Harada, K., 1974, in: "The Origin of Life and Evolutionary Biochemistry," K. Dose, S. W. Fox, G. A. Deborin, and T. E. Pavlovskaya, eds., p. 183, Plenum, New York.
Harada, K., and Iwasaki, T., 1975, Chem. Lett., 185.
Harada, K., Nomoto, M. M., and Gunji, H., 1981, Tetrahedron Lett., 22:769.
Harada, K., and Nomoto, M. M., 1982, unpublished experiments.
Harada, K., Nomoto, S., Suzuki, S., and Yamada, A., 1982, unpublished experiments.
Harada, K., and Suzuki, S., 1977a, Nature, 266:275.
Harada, K., and Suzuki, S., 1977b, Naturwiss., 64:484.
Harada, K., Suzuki, S., and Ishida, H., 1978, Experientia, 34:17.
Harada, K., Suzuki, S.; Ishida, H., Matsuyama, M., and Tamura, M., 1978, in: "Origin of Life," H. Noda, ed., p. 141, Japan Scientific Societies Press, Tokyo.
Harada, K., Suzuki, S., and Ishida, H., 1978, BioSystems, 10:247.
Harada, K., and Terasawa, J., 1980a, Chem. Lett., 441.
Harada, K., and Terasawa, J., 1980b, Chem. Lett., 1545.
Harada, K., Terasawa, J., and Gunji, H., 1981, in: "Origin of Life," Y. Wolman, ed., p. 173, Reidl, Dordrecht, Holland.
Harada, K., Terasawa, J., and Suzuki, S., 1978, Naturwiss., 65:259.
Hickling, A., 1971, in: "Modern Aspects of Electrochemistry," J. O'M. Bockris and B. E. Cornway, eds., No. 6, p. 329.
Hickling, A., and Ingram, M. D., 1964a, J. Chem. Soc., 783.
Hickling, A., and Ingram, M. D., 1964b, J. Electroanal. Chem., 8:65.
Hickling, A., and Newns, G. R., 1959, Proc. Chem. Soc., 368.
Hickling, A., and Newns, G. R., 1961, J. Chem. Soc., 5177; 5186.
Kokufuta, E., Fujii, S., Ishibashi, H., Yokoi, H., Harada, K., and Nakamura, I., 1980, Polymer Bull., 3:173.
Kokufuta, E., Fujii, S., Shibasaki, T., Harada, K., and Nakamura, I., 1981, unpublished results.
Mazzocchin, G. A., Magno, F., and Bontempelli, G., 1973, J. Electroanal. Chem. Interfac. Electrochem., 45:471.
Miller, S. L., 1953, Science, 117:528.
Miller, S. L., 1955, J. Am. Chem. Soc., 77:2351.

Nomoto, M. M., Sakai, F., and Harada, K., 1981, Polymer Bull., 5:451.
Steinman, G. D., 1966, Science, 154:1344.
Steinman, G. D., and Lillevik, H. A., 1964, Arch. Biochem. Biophys., 105:303.
Suzuki, S., Tamura, M., Terasawa, J., and Harada, K., 1978, Bioorg. Chem., 7:111.
Terasawa, J., and Harada, K., 1980, Chem. Lett., 73.

CHARACTERIZATION OF IONIZABLE GROUPS IN THERMALLY PREPARED POLYAMINO ACIDS

Etsuo Kokufuta and Kaoru Harada†

Institute of Applied Biochemistry (Department of Chemistry†), University of Tsukuba
Sakura-mura, Ibaraki 305, Japan

INTRODUCTION

In an attempt to clarify the process of evolution of small molecules to macromolecules under prebiotic conditions, attention should be paid to the formation of polymers of amino acids. Only the thermal method successfully yields polyamino acids under simulated geological conditions (e.g., Fox and Dose, 1977). The thermally prepared polyamino acids possess some proteinlike properties; therefore, they have drawn attention as possible abiotic proteins (Fox, 1965).

Thermal polycondensation of amino acids gives rise to polymers ranging from homopolymers of aspartic acid (Kovács et al., 1953; Vegotsky et al., 1958), glycine (Meggy, 1956), or lysine (Harada, 1959) to copolymers (proteinoids) containing the 18 common amino acids in contemporary protein (Fox and Harada, 1958, 1960; Fox et al., 1962, 1963; Fox, 1965; Rohlfing, 1967). There are numerous studies on the physicochemical and biological properties of thermal polyamino acids and/or proteinoids (Fox and Dose, 1977): (1) amino acid composition; (2) heterogeneity; (3) molecular weight; (4) solubility and precipitability; (5) configuration of residues; (6) linkages; (7) infrared absorption spectra; (8) hydrolytic behavior; (9) nutritive quality; (10) enzymelike activities; (11) an origin of metabolism; (12) hormonal activity. In the previous studies, however, the main disadvantages was the lack of the fundamental data which are required for understanding the details of these proteinoid properties. For example, the dissociation constant, which is an important parameter to illustrate the enzymelike activity of proteinoid, has not yet been studied. Such an investigation could also provide useful information about the α- and ω-linkages of acidic and basic amino acid residues in thermal polypeptide and proteinoid (Kokufuta et al., 1977, 1978, 1979; Kokufuta and Harada, 1981; Harada et al., 1978).

Potentiometric titration is a suitable method for determining dissociation constants, not only for monomeric acids and bases but also for polymers. The dissociation constant of each ionizable group in protein can be estimated by analyzing the titration data with an electronic computer (Shaw and Hartzell, 1976). However, this technique is not directly applicable to proteinoid, because the identity of the ionizable groups is not completely known (Fox and Dose, 1977). In such a case, one must then investigate in a little more fundamental way the nature of thermal homo- and copoly-amino acids.

In this chapter, the analytical method for evaluating the dissociation constant of ionizable groups in the polymer is first described, followed by the results obtained from some thermal poly-amino acids. These results are then used for the discussion about ionic properties and linkages of proteinoids. The formation mechanisms of aspartyl, glutamyl, and lysyl residues in proteinoids are also discussed on the basis of the analytical results.

ANALYTICAL METHODS

One Species of Ionizable Groups

The equilibrium which exists between protons and acidic (~AH) or basic (~BH^+) groups attached to the polymer chain may be described as:

$$\sim AH \rightleftarrows \sim A^- + H^+ \quad (1)$$

and

$$\sim BH^+ \rightleftarrows B + H^+. \quad (2)$$

The apparent dissociation constant (pK_a) can be defined by:

$$pK_a = pH + \log(1-\alpha_d)/\alpha_d \quad (3)$$

$$\alpha_d = \alpha_n + (C_{H^+} - C_{OH^-})/C_I, \quad (4)$$

where C_{H^+} and C_{OH^-} represent the molarity of protons and hydroxyl ions, respectively, C_I is the molar concentration of the ionizable groups, α_d is the degree of dissociation and α_n is the degree of neutralization. Eqs. (3) and (4) are the formal equivalent of the representations which deal with the dissociation equilibrium for monomeric acids and bases. In the case of polymers, however, the equilibrium is strongly influenced by polyion charges (e.g., Tanford, 1969); thus, pK_a is expressed as a function of free energy change (ΔG_e) due to an electrostatic interaction between protons and polyions as follows:

$$pK_a = pK_{int} + 0.434(\Delta G_e/RT), \quad (5)$$

where R denotes gas constant, T is absolute temperature, and pK_{int} is the intrinsic dissociation constant which may be taken as pK_a under imaginary conditions where each ionizable group linked to the chain backbone behaves independently. Therefore, it is preferable to evaluate the ionic property of polyamino acid by using the value of pK_{int}. For this purpose, the titration is carried out frequently by adding neutral salts so as to adequately eliminate the effect of the polyion charges. The pK_{int} value is then estimated by two different approaches: One is graphical extrapolation of pK_a to $\alpha_d \to 0$, and the other is by means of the Henderson-Hasselbalch equation (e.g., Kokufuta et al., 1977):

$$pH = pK_a + n \log (1-\alpha_d)/\alpha_d . \quad (6)$$

Here, n is the empirical constant which denotes the magnitude of the interaction between protons and polyions as the deviation of n from unity*; thus, pK_{int} corresponds to pK_a at n = 1.

Usually the pK_{int} value varies slightly depending on the ionic strength even when the other conditions are held at constant. This is related to the effect of the ionic atmosphere which is constructed around an ionizable group. In this chapter, an intrinsic dissociation constant (pK^0_{int}), which is not affected by ionic strength, is thus obtained by the following equations (Harris and Rice, 1955); i.e., for polyacids;

$$pK^0_{int} = pK_{int} + 0.434(e^2\kappa/3DkT) \quad (7)$$

and for polybases,

$$pK^0_{int} = pK_{int} - 0.434(e^2\kappa/3DkT) \quad (8)$$

Here, the term $0.434(e^2\kappa/3D\,T)$ (e, electronic charge; κ, Debye-Hückel parameter; D, the dielectric constant of the solvent; k, Bolzmann constant) represents the free energy change for building up the ionic atmosphere.

---

*Monomeric acids and bases give n = 1 when the titration data are analyzed by Eq. (6).

Two Species of Ionizable Groups

We consider a polymeric acid or base which bears two species of ionizable groups ($\sim$AH and $\sim$BH). When the dissociation of each ionizable group takes place independently in different pH ranges, their pK^0_{int} values can be estimated by the methods described above. However, when the dissociation of the two ionizable groups overlap with each other, the analysis of the titration data becomes very complex. In such a case, an analytical method has been developed by Kokufuta et al. (1978, 1979).

Assuming that the activity coefficients are unity,* the dissociation constant of each ionizable group can be expressed as

$$K_1 = [H^+][\sim A^-]/[\sim AH] \quad (10)$$

$$K_2 = [H^+][\sim B^-]/[\sim BH],$$

where brackets signify the concentration of the species. The total concentration of each ionizable group is then given by:

$$C_1 = [\sim A^-] + [\sim AH] \quad (11)$$

and

$$C_2 = [\sim B^-] + [\sim BH]. \quad (12)$$

When a strong monoacidic base, which may be completely dissociated, has been added to give a molar concentration (M), the requirement of "the electroneutrality" is defined by

$$M + [H^+] = [\sim A^-] + [\sim B^-] + [OH^-]. \quad (13)$$

From the combination of Eq. (9) with Eq. (11) and that of Eq. (10) with Eq. (12), we obtain

$$[\sim A^-] = C_1K_1/([H^+] + K_1) \quad (14)$$

and

$$[\sim B^-] + C_2K_2/([H^+] + K_2). \quad (15)$$

Introducing the terms of [$\sim A^-$] and[$\sim B^-$] into Eq. (13), the following expression can be obtained

*This assumption may be permissible under the conditions where polyion charges are adequately reduced; i.e., $\Delta G_e \simeq 0$.

$$(1/K_i)(M + [H^+] - [OH^-])([H^+] + K_1) =$$

$$C_1 + \frac{K_2([H^+] + K_1)}{K_1([H^+] + K_2)} C_2 . \tag{16}$$

Taking Eq. (4) into account, Eq. (16) can be converted into

$$Y = C_1 + C_2 X$$

$$Y = (1/K_1)(K_1 + C_{H^+})\alpha_d^t \, C_I^t \tag{17}$$

$$X = K_2(K_1 + C_{H^+})/K_1(K_2 + C_{H^+}) ,$$

where α_d^t and C_I^t (= $C_1 + C_2$) represent the degree of dissociation and the molar concentration for all the ionizable groups, respectively. Eq. (17) is considered to be the formal resemblance of Speakman's equation (1940) which has been used to determine the dissociation constant for weakly dibasic acids such as adipic acid. The values of pK_1 (= $-\log K_1$) and pK_2 (= $-\log K_2$) can be estimated from the titration data by Eq. (17), if both C_1 and C_2 are known and the data can be obtained under the condition where $\Delta G_e \to 0$. When C_1 and C_2 are unknown, the pK values can also be estimated by using some C_1 and C_2 values so as to best fit the X-Y plots to a linear relationship (data-fitting method). The fitness is easily determined by evaluating the correlation coefficient (r);

$$r = \frac{\frac{1}{N}\sum_{i=1}^{N} (X_i - \bar{X})(Y_i - \bar{Y})}{\sigma_X \, \sigma_Y} , \tag{18}$$

where $\bar{X}$ and $\bar{Y}$ are average values of X_i and Y_i, respectively, σ_x and σ_y are standard deviations of X_i and Y_i, respectively, and N is the data number. An electronic computer can facilitate these calculations.

The validity of Eq. (17) has been confirmed by comparing the analytical results for poly(α- and β-DL-aspartic acids), which are obtained by the titration and H^1-NMR methods (Matsuyama et al., 1980). Three samples were prepared by different methods: α-PAA(H) by the hydrogenolysis of poly(β-benzyl DL-aspartic acid)*; β-PAA(H) by the hydrogenolysis of poly(α-benzyl DL-aspartic acid)**;

*Synthesized by the Leuchs method.

**Synthesized by the active ester method.

β-PAA(OH) by the alkaline hydrolysis of poly(α-methyl DL-aspartic acid).* The titration was carried out at ionic strength 0.5, where the polyion charges are sufficiently eliminated. The results obtained by the data-fitting method are shown in Table 1. It is found that β-PAA samples contain both α-carboxyl groups (pK^0_{int} = 3.43±0.1) and β-carboxyl groups (pK^0_{int} = 4.44±0.1), where α-PAA(H) has only β-carboxyl groups (pK^0_{int} = 4.44±0.1). On the other hand, the NMR spectrum of α-PAA(H) shows one signal (4.1 ppm) assigned to the methine groups, but the methine signal for both β-PAA(H) and β-PAA (OH) splits into two peaks (3.7 and 4.0 ppm)(see Fig. 1). The content of α- or β-carboxyl groups in β-PAA samples, determined from the integral ratio of the two peaks, agrees with that obtained by the titration (Table 1). Thus, it may be concluded that the dissociation constants of two ionizable groups existing in a polypeptide chain can be determined from the titration data by using Eq. (17).

RESULTS FOR SOME POLYAMINO ACIDS

The major ionizable groups in proteinoids were found to be carboxyl and amino groups by means of amino acid analysis (Fox et al., 1963; Waehneldt and Fox, 1968). The carboxyl group exists in both aspartyl and glutamyl residues, while the amino group is mainly in lysyl residue. Therefore, the thermal polymers containing such amino acid residues can be useful for obtaining information about the ionic properties of proteinoids. This section describes the characterization of the ionizable groups in thermal polymers containing aspartyl, glutamyl, or lysyl residues.

Poly(Aspartic Acid)

Poly(aspartic acid) (PAA) can be obtained by alkaline hydrolysis of thermal poly(succinimide) or poly(anhydroaspartic acid) (e.g., Fox and Harada, 1966). Kokufuta et al. (1977) have studied the potentiometric titration behavior of thermal PAA. As shown in Fig. 2, the titration curve at ionic strength 0 has two inflection points at α_n = 0.29 and 1.0. The first inflection point disappears when the titration was made at C_I below 10^{-3} mol/l and/or ionic strengths above 0.5. Thus, they assumed that thermal PAA contains both α- and β-carboxyl groups which dissociate in the ranges of α_n below and above 0.29, respectively. If this assumption is justified, their dissociation constants can be estimated from the titration data at the corresponding α_n regions by means of the graphical extrapolation (see Fig. 3) and of the Henderson-Hasselbalch equation. The results obtained are listed in Table 2. The pK^0_{int} values of α- and β-carboxyl groups could be compared favorably with the pK_1

*Synthesized by the Leuchs method.

Table 1. Comparison of potentiometric titration method with NMR method for determining the contents of α- and β-carboxyl groups in poly(α- and β-DL-aspartic acids)

		Titration method[a]				NMR method[b]	
		α-COOH		β-COOH		α-COOH (β-CH)	β-COOH (α-CH)
	n^c	Content (%)	pK^0_{int}	Content (%)	pK^0_{int}	Content (%)	Content (%)
α-PAA(H)	1.02	2	3.34±0.1	98	4.44±0.1	0	100
β-PAA(H)	1.01	43	3.34±0.1	57	4.44±0.1	60	40
β-PAA(OH)	1.02	34	3.34±0.1	66	4.44±0.1	33	66

Source: Matsuyama et al., 1980.
[a] Carried out at ionic strength 0.5 and 25 °C.
[b] NMR spectra are shown in Fig. 1.
[c] Determined by Eq. (6).

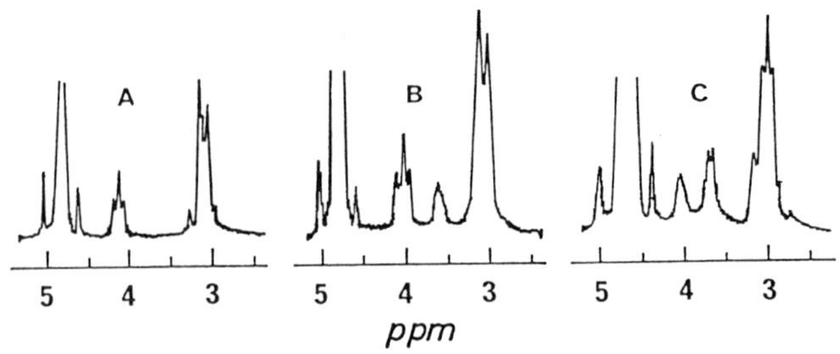

Fig. 1. ^{1}H-NMR spectra of three synthetic poly(aspartic acids): (A) α-PAA(H) at pD 3.31; (B) β-PAA(OH) at pD 4.16; (C) β-PAA(H) at pD 3.90. The measurements were made in D_2O at 25°C. Source: Matsuyama et al. (1980).

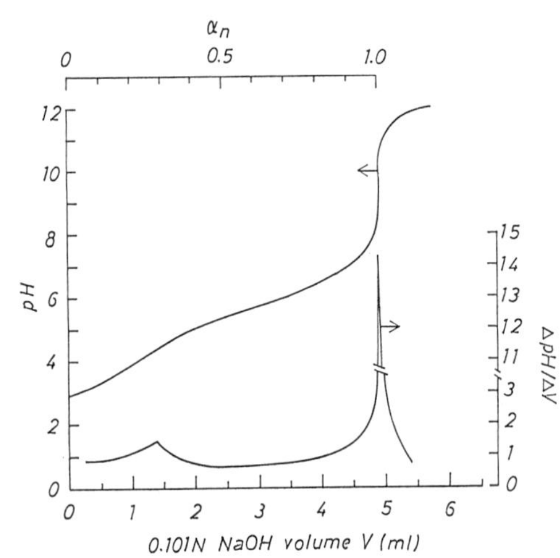

Fig. 2. Potentiometric titration curve of thermal PAA prepared from DL-aspartic acid (ionic strength = 0). The differential titration curve of ΔpH/ΔV vs. V was obtained by the graphical differentiation of the curve of pH vs. V. Source: Kokufuta et al., 1977.

of N-acetylaspartic acid (pK_1 3.37 and pK_2 4.91) and the pK_2 of glycylaspartic acid (pK_1 2.81, pK_2 4.45, and pK_3 8.60), respectively. These aspartic acid derivatives are suitable as monomeric models for PAA for the following reasons: (a) The α-carboxyl group of N-acetylaspartic acid is not affected by the charged amino group;

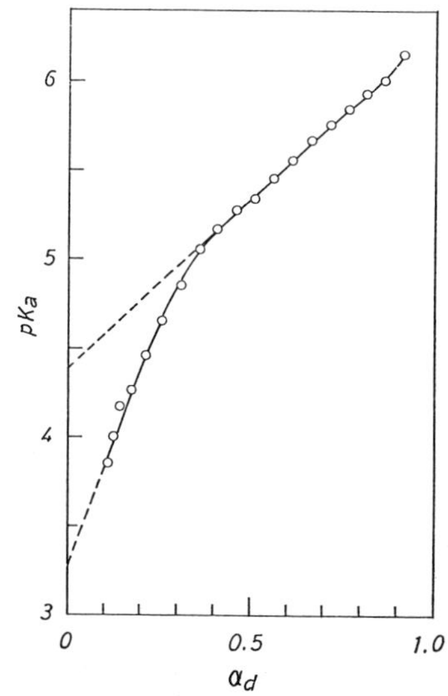

Fig. 3. Dependence of pK_a on α_d obtained from the titration curve in Fig. 2. Source: Kokufuta et al., 1977.

Table 2. Analytical results of the titration data at 25°C for thermal PAA

Ionic Strength	n	α-COOH pK^0_{int}	α-COOH Content (%)	β-COOH pK^0_{int}	β-COOH Content (%)	Analytical method
0		3.26	29	4.38	71	Graphical extra-poration
0.1	1.0[a]	3.23		4.78		(Henderson-(Hasselbalch
1.0	1.0			4.32		(equation
1.0	1.0	3.25	26	4.38	74	Data-fitting

Source: Kokufuta et al., 1977; Kokufuta and Harada, 1981.
[a]Value of n for β-COOH is 1.22.

(b) the β-carboxyl group of glycylaspartic acid is not influenced by the dissociated α-carboxylate ion because of a Zwitter ion structure with the amino group; (c) both model compounds have amide linkages similar to those in PAA. Therefore, the agreement between the pK values of the sample and models could justify the analytical results.

In order to confirm further the results mentioned above, the titration data at ionic strength 1.0 was analyzed by the data-fitting method (Kokufuta and Harada, 1981). The pK^0_{int} values estimated are shown in Table 2. A good agreement is observed in the results determined by three different analytical methods .

PAA can also be prepared by aqueous thermal condensation of asparagine (Kovács and Nagy, 1961). Harada et al. (1978) have studied the ionic behavior of PAA obtained by refluxing an aqueous solution of asparagine or isoasparagine. The results obtained by the data-fitting method are summarized in Table 3. It was found that the values of pK^0_{int} for α- and β-carboxyl groups are similar to those of the thermal PAA prepared in the solid state, while a difference in the ratio of α- and β-carboxyl groups exists between both thermal polymers.

Copolymer of Glutamic Acid and Alanine

Glutamic acid does not form a homopolymer by heating, but does react with other amino acids to yield the corresponding copolymers (Fox and Harada, 1966). The copolymer of glutamic acid and alanine, copoly(glu, ala) (Glu:Ala = 1:2.8),* shows a monotonous titration curve at different ionic strengths, which is characterized by an inflection point at $\alpha_n = 1.0$ (Kokufuta et al., 1978). The values of pK^0_{int}, which are evaluated by means of the graphical extrapolation and Henderson-Hasselbalch equation, are listed in Table 4, together with those of various poly(glutamic acids) (Waley, 1955; Edelhoch and Bateman, 1957; Ciferri et al., 1968). The results for copoly(glu, ala) are in fair agreement with those for the γ-linked polymers but different from the α-linked ones; thus suggesting that copoly(glu, ala) mostly or entirely contains α-carboxyl groups. To confirm this result, the titration data were analyzed with Eq. (17). The values of K_1 (1.58×10^{-4}) and K_2 (1.86×10^{-5}), corresponding to the average values of pK^0_{int} for poly(γ-D-glutamic acid) and poly(α-L-glutamic acid), respectively, were used in the calculation. The plots of Y vs. X obtained are shown in Fig. 4. It was found that Y is independent of X; therefore, clearly indicating that the glutamyl residues in copoly(glu, ala) bear only α-carboxyl groups which are characterized by $pK^0_{int} = 3.89 - 3.98$.

Copolymer of Aspartic Acid and Glutamic Acid

The thermally prepared PAA has been found to contain both α- and β-carboxyl groups, whereas only α-carboxyl groups exist in

*Obtained by alkaline treatment of the thermal polymer containing about 4% of 6-membered imide residues.

Table 3. Analytical results of the titration data for PAA prepared by aqueous thermal polycondensation[a]

Preparation condition			α-COOH		β-COOH	
Monomer	Solvent	pH	pK^0_{int}	Content (%)	pK^0_{int}	Content (%)
$AspNH_2$	1N NaCl	7.0	3.39±0.2	25±6	4.59±0.1	75±6
$IsoAspNH_2$	1N NaCl	7.0	3.39±0.2	15±8	4.59±0.1	85±8
$AspNH_2$	H_2O	4.2	3.39±0.2	60±5	4.59±0.1	40±5

Source: Harada et al., 1978.
[a]The titration data (25°C, ionic strength 0.5) were analyzed by the data-fitting method. The values of n in Eq. (6) were 1.03±0.01 under the titration conditions used.

Table 4. Values of pK^0_{int} of thermal copoly(glu, ala) and various poly(glutamic acids) at 25°C

Polypeptide	Ionic strength	n	pK^0_{int}[a] A	B	Reference
Thermal copoly-(glu,ala)	0	1.41	3.93	3.89	Kokufuta et al., 1978
	0.5	1.07	3.96	3.93	
	1.0	1.00	3.98	3.98	
Poly(γ-L-glutamic acid)[b]	1.0		3.94±0.1		Waley, 1955
Poly(γ-D-glutamic acid)[c]	0.01			3.76	Edelhoch and Bateman, 1957
	0.1			3.80	
	1.0			3.70	
	1.0		3.94±0.1		Waley, 1955
Poly(α-L-glutamic acid)[b]	0.1			4.71	Ciferri et al., 1968
	0.4			4.69	
	0.7			4.74	
	1.2			4.78	

Source: Kokufuta et al., 1978.
[a]Estimated by graphical extrapolation (A) and Henderson-Hasselbalch equation Eq. (B).
[b]Synthesized.
[c]Obtained from the organism B. anthracis.

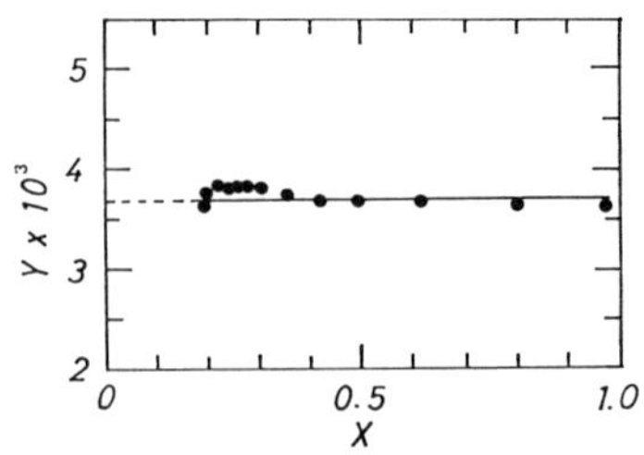

Fig. 4. The plots of Y vs. X for thermal copoly(glu, ala). The values of X and Y were calculated from the titration data at ionic strength 1.0 by Eq. (17). Source: Kokufuta et al., 1978.

copoly(glu, ala). The thermal copolymer of aspartic acid and glutamic acid, copoly(asp, glu), is of interest in the comparison with PAA and copoly(glu, ala). Fig. 5 shows the comparison of the titration curve of copoly(asp, glu) (Asp:Glu=8.8:1.2) with those of PAA and copoly(glu, ala). It is observed that, in the α_n region below 0.5, the pH values for copoly(asp, glu) are lower than those for PAA. Furthermore, the first inflection point in the curve of PAA is shifted to the high α_n range. Taking into account that copoly(asp, glu) contains 12% of the glutamyl residues, the shifting (about 10%) of α_n could be imputed to the glutamyl residues in the copolymer. Therefore, it is reasonable to consider that the glytamyl residues have α-carboxyl groups even in copoly(asp, glu).

Poly(Lysine)

The potentiometric titration behavior of thermal poly(lysine) (PL) has been studied at different ionic strengths (Kokufuta et al., 1979). The curves of pK_a against α_d show monotonous change (see Fig. 6), whereas that of Leuchs poly(α-L-lysine) depicts a characteristic pattern due to the helix-coil transition (Ciferri et al., 1968). Another remarkable feature for PL is that the n value in Eq. (6) is not close to one even when the titration was made under high ionic strength; that is, ΔG_e is not reduced by the electrostatic screening of the polyion charges with small anions. This is similar to the ionic behavior of branched poly(ethyleneimine) (Shepherd and Kitchener, 1956; Thiele and Gronau, 1963; Treslong and Staverman, 1974), which is governed by the nearest neighbor interaction* originally proposed by Katchalsky and co-workers (1957). From these results, it is considered that thermal PL behaves as a branched random-chain polybase.

*This refers to a possible strong interaction of a charged group with the neighboring uncharged groups along the polymer chain.

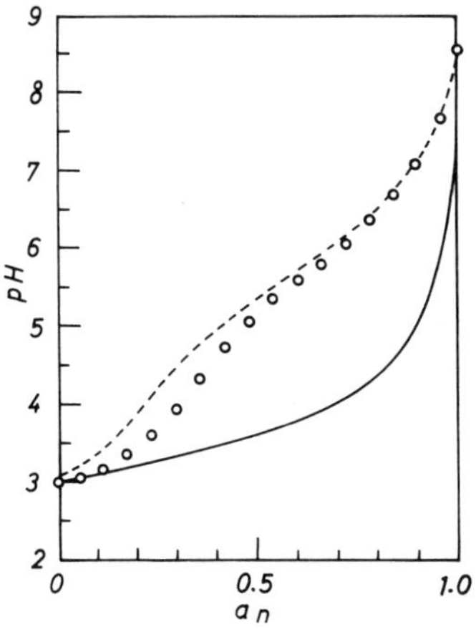

Fig. 5. Comparison of the titration curve for thermal copoly(asp, glu) (plots) with those of thermal PAA (dashed line) and copoly(glu, ala) (full line). The contents of glutamyl residues are 12% for copoly(asp, glu) and 26% for copoly-(glu, ala). The titration was made at C_I=0.05 mol/l and ionic strength = 0.

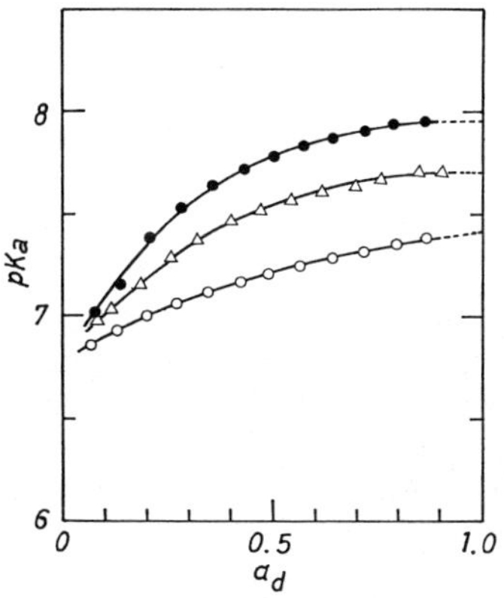

Fig. 6. Dependence of pK_a on α_d for thermal PL at different ionic strengths: (○) 0.1; (△) 0.5; (●) 1.0. Source: Kokufuta et al., 1979.

As was described above, the titration data for PL cannot be obtained under the conditions where $\Delta G_e \rightarrow 0$. However, the analyses by the data-fitting method could give reasonable pK^o_{int} values of α- and ε-amino groups (see Table 5), because they are comparable with the pK values of ε-amino groups (10.2 - 9.8) and α-amino groups (7.8) in proteins such as ribonuclease and bovine plasma

Table 5. Analytical results of the titration data at 25°C for thermal PL and Copoly(lys, ala)[a]

	Ionic strength	n	α-NH_2 pK^0_{int}	α-NH_2 Content (%)	ε-NH_2 pK^0_{int}	ε-NH_2 Content (%)
PL	0.1	1.26	7.1	96	10.4	4
	0.5	1.40	7.3	93	10.6	7
	1.0	1.49	7.3	94	10.6	6
Copoly-(lys,ala)	0.1	1.48	7.3	93	10.6	7
	0.5	1.53	7.4	95	10.7	5
	1.0	1.59	7.4	94	10.7	6

Source: Kokufuta et al., 1979.
[a]Obtained by the data-fitting method.

albumin (Tanford, 1969). Thus it may be concluded that the titratable groups* in thermal PL are of about 95% α-amino groups (pK^0_{int}= 7.1 - 7.3) and about 5% ε-amino groups (pK^0_{int}=10.4 - 10.6).

Copolymer of Lysine and Alanine

The heating of a mixture of lysine and alanine yields copoly-(lys, ala) (Fox and Harada, 1966). This copolymer was expected to be unaffected by the nearest neighbor interaction, because of the decrease in the ionizable group content caused by the copolymerization of the neutral amino acid. However, the ionic behavior of copoly(lys, ala) (Lys:Ala=1:0.83) was very similar to that of PL (Kokufuta et al., 1979), which could be explained by assuming that a part of the lysyl residue forms branching points even in the copolymer with alanine. The analytical results of the titration data by the data-fitting method are listed in Table 5, together with those of PL. The pK^0_{int} values and the contents of α- and ε-amino groups were in good agreement with those for PL.

*It has been found that all the amino groups in branched poly-(ethyleneimine) cannot titrate with hydrochloric acid because of the nearest neighbor interaction (Kokufuta, 1979). Thus the amino groups in PL may not be titrated completely, although no remarkable difference was observed in the contents of the ionizable groups determined by means of titration and amino acid analysis (Kokufuta et al., 1979).

APPLICATION OF THE RESULTS FOR UNDERSTANDING SOME PROPERTIES OF PROTEINOIDS

Dissociation Properties of Acidic and Basic Groups in Proteinoids

Since the thermal polymers described above were obtained under the conditions analogous to proteinoid preparations, they should serve as good models for proteinoids. The ionic properties of proteinoid can thus be inferred from the results for the thermal polymers. The values of pK^0_{int} for each thermal polymer are summarized as follows: 3.23 - 3.39 for α-COOH (aspartyl); 3.89 - 3.98 for α-COOH (glutamyl); 4.32 - 4.59 for β-COOH (aspartyl); 7.1 - 7.4 for α-NH_2 (lysyl); 10.4 - 10.7 for ε-NH_2 (lysyl).

The dissociation constants of ionizable groups in various pro-proteins, summarized by Hayashi (1977), are listed in Table 6. The pK values for the carboxyl and amino groups in thermal polyamino acids are found to be similar with those in proteins. It has been bell known that the pK of an ionizable group in a protein varies much depending on several conditions in the neighborhood of the ionizable groups. From the results described above, therefore, it could be considered that there is no distinguishable difference between physicochemical conditions around the carboxyl or amino group in proteinoid and protein, although configurational and conformational differences exist between both polymers. The high

Table 6. Summarized results for the dissociation constants of ionizable groups in proteins

Ionizable group	Dissociation constant
α-COOH	3.0 - 5.8
ε-COOH	3.5 - 5.0
γ-COOH	4.5
imidazolyl	5.5 - 10.8
α-NH_2	7.5 - 8.5
ε-NH_2	9.5 - 10.5
guanidyl	11.5 - 12.5
phenolic OH	10 - 13
mercapto	9 - 11

Source: Hayashi, 1977.

amounts of γ-glutamyl and ε-amino linkages found in the thermal polymers are of course not common in proteins.

Linkages of Aspartyl, Glutamyl, and Lysyl Residues in Proteinoids

Aspartyl residue. The determination of α- and ω-linkages is important in understanding the structure of polyamino acids and proteinoids. Kovács et al. (1961) have studied the proportions of α- and β-linkages in thermal poly(aspartic acid) by means of chemical degradation, and reported a ratio of α:β = 1.0:1.3. However, the results obtained by the chemical method should only be accepted in such a case where a stoichiometric relationship is confirmed in all the reactions used for the analyses. On the other hand, the physical measurements, e.g., NMR method, could be suitable for determining the ration of α- and ω-linkages. By ^{1}H-NMR spectroscopy, Andini et al. (1975) and Temussi et al. (1976). claimed that the thermal poly(aspartic acid) contains no α-linked aspartyl residues. The disagreement between both results was explained by assuming a difference in the samples used (Fox, 1976). However, the conclusion from Andini et al. is based only on a difference in the chemical shift of the CH_2 groups in thermal poly(aspartic acid) and Leuchs poly(α-L-aspartic acid), but not on the split signals of the CH_2 in the α- and β-linked residues (see Fig. 1). Recently, ^{13}C-NMR method was applied to the complexes of Co(II) with poly(aspartic acids) prepared by hydrolyzing poly(succinimide) at different pH values (Pivcová et al., 1982).* The signal of the CH_2 and CH in α- and β-aspartyl residues successfully split into two peaks, and their integral ratio showed α:β = 2.5:7.5 to 4.5:5.5, which varies depending on the pH values of the hydrolysis.

The existence of the α- and β-linkages in thermal poly(aspartic acid) was also confirmed from the titration data (Tables 2 and 3). Thus, there has been no doubt that both α- and β-peptide bonds exist in thermal poly(aspartic acid) and also in proteinoid. However, agreement between the ratio of α- and β-linkages, estimated by the titration and NMR methods, has not been obtained yet for thermal poly(aspartic acid), although no difference between both analytical results was observed in the synthetic poly(aspartic acids) (see Table 1).

*It should be noted that the sample, subjected to ^{13}C-NMR, is not obtained by stoichiometric complexing between Co(II) and the polymer, although such complex is expected to be unsuitable for the measurement because of the precipitation in aqueous solvents.

Glutamyl residue. The titration data for copoly(glu, ala) reveal that the copolymer is composed of γ-linked glutamyl residues (Table 4, and Fig. 4). The glutamyl residues are also γ-linked even in thermal copoly(asp, glu), whereas both α- and β-linked aspartyl residues exist in the copolymer (Fig. 5). Therefore, it is reasonable to consider that glutamic acid residues tend to be linked through the γ-carboxyl groups in thermal copolymers and also in proteinoids.

Lysyl residue. The ratio of α- and ε-linked lysyl residues has been estimated by hydrolyzing the dinitrophenylated thermal poly(lysine) and copoly(lys, ala) (Heinrich et al., 1969; Kokufuta et al., 1979). The results obtained are summarized as follows: α:ε = 2.0:8.0 to 3.3:6.7. In contrast to these results, the potentiometric studies* indicated that both poly(lysine) and copoly(lys, ala) consist mostly of ε-linked lysyl residues (see Table 5).

To clarify this disagreement, the titration data for poly(lysine) was compared with the results of the chemical method (Kokufuta et al., 1979). In Fig. 7, the broken dashed line represents the relationship of Eq. (17) which agrees with the results (α-NH_2:ε-NH_2= 3.3:6.7) of the chemical method. The black circles are the X-Y plots of the titration data which closely match the dashed line. To fit the titration data to the dashed line, it is necessary to introduce unreasonable pK^0_{int} values into Eq. (17); that is, 7.5 for α-NH_2 and 6.9 for ε-NH_2 (the former is particularly lower than the pK values for usual polypeptides and proteins). Even though these pK^0_{int} values are used in Eq. (17), the fitness [0.799 as r in Eq. (18)] between the black circles and the dashed line is much smaller than that (r = 0.997) between the open circles and the full line which give the analytical results in Table 5. Thus, the disagreement between the titration and chemical methods is possibly due to incomplete dinitrophenylation and/or hydrolysis during the analytical process.

General Discussion

It has become apparent that thermal polyamino acids contain β-, γ- and ε-peptide bonds which are rarely found in contemporary proteins. Some authors (Andini et al., 1975; Temusii et al., 1976; Pivcová et al., 1982) have discussed these results in terms of the evolutional theory. However, we should not forget that the thermal polymers including proteinoids are "precursor" of contemporary proteins. Therefore, the determination of α- and ω-linkages is considered to be important mainly for learning the structure of thermal

*There are a few questions about whether all the ionizable groups are completely titratable or not (see above).

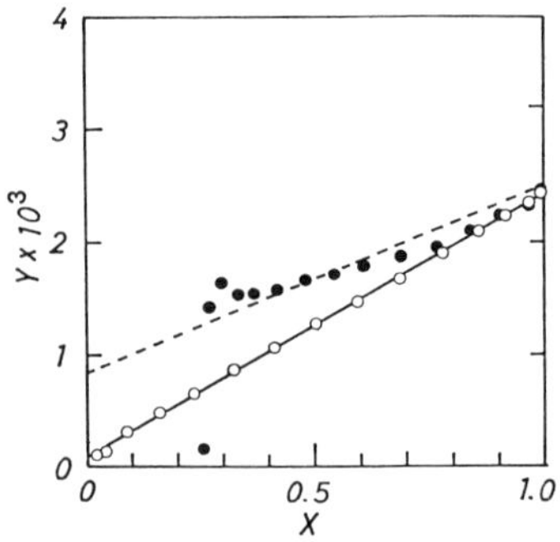

Fig. 7. The plots of Y vs. X for thermal PL at ionic strength 0.1. Open circles are the plots at maximal r value, i.e., the case in which K_1 and K_2 in Eq. (17) were selected to obtain the best fit to a linear relationship (full line) by the data-fitting method. Dashed line presents the relationship of Eq. (17) expressed by the values of C_1 and C_2 obtained from the results of the chemical degradation method; Y = 0.00083 + 0.0016X. Black circles indicate the case in which the values of K_1 and K_2 were selected to give plots which would best fit the dashed line. Source: Kokufuta et al., 1979.

polymers (protoprotein) which can be formed under assumed prebiotic conditions. It is a necessary step in understanding how the thermal proteinoids possess the many enzymic and structural features which could have been extremely important in the prebiotic era.

On the Formation Mechanism of Aspartyl, Glutamyl, and Lysyl Residues in Proteinoids

Aspartyl residue. It has been well known that the heating of aspartic acid yields poly(succinimide) which is readily converted to poly(aspartic acid) by treatment with dilute alkali or by warming in aqueous solution (Vegotsky et al., 1958; Kovács et al., 1961; Hoagland and Fox, 1967). As was mentioned previously, the aspartyl residues have α- and β-carboxylic groups. Therefore, the 5-membered imide structure in poly(succinimide) is converted to both α- and β- peptide bonds:

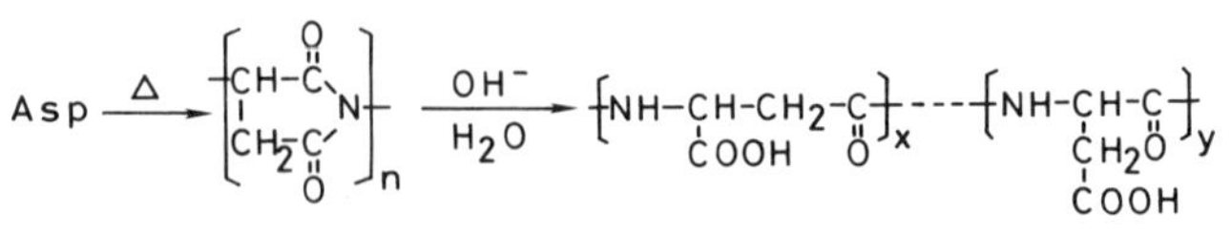

The aqueous thermal polycondensation of asparagine was assumed to form a β-peptide bond because of the intermolecular transamidation of the β-amide group. However, poly(aspartic acid) obtained by refluxing aqueous solution of asparagine or isoasparagine, contains both α- and β-peptide bonds (see Table 3). From these results, Harada et al. (1978) proposed that the 5-membered imide structure is a probable intermediate in the interconversion of α- and β-aspartyl residues during the aqueous thermal condensation:

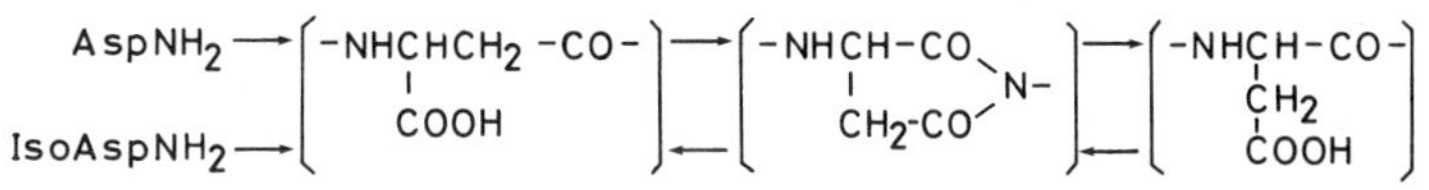

If this is the case, the ratio of α- and β-linkages might be determined thermodynamically.

Glutamyl residue. Glutamic acid is converted by heating to pyrrolidone-2-carboxylic acid, which does not polymerize but reacts with other amino acids to yield the corresponding copolymers. In the formation of the copolymer of glutamic acid, two possible mechanisms could be considered: (a) the condensation between the carboxyl group of the lactam and the amino group of the amino acid, and (b) the reaction of the lactam ring with the amino group of the amino acid. The titration data have indicated that no α-linked glutamyl residue is present in copoly(glu, ala) (Table 4). Therefore, it becomes apparent that the lactam ring reacts with other amino acid to form γ-peptide bonds by a transamidation reaction (mechanism b):

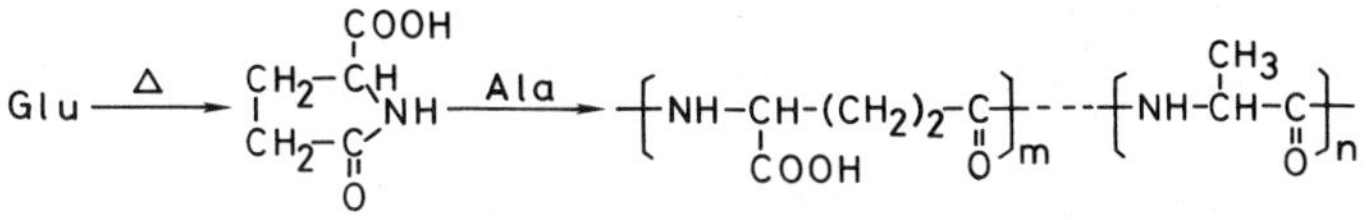

This is supported by the fact that the amount of the imide residues in copoly(glu, ala) is very small (Kokufuta et al., 1978).

Lysyl residue. It has been found that the thermal polymers of lysine are composed mainly of ε-linked residue (Table 5). This could indicate the following peptide bond propagation mechanism. Lysine is first converted by heating to the 6-membered α-amino lactam.* The lactam ring then reacts with the amino group of the

*If the lactam and free lysine coexist in the initial state during the course of heating, the α-amino group of free lysine may react

other lactam or amino acid to form ε-lysyl residues by a transamidation reaction. The reaction of the lactam ring with amino groups in lysine peptide could form a branching point.

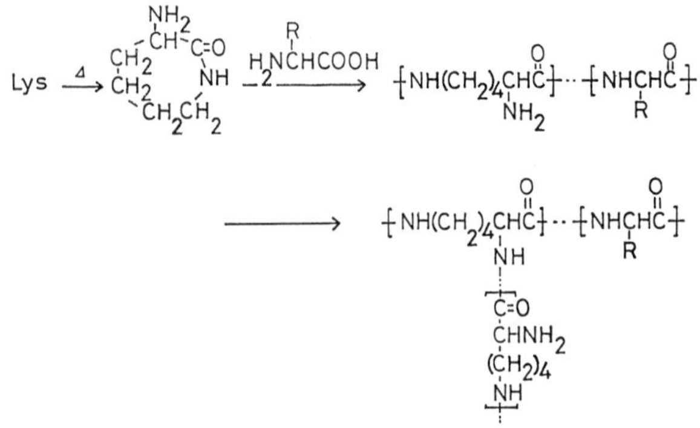

General discussion. The α- and ω-linkage formation mechanisms of aspartyl, glutamyl, and lysyl residues have been described by investigating the titration behavior of the thermal homo- and copolymers containing such amino acid residues. As was mentioned previously, these thermal polymers could be regarded as models for proteinoids. Thus, the mechanism proposed here could provide information about the proteinoid formation under prebiotic conditions.

REFERENCES

Andini, S., Benedetti, E., Ferrara, L., Paolillo, L., and Temussi, P. A., 1975, Origins Life, 6:147.
Ciferri, A., Puett, D., Rajagh, L., and Hermans, J., 1968, Biopolymers, 6:1091.
Edelhoch, H., and Bateman, J. B., 1957, J. Amer. Chem. Soc., 79: 6093.
Fox, S. W., 1965, Nature, 205:328.
Fox, S. W., 1976, J. Mol. Evol., 8:301.
Fox, S. W., and Dose, K., 1977, "Molecular Evolution and the Origin of Life," 2nd edn., pp. 138-200, Marcel Dekker, New York.
Fox, S. W., and Harada, K., 1958, Science, 128:1214.
Fox, S. W., and Harada, K., 1960, J. Amer. Chem. Soc., 82:3745.
Fox, S. W., and Harada, K., 1966, in: "A Laboratory Manual of Analytical Methods of Protein Chemistry," P. Alexander and H.P. Lundgren, eds., Vol. 4, pp. 129-151, Pergamon Press, New York.

with the lactam ring to form an α-peptide bond. However, the titration data show that this is not the main pathway in the peptide bond propagation.

Fox, S. W., Harada, K., and Rohlfing, D. L., 1962, in: "Polyamino Acids, Polypeptides, and Proteins," M. Stahmann, ed., p. 47, University of Wisconsin Press, Madison.
Fox, S. W., Harada, K., Woods, K. R., and Windsor, C. R., 1963, Arch. Biochem. Biophys., 102:439.
Harada, K., 1959, Bull. Chem. Soc. Jpn., 32:1008.
Harada, K., Matsuyama, M., and Kokufuta, E., 1978, Polym. Bull., 1:177.
Harris, F. E., and Rice, S. A., 1955, J. Polym. Sci., 15:151.
Hayashi, K., 1977, "Tanpakushitsu no Denkiteki-seishitsu (in Japanese) (Electrochemical Properties of Proteins)," p. 53, University of Tokyo Press, Tokyo.
Heinrich, M. R., Rohlfing, D. L., and Bugna, E., 1969, Arch. Biochem. Biophys., 130:441.
Hoagland, P. D., and Fox, S. W., 1967, J. Amer. Chem. Soc., 89:1389.
Katchalsky, A., Mazur, J., and Spitnik, P., 1957, J. Polym. Sci., 23:513.
Kokufuta, E., 1979, Macromolecules, 12:350.
Kokufuta, E., and Harada, K., 1981, in: "Origin of Life," Y. Wolman, ed., pp. 277-284, Dordrecht, Holland.
Kokufuta, E., Suzuki, S., and Harada, K., 1977, BioSystems, 9:211.
Kokufuta, E., Terada, T., Suzuki, S., and Harada, K., 1978, BioSystems, 10:299.
Kokufuta, E., Terada, T., Tamura, M., Suzuki, S., and Harada, K., 1979, Arch. Biochem. Biophys., 196:23.
Kovács, J., and Nagy, H., 1961, Nature, 191:531.
Kovács, J., Könyves, I., and Pusztai, A., 1953, Experientia, 9:459.
Kovács, J., Kovács, H. N., Könyves, F., Csaszar, J., Vajda, T., and Mix, H., 1961, J. Org. Chem., 26:1084.
Matsuyama, M., Kokufuta, E., Kusumi, T., and Harada, K., 1980, Macromolecules, 13:196.
Meggy, A. B., 1956, J. Chem. Soc., 1444.
Pivcová, H., Saudek, V., and Drobnik, H., 1982, Polymer, 23:1237.
Rohlfing, D. L., 1967, Arch. Biochem. Biophys., 118:468.
Shaw, R. W., and Hartzell, C. R., 1976, Biochemistry, 15:1909.
Shepherd, E. J., and Kitchener, J. A., 1956, J. Chem. Soc., 2448.
Speakman, J. C., 1940, J. Chem. Soc., 855.
Tanford, C., 1969, "Physical Chemistry of Macromolecules," Chap. 8, Wiley, New York.
Temussi, P. A., Paolilio, L., Ferrara, L., Benedetti, E., and Andini, S., 1976, J. Mol. Evol., 7:105.
Thiele, H., and Gronau, K. H., 1963, Makromol. Chem., 59:207.
Treslong, C. J. B., and Staverman, A. J., 1974, Recl. Trav. Chim. Pays-Bas, 93:171.
Vegotsky, A., Harada, K., and Fox, S. W., 1958, J. Amer. Chem. Soc., 80:3361.
Waehneldt, T. V., and Fox, S. W., 1968, Biochim. Biophys. Acta, 160:239.
Waley, S. G., 1955, J. Chem. Soc., 517.

ON PROTEINOID STRUCTURE

Paul Melius and Vasilios Nicolaou

Department of Chemistry
Auburn University
Auburn, AL 36849

The structure of proteinoids should be of special interest to the peptide chemist and those studying the prebiotic chemical evolution preceding the origin of life. Fox and his colleagues (1980) have experimentally discovered a number of interesting properties of the proteinoids (thermal polyamino acids) such as their ability to form microspheres (protocells) in aqueous solutions and specific catalytic activities which would be necessary to form polymers. A number of parameters can influence the polymerization of the amino acids such as: reactivities, steric constraints of the amino acid side chains as well as the growing polymer chain, stability of intermediates and products, hydrogen-bonding interactions and amino acid composition of the polymerization mixture. A great deal of information is necessary on structure of the polymers in order to understand their ability to coalesce into microspheres and ability to act as specific catalyts.

The formation of amino acids can occur abiotically as analysis of meteorites and laboratory experiments such as Miller's (1955) have demonstrated. Thus small molecules such as hydrogen cyanide, methane, ammonia, ammonium cyanide, etc., are able to react under appropriate conditions to form a large number of amino acids. Fox (1968) and his colleagues have demonstrated that mixtures of amino acids containing glutamic acid and/or lysine and aspartic acid can be polymerized by heat to produce peptide polymers called proteinoids to distinguish them from biotically synthesized proteins. Contrary to predictions, the experimental evidence indicates that these polymers are nonrandom with regard to their composition and their amino acid sequence.

THERMAL PROTEINOIDS

The most striking characteristic in the thermal proteinoids as prepared by the Fox and Harada (1960) procedures is the pyroglutamic (pyroGlu) N-terminal group (Phillips and Melius, 1974) when the polymerization mixture contains glutamic acid. The N-terminal pyroGlu group was identified by Hubbard (1973) and Sheng (1974) using pyrrolidine carboxyl peptidase to remove the end group and dilute NaOH at 60°C to convert it to a Glu group.

Other generalizations that we can make are that acidic amino acids produce smaller polymers in the range of 4,000 to 12,000 molecular weight. When Lys is present in significant concentrations, polymers of 100,000 (Heinrich et al., 1969) molecular weight were observed by gel electrophoresis and untracentrifugation. Asquith et al. (1970) found that enzymatic digests of wool keratin contained ε-(γ-L-glutamyl)-lysine and that this peptide increased when the wool was pretreated by heating. So we might expect the same reactions in lysine thermal copolymers to lead to cross-linking and thus larger sized polymers than when glutamic acid is the only trifunctional amino acid present in the polymerized mixture. Fox and coworkers (Melius, 1982) demonstrated that the composition of amino acids in the polymerization mixture had a profound effect on the amino acids incorporated in the polymer. This fact has been experimentally verified by other investigators. The different amino acid composition of the polymers as compared to the amino acid composition of the polymerization mixture, suggested an internal nonrandomness in the amino acid distribution (Fox, 1976). Thus the evidence of limited heterogeneity or the nonrandom character in the proteinoids is; N-terminal and C-terminal amino acid composition, limited number of species by electrophoresis, ultracentrifugation, DEAE-cellulose chromatography, paper chromatography and amino acid composition of polymers as compared to the polymerization mixture (Melius, 1977).

CHEMICALLY ACTIVATED PROTEINOIDS

Many techniques have been used for activating amino acids for peptide bond formation from cyanide and nitrite compounds to diimides, N-carboxyanhydrides and adenylates which come closer to biological polypeptide formation. Steinman and Cole (1967) polymerized amino acids in water using dicyanamide (DCA) and found that the nature of the side group of the amino acid, at least for nonpolar residues, appeared to be a major factor in the type of dipeptides produced. They found a striking similarity of dipeptide frequency in the synthetic dipeptides as compared to egg lysozyme, ribonuclease and sheep insulin. A further interesting finding in Steinman and Cole's work, was that an increase in size of peptide in case of Gly caused an increase in adsorption to montmorillonite which was used to augment the DCA-mediated peptide synthesis. The peptides were also more reactive in condensing with the monomer. It was suggested that the kinetic acceleration of chain growth was autocatalytic. It was further suggested that one peptide could serve to catalyze the synthesis of a second, the second that of a third and so forth. As a test, the dimerization of Ala and Leu was studied in the absence and presence of polyglycine (0.25% w/r). The yields of the two peptides were increased an average of 38 per cent upon addition of the polymer. Thus the size of the side chain in amino acids, at least of nonpolar residues, appears to be an important factor in the polymerization of amino acids under the influence of DCA. Ponnamperuma and Peterson (1965) used cyanamide (CA) to form peptides from glycine and leucine in water. They were able to identify Gly-Gly, Gly-Leu, Leu-Gly, Leu-Leu, and Gly-Gly-Gly by chromatography.

Amino acid adenylates were used by Banda and Ponnamperuma (1971) to form polypeptides. Katchalski (1967) earlier had shown that the amino acid adenylates spontaneously polymerize at alkaline pH in water. Proline and alanine each were found to yield oligopeptides up to the tetramers. Fox and coworkers (1969) reported the polymerization of a mixture of adenylates where all the protein amino acids were incorporated into the polymer. Banda and Ponnamperuma (1971) found that peptide fragments were formed during the adenylate stage, but an even greater amount resulted from the alkaline polymerization of the adenylates. They used a mixture of sixteen amino acids common to proteins except for β-Ala. They found that Ala, Gly, Ser, Asp, Pro were all incorporated more extensively in the polymer that was formed.

The polymer obtained was nonrandom as only two distinct fractions were obtained and their amino acid composition did not reflect that of the starting mixture. Also electrophoretic separation resulted in a small number of distinct bands. It is not known where the selectivity is effected; at the adenylate formation step or the polymerization step. However adenylate formation of Ala, Ser, and Leu are nearly the same.

Both Fox and Ponnamperuma found that Ala, Gly, Glu, Asp, and Leu are incorporated preferentially into the polymers formed from adenylate activated amino acids. Weber and Orgel (1979) prepared dipeptides by the reaction of 2'(3') O-(glycyl)-adenosine-5'-(O-methyl phosphate) and amino acid at pH 8.2. Gly-Ser was obtained in 35.3% yield, whereas Gly-Cys in 11.8% and Gly-Thr (5.4%). The other 12 amino acids produced dipeptides in yields less than 2%. So there is a decided difference in the incorporation of amino acids into polymers under various reaction conditions.

In just examining the amino acid composition of thermally polymerized amino acids or polymers obtained by amino acid activation such as use of acyladenylates, carbodiimides or cyanamides, we can see the profound effects of the reaction medium on specific incorporation of amino acids. Thus a wide variety of amino acid sequences must occur in these amino acid polymers.

PROTEINOIDS OF LIMITED AMINO ACID COMPOSITION

As an example of the reaction conditions on incorporation of amino acids into polymer, Nakashima et al. (1970) found that magnesium ions caused a decrease in the incorporation of basic amino acids when the amino acyl adenylates were polymerized. When formyl methionine, acetylalanine and acetylserine were added to the reaction mixture, there was an increased uptake of those amino acids with complete elimination of cystine from the polymer formed and with decreases in many of the other amino acids.

Incorporation of amino acids into the thermally produced polymers has been reviewed by Fox (1968) and Melius (1977). The effects of amino acid composition of the mixture being polymerized on the composition of the polymer have been extensively studied. Usually if an amino acid is present in a high ratio it will be incorporated effectively into the polymer that is formed. However significant quantities of a trifunctional amino acid such as Glu, Asp or Lys is necessary for polymerization and not just thermal pyrolysis of the amino acids.

In order to reduce the complexity of the chemistry taking place, mixtures of two to eight amino acids have been investigated. Inoue and Yamada (1979) have obtained pyrogultamyl dipeptides by heating glutamic acid with another amino acid on an equimolar basis at 175°C. Glu and Ala gave a 26% yield of pyroGlu-Ala. About 3% pyroGlu-Ala-Ala was obtained and 2% pyroGlu-Glu. Yields of pyroGlu dipeptide ranged from 10% for Tyr to 80% for Lys. When Glu, Gly, and Ala were used, pyroGlu-Gly-Ala, pyroGlu-Ala and pyroGlu-Gly were all produced. A large proportion of pyroGlu remained unreacted.

Nakashima et al. (1977) thermally polymerized a mixture of Glu, Gly and Tyr, and isolated two tripeptides, which were identified by mass spectrometry after permethylation of the peptides. The spectrometric analysis indicated the two tripeptides to be: pyroGlu-Gly-Tyr and pyroGlu-Tyr-Gly. Thus two tripeptides were formed out of a possible 36 (Fox, 1980) different peptides utilizing the three amino acids present in the polymerization. Mixtures of amino acids such as those used by Philips and Melius (1974), Nakashima et al. (1977) and Inoue and Yamada (1979), which do not contain Lys, produce ninhydrin negative peptides. This has been demonstrated to be due to the N-terminal pyroGlu group. The pyroGlu may be an initiating or terminating group for the peptide formation. This does limit the number of possible peptides that can be formed. For example, where dipeptides are major products: (polymerization mixture containing Glu and Gly) pryoGlu-Gly and pyroGlu-Glu would be the products, however we can combine Glu and Gly into four dipeptides; Glu-Glu; Glu-Gly; Gly-Glu; and Gly-Gly.

In case of three amino acids Glu, Gly, and Tyr used by Fox (1980) there is a possibility of 36 tripeptides when α and γ peptide bonds to Glu are considered. Experimentally only two tripeptides were isolated as previously stated. This work by Nakashima et al. (1977) was confirmed by Hartmann et al. (1981). Hartmann et al. (1981) obtained experimental evidence for the rapid conversion of Glu to pyroGlu which had earlier been reported by Phillips and Melius (1974). The pyroGlu then reacted with cyclic Gly-Tyr to produce pyroGlu-Tyr-Gly and pyroGlu-Gly-Tyr.

GENERAL CHARACTERISTICS

The formation of pyroglutamic acid limits the number of potential peptides that can be formed because the α-amino group of pyroGlu is internally blocked, allowing peptide bond formation only at the α-carboxyl group. However in large polymers containing more than one residue of Glu, as determined by acid hydrolysis of the polymer, and subsequent amino acid analysis by acid hydrolysis of the polymer, and subsequent amino acid analysis by the ninhydrin reaction, some of the glutamic acid residues have to be non-pyroGlu for example where the polymer is linear or does not contain any R groups allowing branched structure (Phillips and Melius, 1974).

Another striking characteristic of the polymers obtained by the thermal polymerization process is the limited number of fractions obtained when the various chromatographic and electrophoretic separation procedures are used (Phillips and Melius, 1974; Melius and Sheng, 1975; Melius, 1977; Dose, 1971). In many instances such as the thermal polymerization of Glu with two to six other amino acids such as Ala, Phe, Gly, Met, Leu, Val, two to three paper chromatographic fractions have been obtained with a fraction containing all the amino acids and other fractions containing Glu and some but not all the other amino acids included in the polymerization mixture. A mixture of Glu, Met, Phe, and Ala produced two chromatographic fractions, one containing all four amino acids and a second with Glu and Ala in a ratio of 1:6 (Melius, 1983). The fraction was ninhydrin negative so we can propose a tentative structure:

$$\text{pyroGlu-(Ala)}_x\text{-Ala}$$

Ala was found to incorporate very readily into thermal polymers as well as polymers where activated amino acids were used (Fox, 1968; Melius, 1977; Krampitz and Fox, 1969). Also a number of reports of -Ala-Ala- sequences from thermal amino acid polymers have been made.

Melius and Sheng (1975) thermally polymerized a mixture of six amino acids: Glu, Ala, Gly, Leu, Pro, and Phe. When the polymerized reaction products were separated by paper chromatography, three peptide fractions were observed by the Rydon-Smyth procedure as the peptides are ninhydrin negative. Also all the free amino acids were detected by ninhydrin reagent. When the pyroGlu N-terminal end groups were converted to Glu with trifluoroacetic acid treatment, they were detected by dansylation and chromatographic identification of dansyl-Glu. Hydrazinolysis was used to determine the C-terminal amino acids which were Gly for peptide fraction 1, Ala for fraction 2 and Leu for fraction 3. Thus the major peptide in each fraction was:

Fraction 1: pyroGlu-----------Gly
Fraction 2: pyroGlu-----------Ala
Fraction 3: pyroGlu-----------Leu

In fractions 1 and 3, Ala content was very low 7.1 mole % and 2.8 mole % respectively. Fraction 2 with an Ala content of 24.4 mole % did have Ala as the major C-terminal amino acid. In fractions 1 and 3 the Gly + Leu content was 49.5 mole % and 44.1 mole % respectively so the Gly and Leu C-terminal amino acids in those two peptides is not surprising. It is of interest that fraction 1 did not contain any Phe or Pro whereas fraction 2 had 8.5 moles % of the two and fraction 3 had 16.1 mole % of the two. Fractions 2 and 3 had approximately twice the molecular weight of fraction 1. An interesting question is whether one of the polymers is a precursor for the others or if all three types polymerize simultaneously. From a comparison of the composition of the three types of polymer it appears most probable that they are all formed simultaneously.

The preceding discussion has indicated the nature of the experimental evidence for the nonrandom arrangement of amino acids in thermal polymers of amino acids. It is also known (Melius, 1977) that Asp forms α- and β-amide as well as imide links, Glu forms α- and γ-links and Glu can form cross-links with Lys. Many nonprotein amino acids will also polymerize into the proteinoids. So, in this regard, proteinoid structure can be more complex than protein structure. However, in regard to ordered amino acid sequences, the proteins are much more ordered from all the currently available experimental information.

The experimental information discussed here would tend to leave us with the impression that a mix of amino acids polymerizes into a static mix of proteinoids. That is not correct as we know from such work as Johnstone and Povall (1975) in their work on conversion of oligopeptides into cyclic dipeptides as a method of peptide sequencing. When oligopeptides are heated or acid catalyzed, the N-terminal amino group attacks the penultimate peptide carbonyl group to split out a cyclic dipeptide, and the reaction can proceed down the chain. Interestingly if the N-terminal group is a pyroGlu residue no cyclic dipeptide is formed. Of course transpeptidation and acid and alkaline catalyzed hydrolysis of proteins are well known. The time and temperature of polymerization and other conditions control the size of the polymers that are formed. Thus the formation and modification of proteinoids results in a very dynamic state of affairs in regard to the type and content of any proteinoid mixture at any specific time. A protocell, of the proteinoid microsphere type, would thus have a wide variety of molecules to select from and would also have the possibility of modifying those molecules when taken up from the environment. This implies a very complex, dynamic, chemical system forming a variety of proteinoid molecules which could undergo selection processes which could lead to more ordered proteinoid systems.

REFERENCES

Asquith, R. S., Otterburn, M. S., Buchanan, J. H., Cole, M., Fletcher, J. C., and Gardner, K. L., 1970, Biochim. Biophys. Acta, 221:342.
Banda, P. W., and Ponnamperuma, C., 1971, Space Life Sciences, 3:54.
Dose, K., and Zaki, L., 1971, Z. Naturforsch., 26b:144.
Fox, S. W., and Harada, K., 1960, J. Am. Chem. Soc., 82:3745.
Fox, S. W., 1968, Encycl. Polym. Sci. Technol., 9:294.
Fox, S. W., 1976, in: "Protein Structure and Function," J. L. Fox, L. Deyl, and A. Blazej, eds., p. 125, Marcel Dekker, New York.
Fox, S. W., and Nakashima, T., 1980, BioSystems, 12:155.
Fox, S. W., 1980, Naturwiss., 67:576.
Hartmann, J., Brand, M. C., and Dose, K., 1981, BioSystems, 13:141.
Heinrich, M. R., Rohlfing, D. L., and Bugna, E., 1969, Arch. Biochem. Biophys., 130:441.
Hubbard, W. L., 1973, "Thermal peptides: An attempt to remove the pyrrolidone carboxyl residues from the N-terminal ends," M.S. thesis, Auburn University, Auburn, Alabama.
Inoue, C., and Yamada, T., 1979, "A synthetic method for pyroglutamyl peptides," Jpn. Kokai Tokkyo Koho, 79:19,967.
Johnstone, R. A. W., and Povall, T. Jeffrey, 1975, J. Chem. Soc. Perkin Trans., 1:1297.
Katchalski, A., and Ailam, G., 1967, Biochim. Biophys. Acta., 140:1.
Krampitz, G., Baars-Diehl, S., Haas, W., and Nakashima, T., 1968, Experientia, 24:140.
Krampitz, G., and Fox, S. W., 1969, Proc. Nat. Acad. Sci. U.S., 62:399.
Melius, P., and Sheng, J.Y.P., 1975, Bioorg. Chem., 4:385.
Melius, P., 1977, in: "Bioorganic Chemistry, Vol. III, Macro- and Multimolecular Systems," E. van Tamelin, ed., Academic Press, New York.
Melius, P., 1982, BioSystems, 15:275.

Melius, P., 1983, unpublished work.
Miller, S. L., 1955, J. Am. Chem. Soc., 77:2351.
Nakashima, T., Lacey, J. C., Jungck, J., and Fox, S. W., 1970, Naturwiss., 57:1.
Nakashima, T., Jungck, J. R., Fox, S. W., Lederer, E., and Das, B. C., 1977, Int. J. Quant. Chem., QBS4:65.
Phillips, R. D., and Melius, P., 1974, Int. J. Peptide Protein Res., 6:309.
Ponnamperuma, C., and Peterson, E., 1965, Science, 147:1572.
Sheng, J. Y. P., 1974, "Studies on thermal polymerization of six amino acids," M.S. thesis, Auburn University, Auburn, Alabama.
Steinman, G., and Cole, M. N., 1967, Proc. Nat. Acad. Sci. U.S., 58:735.
Weber, A. L., and Orgel, L. E., 1979, J. Mol. Evol., 13:185.

PYROPHOSPHATE AS A POSSIBLE PRECURSOR OF ATP

Fritz Lipmann

The Rockefeller University

New York, New York 10021

I am very pleased to participate in honoring Dr. Sidney Fox in this anniversary volume. I am ever so grateful that he invited me to the memorable 1965 conference on the Origins of Prebiological Systems. I proposed there the search for biochemical fossils in present-day organisms (Lipmann, 1965). As a rare example, I mentioned then the pyrophosphate (PP_i) linked reaction of Siu and Wood (1962):

$$\text{Oxaloacetate} + PP_i \rightarrow \text{Phosphoenolpyruvate} + CO_2 + P_i$$

where PP_i appears to replace the ATP. It was proposed that PP_i may be considered the precursor of ATP in evolution. Such a possibility has recently been re-emphasized in various laboratories, particularly by Harland Wood et al. (1977) and by R. Reeves (1982). Reeves has worked predominantly with entamoeba histolytica, a parasitic anaerobe and emphasizes a PP_i-linked phosphorylation in glycolysis in the case of 6-P fructophosphokinase. Included here are other interesting reactions such as phosphorylation of serine by PP_i.

Harland Wood's review (Wood et al., 1977) deals thoroughly with many cases of pyrophosphate linked transfer of energy rich phosphate substituting for the use of ATP. They were found in his laboratory, using largely propionic acid bacteria. Recently, he published a fascinating report (Wood et al., 1972) on the analysis of metabolic pathways found with an acetogenic autotroph (Clostridium thermoaceticum) which grows on CO/CO_2, and H_2, and may convert glucose into three moles of acetate (Wood et al., 1982). He proposed that the conditions in prebiological time may have favored this form of life using CO_2 and H_2 as the source of carbon and some forms of cobalt, nickel, selenium, and tungsten as catalysts: They may likely have included methane and mercaptans (Shapiro and Wolfe, 1980) together with a methyl-coenzyme M, CH_3-S-$(CH_2)_2$-SO_3 as methyl donor and PP_i or polyphosphates instead of ATP. This parallels with recent work in Harry Peck's laboratory (Liu et al., 1982) on a sulfate reducing bacterium, desulfovibrio, that in the activation of sulfate for its reduction uses:

$$ATP + SO_4^{2-} \rightarrow APS \text{ (adenylsulfate)} + PP_i .$$

In this reaction the PP_i-energy remains unused and the organism has to use electron transfer to generate ATP. However, they mention that an anaerobe of this genus uses:

$$\text{acetate} + PP + ADP \rightarrow \text{acetate} + P_i + ATP$$

and found that desulfomaculatum nigrificans would grow anaerobically in presence of yeast extract, acetate, and PP_i. Since in the marine environment presumably PP_i may be absent, they expect polyphosphate formation and breakdown in nature as had been discussed (Harold, 1966). The just reviewed recent work deals with anaerobic bacteria of a kind not far removed from the earliest representatives of life found on earth. This, one might assume, to be illustrated by a plot of evolutionary change against the age of the earth by Schopf (1970), modified by Lipmann (1976). There, the prebiotic period is placed around the first one-half billion of years of the earth: the cooling period. This development is generally considered to have taken place anaerobically due to an earth-atmosphere devoid of oxygen (Oparin, 1970). These recent results support the proposition to further search for parallels in present organisms replacing the more complex molecules available now.

For quite a while we have in this laboratory been following up reports by Kurahashi et al. (Tomino et al., 1967), concerning the mechanism of the biosynthesis of antibiotic peptides beginning with gramicidin S. and extending to tyrocidin and the linear of gramicidin in extracts of a B. brevis. These antibiotics are synthesized by an enzymatic method differing from the ribosomal one and less complex, indicated already by the work of Kurahashi et al. (Tomino et al., 1967). Their synthesis had been observed by an ATP linked reaction in extracts of a B. brevis treated exhaustively with ribonuclease. We found the ATP linked with amino acid activation to result in their thioesterification on polyenzymes (Gevers et al., 1969) followed by a polymerization through the swinging arm of a pantethein-peptide as seen early in fatty acid synthesis (Lipmann, 1973). I am not anymore sure of this to be a model for an early mechanism of amino acid activation. However, I am encouraged by a letter I received some years ago from Dr. Lars Onsager who, after retiring from Yale, went to Miami and became interested in the origin of life. He mentioned in this letter that in a lecture at an earlier Lindau meeting (unpublished) he had reflected on the way by which life might have originated. Among other considerations, he had discussed an amino acid activation by way of thioesterification without knowing about our recent finding of such a pathway. The similarity between this type of amino acid activation and polymerization and acylate activation and polymerization in long chain fatty acid synthesis I have discussed somewhat daringly in a review, "Attempts to Map a Process Evolution of Peptide Biosynthesis" (Lipmann, 1971).

REFERENCES

Gevers, W., Kleinkauf, H., and Lipmann, F., 1969, Proc. Nat. Acad. Sci. U.S., 63:1335.

Harold, F. M., 1966, Bact. Rev., 30:772.

Lipmann, F., 1965, in: "Origins of Prebiological Systems," S. W. Fox, ed., p. 259, Academic Press, New York.

Lipmann, F., 1971, Science, 173:875.

Lipmann, F., 1973, Accounts of Chem. Res., 6:361.

Lipmann, F., 1976, in: "Reflections on Biochemistry," A. Kornberg, G. L. Horecker, L. Cornudella, and J. Oro, eds., p. 34, Pergamon, New York.

Liu, C. L., Hart, N., and Peck, Jr., H., 1982, Science, 217:363.

Oparin, A. I., 1970, Biol. Rev., 45:91.

Reeves, R. E., 1976, TIBS, March Issue, p. 53.

Shapiro, S., and Wolfe, R. S., 1980, J. Bacteriology, 141:728.

Schopf, T. W., 1970, Biol. Rev., 45:319.

Siu, P. M. L., and Wood, H. G., 1962, J. Biol. Chem., 237:2164.

Tomino, S., Yamada, M., Itoh, H., and Kurahashi, K., 1967, Biochemistry, 6:2552.

Wood, H. G., O'Brien, W. E., and Michaelis, G., 1977, Adv. Enzymol. Vol. 45, Wiley, New York.

Wood, H. G., Drake, H. L., and Hu, S., 1982, Proc. Biochem. Symposia 29, pub. by Editors Annual Review, Inc.

SELF-ORGANIZING MOLECULAR SYSTEMS

Jerome J. Wolken

Mellon Institute
Carnegie-Mellon University
Pittsburgh, Pennsylvania 15213

INTRODUCTION

Pasteur in 1860 showed by an ingenious set of experiments that life could only come from existing life. This dispelled for a time belief in the spontaneous generation of life, but, in 1878, Pasteur wrote: "I have been looking for it (spontaneous generation [author's italics]) for 20 years but I have not yet found it, although I do not think it is an impossibility." In this same context, Charles Darwin, in a letter to Joseph Hooker in 1871 that is widely quoted wrote:

> It is often said that all the conditions for the first production of living organisms are now present, which could ever have been present. But if (and- oh, what a big if) we could conceive in some warm little pond, with all sorts of ammonia and phosphoric salts, heat, electricity, etc., present, that a protein compound was chemically formed ready to undergo still more complex changes, . . .

Darwin and Pasteur in their own ways dwelled on the possibility that life occurred as a spontaneous event when the physical and chemical conditions were right.

A new impetus for seeking the origin of life emerged in the 1930s with the writings of Bernal (1933, 1951, 1967), Haldane (1929, 1966) and Oparin (1938, 1968), but remained dormant until the 1950s when advances in biochemistry suggested new approaches to the search for the origin of life. These experiments and various resulting hypotheses have been summarized in Blum (1968), Calvin (1969, 1975),

Dauvillier (1965), Fox (1965a,b; 1973, 1980a,b), Fox and Dose (1972), Kenyon and Steinman (1969), Keosian (1964), Ponnamperuma (1965, 1972), Wald (1952, 1964), by others and in numerous symposia. More than a century after Darwin's death the subject of the origin and evolution of life is of greater interest now among scientists than it has ever been.

To briefly review, Oparin (1938, 1968) proposed that the prebiological atmosphere was reducing and consisted of hydrogen, methane, ammonia, nitrogen, and water vapor, and that the first carbon compounds were hydrocarbons. Haldane (1928, 1966) assumed that the primitive atmosphere contained CO_2, NH_3, and water, but no oxygen, and he claimed that such a mixture exposed to ultraviolet radiation would give rise to a vast variety of organic compounds. Miller (1953, 1955, 1957) was one of the first to test this hypothesis and was able to synthesize amino acids in a prebiological simulation experiment. He applied electric discharges to a mixture of methane, ammonia, hydrogen, and water. Among the amino acids identified were glycine, alanine, β-alanine, aspartic acid, α-amino-n-butyric acid, and other organic compounds, but purines and pyrimidines were absent.

When similar gaseous mixtures were subject to temperatures of 80-90°C, ultraviolet radiation, ionizing radiation, and electric discharge, analysis of the contents of these reactions chambers showed that small quantities of a great variety of organic molecules had been produced. These included many organic acids of low molecular weight, amino acids, urea and simple sugars including ribose. Therefore, many organic molecules of considerable biological importance were synthesized in the prebiological environment and have been synthesized in the laboratory without the agency of any living cells or cellular material.

How did the molecules in the prebiological environment self-order to form an assembly of macromolecules? Bernal (1951) suggested that clays could provide sufficient surfaces for organic molecules to be adsorbed, and, in the presence of a catalyst, the synthesis of macromolecules could have rapidly taken place. As an example, vinyl monomers, either adsorbed on the external surfaces or between the lamellae of clay, polymerize. Rao et al. (1980) also concluded that clays may have played a role in the very early polymerization reactions. The hypothesis of crystalline-organic particles bound by adsorption of organic molecules onto surfaces of mineral clay is supported by recent findings in which amino acids were isolated from meteorites and found to be L-amino acids.

Molecules can be brought together by being built into a crystal, in which case the lattice forces will hold the molecules together. Therefore, crystallization could have served as a template for a self-replicating system. In the process of crystal growth, the progressive accretion of molecules on a crystal could have occurred. Cairns-Smith (1971) developed a model in which the first functional macromolecules, proteins, replicated on templates. These templates may have been microscopic crystals of various kinds of clay containing trace metallic ions that could have acted as catalysts. Cairns-Smith (1971, 1982) further points out that the adsorption of amino acids or protein chains to the surface of a particular template will be specific, depending on the configuration of the organic molecules. That is, aggregates of clay particles and protein chains could have specific properties and could react to changes in the environment.

Another view is that organic molecules separated at different phase boundary layers in the sediment strata during certain temperature stages in the development of the earth's surface. Evidence is found among layers where organic materials in agglomerates contained phospholipids that could have served as boundary membrane layers for the synthesis of complex organic molecules into macromolecular structures. Hydrocarbons were not only present in clays but were probably present on oceanic surfaces. Hydrocarbons on the oceanic surface, subjected to ultraviolet radiation, could have led to a phospholipid-water bilayer, which in turn formed microspheres. Thus, primordial cells could have spontaneously arisen in the environment. Therefore, radiation, temperature variations, and diurnal alterations were important factors for the synthesis of organic molecules, their polymerization and aggregation into macromolecular protocells.

Attempts to create structures that resembled cells from a variety of organic and inorganic molecules go back to the experiments of Leduc (1911). The possibility of building high molecular weight compounds from dilute colloidal solutions was of interest to Bungenberg de Jong (1936). Oparin (1938) applied these studies in his search to find a dynamic colloidal system as a model for cellular behavior. One such model system was originally made by mixing gelatin with gum arabic in water to 42°C until a clear solution was obtained; others were made with gelatin and lecithin and a variety of other substances. The important observation Oparin made was that at a critical pH, microspheres came out of solution; he named these "coacervates." These are colloidal microspheres with osmotic properties that have many features associated with living cells. Oparin and his associates continued to experiment with this model system demonstrating a variety of biochemical reactions.

An approach based on the hypothesis that amino acids were present in the prebiological environment was begun by Fox in the early 1960s to bring about the polymerization of amino acids. Initially, the method consisted of heating a mixture of amino acids to temperatures of 160-200°C for several hours under anhydrous conditions in an atmosphere of nitrogen. The mixture contained aspartic acid, glutamic acid, and lysine (Fox, 1965a,b; Fox et al., 1963).

Using thermal polycondensation they were able to copolymerize amino acids, which they named proteinoids. When dry proteinoid is treated with hot water, microspheres separate out of the cooling clear solution. The proteinoid microspheres vary in size from a few microns to 100μ in diameter, are relatively stable, and, depending on their preparation, exhibit living cell-like behavior. That is, they possess a cell membrane, bud, coalesce, increase in size and divide. Fox (1980a,b) and his collaborators have demonstrated that proteinoid protocells are able to catalyze a considerable number of biochemical reactions.

It is a long way from the synthesis of organic compounds and macromolecular structures, i.e., from chemical evolution, to the origin of life and to biological evolution. Although bits of experimental data suggestive of biochemical origins are being put together, much information and crucial data bearing on the biochemical origins of life remains to be deciphered. Since we do not know at present how life originated on earth, we can continue to search for experimental models that will give us some clues and insight into how life may have originated. While models may be far from the living state, they do though provide a system that simulates many processes which we associate with a living cell.

LIQUID CRYSTALLINE SYSTEMS

Properties and Structures of Liquid Crystals

A model which seems applicable to the question of how molecules became self-ordered and assembled into macromolecular structures is that of liquid crystalline systems (Brown and Wolken, 1979). Such models have not been explored in depth to the problem of biogenesis, but it is of interest to see to what extent they may contribute to our understanding of self-aligning and replicating molecular structures. In order to do so, it will be helpful to describe briefly liquid crystalline structures and their properties before drawing any analogies to the cell.

Freidrich Reinitzer (1888), an Austrian botanist, discovered the "liquid crystalline" state. He prepared the cholesteryl ester, cholesteryl benzoate, and observed that it had two "melting points" with different properties. At 145.5°C, the solid cholesteryl benzoate structure collapsed to form a turbid liquid (now known to be a liquid crystal), and when further heated to 178.5°C, it become transparent. Lehmann (1904) made a systematic study of organic compounds and found that many of them exhibited properties similar to cholesteryl benzoate, in that the compounds behaved both as a liquid in their mechanical properties and as a crystalline solid in their optical properties. Lehmann (1904, 1922) originated the term "liquid crystal" and recognized that such properties exhibited by liquid crystals are analogous to those of the living state. Then Rinne (1933) and Bernal (1933, 1951) pointed out that naturally occurring liquid crystals are intimately connected with life processes.

Although the terms "mesomorphic states," "paracrystals" and "anisotropic liquids" are used to describe the properties of these structures, they are commonly referred to as *liquid crystals*. There are two major types of liquid crystals: the thermotropic and the lyotropic. Thermotropic liquid crystals are formed by heating, and lyotropic liquid crystals are formed by mixing two or more compounds. The thermotropic liquid crystals are divided into two groups described as *nematic* and *smectic*.

Nematic liquid crystals are structurally different from isotropic liquid crystals in the spontaneous orientation of the molecules along their long axis. In the nematic form the molecules maintain a parallel or nearly parallel arrangement to each other. They are mobile in three directions and can rotate about one axis. The smectic structure (except smectic D) is stratified with the molecules arranged in layers. Their long axes lie parallel to each other in the layers, approximately normal to the plane of the layers. The molecules can move in two directions in the plane, and they can rotate about one axis. Within the layers the molecules can be arranged either in rows or can be randomly distributed.

Included with the nematic liquid crystals is a subclass that is considered *cholesteric-nematic*, but is referred to as *cholesteric*. Many of these compounds are derivatives of cholesterol (Fig. 1). In the cholesteric liquid crystals, the molecules pack in layers. Although most of the molecules in the cholesteric state are essentially flat, side chains project upward from the plane of each molecule with some hydrogen atoms extending below. Thus, the direction of the long axis of the molecules in a chosen layer is slightly displaced from the direction of the axis in adjacent layers and produces a helical structure.

A nematic liquid crystal can be converted to a cholesteric liquid crystal by a molecule that has an asymmetric atom. Cholesteric liquid crystals (cholesteryl esters) are miscible with nematic liquid crystals forming a helical structure. Molecules associated with living cells, polysaccharide polymers, polynucleotides, polypeptides, DNA, *t*RNA, assemble in a cholesteric helical structure.

The lyotropic liquid crystals differ from thermotropic liquid crystals for they are obtained by dispersing a compound with another compound, one of which is a solvent. They are strongly birefringent. Most detergents, soaps, and surfactants dispersed in water will form lyotropic liquid crystals (Fig. 2). These compounds are amphiphiles for they possess in their molecular structure an ionic

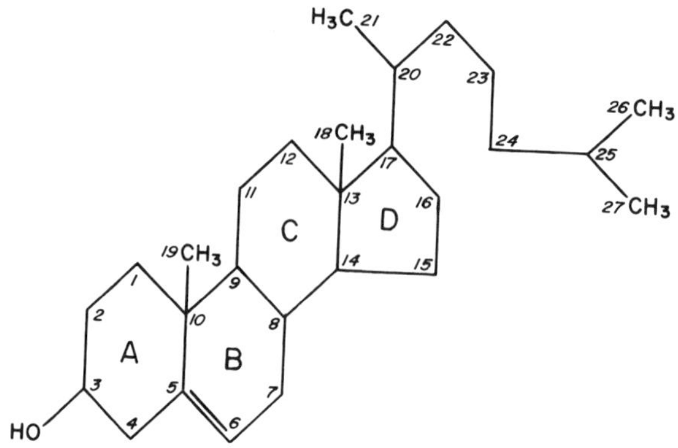

Fig. 1. Cholesterol molecular structure.

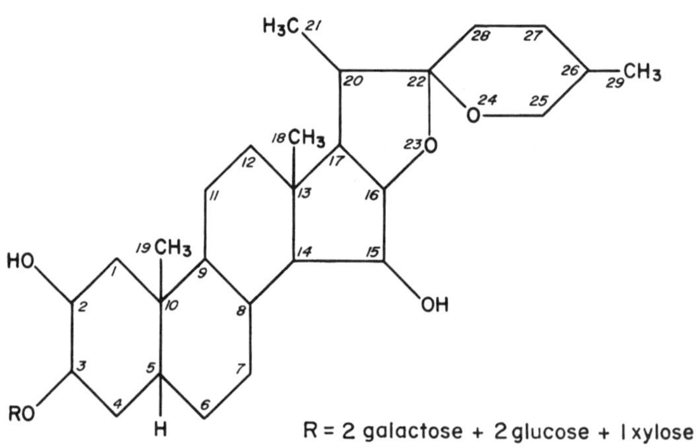

Fig. 2. Digitonin molecular structure, digitalis glycoside, a nonionic detergent.

group that is water soluble and an organic part that is insoluble in water.

Starting with the crystalline form of an amphiphile and water, a series of structures can be generated from the crystal to the true solution. With certain combinations the polymorphic mesophases formed will show molecular lamellar structures, cubic structures, and hexagonal structures. Removing water can reverse the order of mesophase formation. These can be represented schematically as follows:

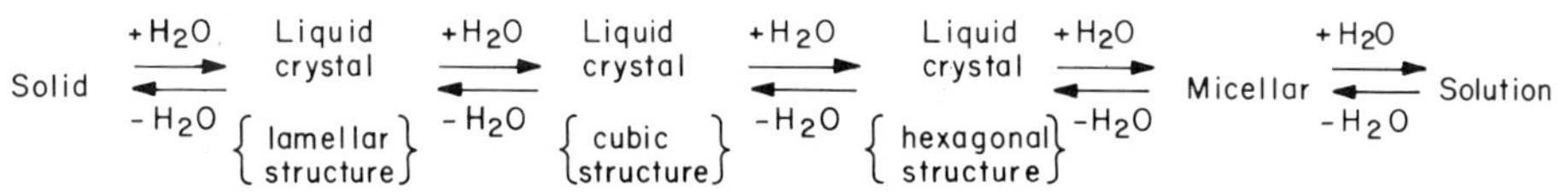

In the cubic structure the molecules pack in a spherical pattern, and the spheres then pack in a cubic design. In the hexagonal structure the molecules pack in a cylindrical pattern, and the cylinders or rods pack hexagonally.

Lyotropic liquid crystals are biologically important for they contain two or more compounds (e.g., lipid-water; lecithin-cholesterol-bile salts-water; lipid-water protein systems) in which water is an integral part of these liquid crystalline structures. Cellular membranes are structurally bilayers of lipid-protein-water systems; they are liquid crystalline structures.

The general classification of liquid crystals is indicated in Fig. 3, and these liquid crystalline structures are schematically illustrated in Fig. 4. In the nematic state the long axes of the molecules lie essentially parallel (Fig. 4a), while in the smectic A structure (Fig. 4b), the molecules show two-dimensional order. Within a layer the molecules are randomly distributed, while between layers the molecular arrangement is equally spaced. The molecules in smectic C liquid crystals are packed in equidistant layers as in the smectic A liquid crystals, but the molecules in a given layer are tilted in relationship to the plane of the layer. The tilt angle is sensitive to temperature and the molecular geometry of the molecules. In the cholesteric structure the molecules are arranged in each layer like those in the nematic structure, but a necessary twist is superimposed on the layers, resulting in a helical structure (Fig. 4c) or in a helical rod (Fig. 4d).

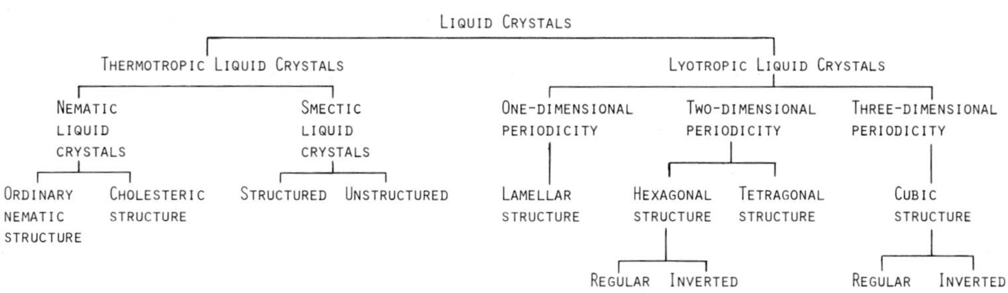

Fig. 3. Classification of Liquid Crystals, Modified from Brown (1977) and Brown and Wolken (1979).

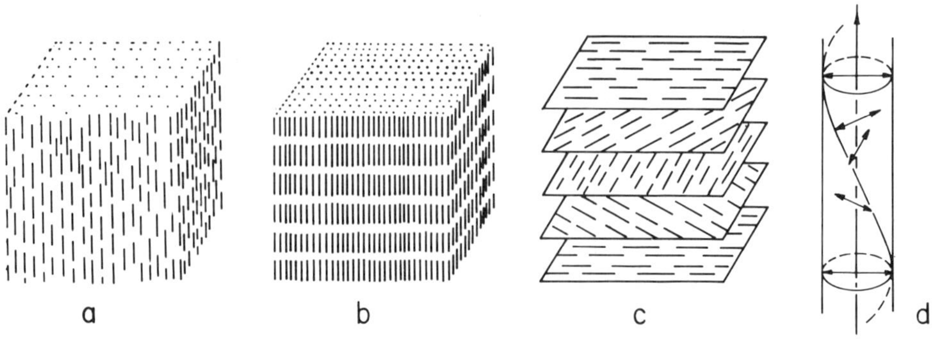

Fig. 4. Structure of the molecular arrangement in three main types of liquid crystals.
a) nematic, the elongated molecules are randomly distributed.
b) smectic, molecules are ordered in layers (lamellae) of equal thickness.
c) cholesteric, the molecules rotate regularly from plane to plane, and (d) helix of a cholestric rod.

The important characteristic of liquid crystals is that they possess both mobility and structural order. Liquid crystals respond to a variety of external stimuli: light, mechanical pressure, temperature, electric and magnetic fields, as well as to changes in the chemical environment. These behavioral properties are those associated with living cells.

Liquid Crystal Cell Model

The properties and structures described for various liquid crystals suggest how a liquid crystalline system may have organized into a protocell. The characteristics of living cells are their stability, their response to environmental stimuli and their ability to reproduce exact copies of themselves. Simple chemistry cannot account for such behavior. The liquid crystal cell should have the structure and behavior of living cells.

The chemical and mechanical properties of liquid crystals are applicable in considering a cell model that has a solid core of nematic liquid crystalline material and is surrounded by a smectic liquid crystal (Fig. 5). The interior of the cell is a nematic liquid crystal that has a number of important features which may be utilized. In the self-ordering nematic structure, the long axes of the molecules are essentially parallel (Fig. 4a), and, as they line up, they generate a "layer" of molecules that serves as a substrate on which chemical reactions can take place. Simple organic reactions such as isomerization and enzymatic oxidation-reduction reactions

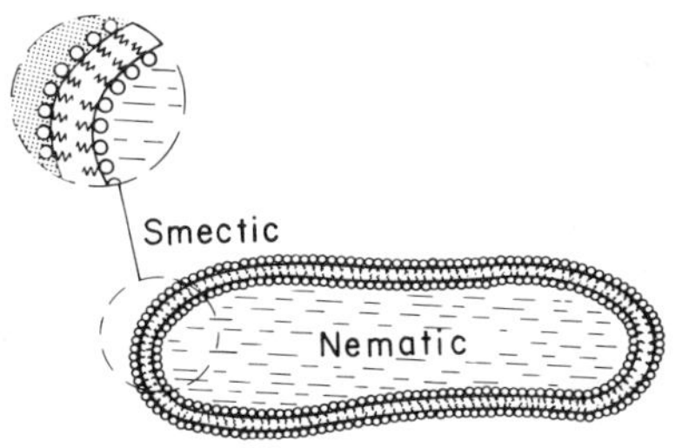

Fig. 5. A schematic of a liquid crystal protocell. (According to Fergason and Brown, 1968, and Brown and Wolken, 1979).

can occur on these liquid crystalline surfaces. The effects of chemical changes will result in changes of shape, and internal distribution of molecules will be accomplished at a much greater rate than would be possible by other states of matter. The surface energy in a droplet of nematic liquid crystal is not a constant but varies as a function of the molecular alignment with the surface.

The smectic liquid crystalline phase has many dynamic properties just like the nematic phase. The cell membrane that has as its principle component a smectic liquid crystal would be able to transfer ions across it (Fig. 4b). If a smectic film contacts a region that lowers its surface energy, it will tend to expand its area of contact. Thus, a smectic film will engulf materials from the environment that tend to lower its surface energy. If a smectic film encloses an isotropic liquid, the shape formed will be determined by the curvature. The shape of liquid crystalline droplets that are uniformly aligned may exhibit a variety of forms from an oblate to a prolate spheriod, depending on the properties of the liquid in which they are immersed. The liquid crystalline structure responds readily to energy changes, and the model could have many other behavioral properties of a living cell (Fergason and Brown, 1968; Brown and Wolken, 1979).

> . . . the oriented molecules in liquid crystals furnish an ideal medium for catalytic action, particularly of the complex type needed to account for growth and reproduction. . . . a liquid crystal has the possibility of its own structure, singular lines, rods and cones, etc. Such structures belong to the liquid crystal as a unit and not to its molecules, which may be replaced by others without destroying them, and they persist in spite of the complete fluidity of the substance. They are just the properties to be required for a degree of organization between that of the continuous substance, liquid or crystalline solid, and even the simplest living cell. (J. B. Bernal, 1933)

It is of interest to examine the kinds of molecules and their properties from which the liquid crystalline cells are structured in comparison to a living cell.

Proteins and Polypeptides

The role of liquid crystals in protein chemistry and structure has only recently become of interest (Brown and Mishra, 1971; Welch et al., 1982). Proteins are composed of amino acids structured together via peptide linkages that form polypeptides. Their structural units consist of about 20 different amino acids that are strung together in chains hundreds to thousands of units long, in different proportions, in all types of sequences, and with a great variety of branching and folding. An infinite number of different proteins is possible for no two species possess exactly the same proteins.

Proteins are generally divided into the fibrous, which are composed of elongated filamentous chains, and the globular, with a considerable amount of folding of the long polypeptide chain. X-ray studies of native proteins and synthetic polypeptides led Linus Pauling (1960) to propose a structure of greatest stability, the α-helix. The α-helix has a spiral chain of repeating amino acids held together by hydrogen bonds (Fig. 6). The helix contains about 3.6 amino acid residues for each full turn of the spiral. Another type of structure is the β-configuration, or pleated sheet, in which two or more peptide chains are tied together laterally by hydrogen bonding. Wherever hydrogen bonding occurs, a crystalline structure is observed. This hydrogen bonding exists when the crystal is dissolved in water to form the liquid crystal. Low molecular weight proteins with water can generate their own liquid crystalline structures, especially if the protein possesses a high degree of polarity.

Turning to the polypeptides from which proteins are structured, the synthetic polypeptides have been the most extensively studied class of macromolecular liquid crystals (Dupré and Samulski, 1979). Robinson (1956, 1958, 1961, 1966) carried out a series of classical studies of the synthetic polypeptide, poly-ν-benzyl-L-glutamate (PBLG). PBLG is a homopolymer of the L-enantiomorph of a single amino acid with the basic repeat unit as shown below:

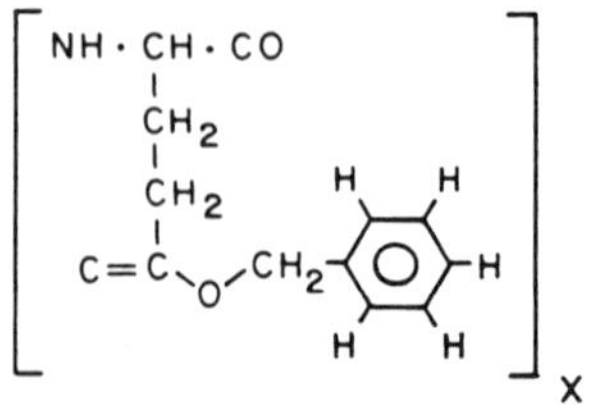

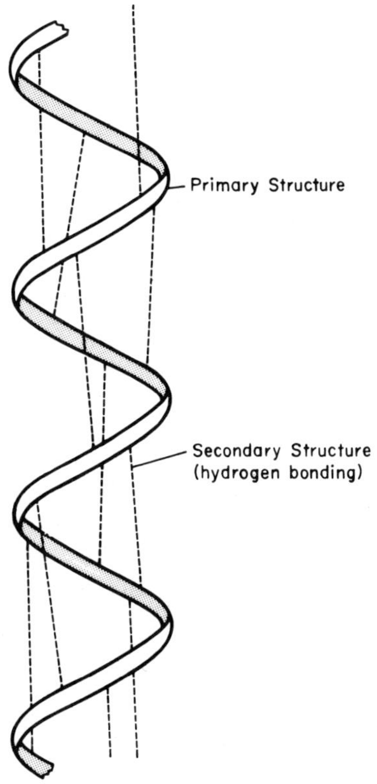

Fig. 6. Schematic model of a right-handed α-helix, as in the structure of proteins.

In concentrated solutions, the polypeptide PBLG is birefringent. The solutions which show birefringence develop parallel and equally-spaced narrow bands which are alternately bright and dark. The spacing between adjacent dark or bright bands is periodic and varies from 2 to 100 μm, depending on the concentration, the solvent, and the temperature. The observed periodicity is relatively independent of the molecular weight of the polymer.

Microscopic observations show that the solute has a twisted structrure with a pitch that can be observed optically in the visible region of the spectrum. If the concentration of the glutamate PBLG is increased, the pitch of the helix decreases. This decrease in pitch can continue until the system looks like a cholesteric liquid crystal (Fig. 4c,d).

Solutions containing equal concentrations of the laevo PBLG and the dextro enantiomorph (mirror image form) show no optical activity and have all the properties of a nematic liquid crystal. Mixtures of solutions of equal concentrations of PBLG in methylene chloride and 1,4-dioxane also give a nematic liquid crystal structure (Fig. 4a).

The polypeptide chain is wound into the α-helix structure. (Note the comparison between Fig. 6 and Fig. 7.) The structure has adjacent coils bound together by hydrogen bonds formed between neighboring -C=O- and NH groups that are four peptide bonds apart. The side chains, $-CH_2CH_2-COOCH_2C_6H_5$, project radially from the helix, as illustrated in Fig. 7. The polymer that is formed behaves as a stiff rod with projecting side units. Other polypeptides also show these properties.

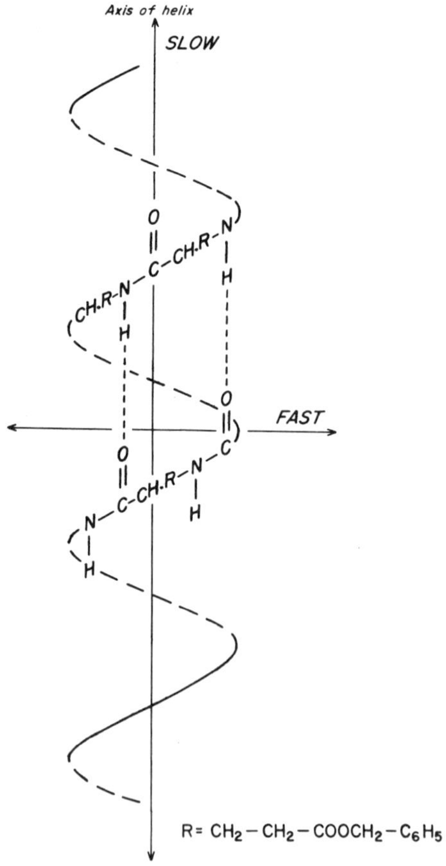

Fig. 7. The α-helix of the synthetic polypeptide, poly-ν-benzyl-L-glutamate (PBLG) showing hydrogen bonds between adjacent coils.

The search for self-ordering and replicating structures has suggested from these studies of Robinson (1961, 1966) that liquid crystals provide a model system for investigating self-assembly and replicating structures. For example, poly-ν-benzyl-L-glutamate in chloroform forms a large number of spherulites upon evaporation. The spherulites coalesce and result in a birefringent material that spontaneously orients itself at the air interface. The polypeptide-rich phase is birefringent and departs from the more dilute medium in the form of spherical liquid droplets or spherulites. If the two-phase solution is cooled or the polypeptide concentration further increases, the droplets grow in size and coalesce, forming a birefringent fluid, a lyotropic liquid crystal. The birefringent structure shows microscopic visible periodicities and indicates a helical arrangement. Spherulites (Fig. 8), like proteinoids, are enclosed by a membrane and divide by passing through an oblate spheroid phase.

Many molecules, particularly elongated molecules in solution, will self-organize and orient. The orientation is such that the long axes of the molecules align parallel to one another. This alignment can take place in the liquid or liquid crystalline state.

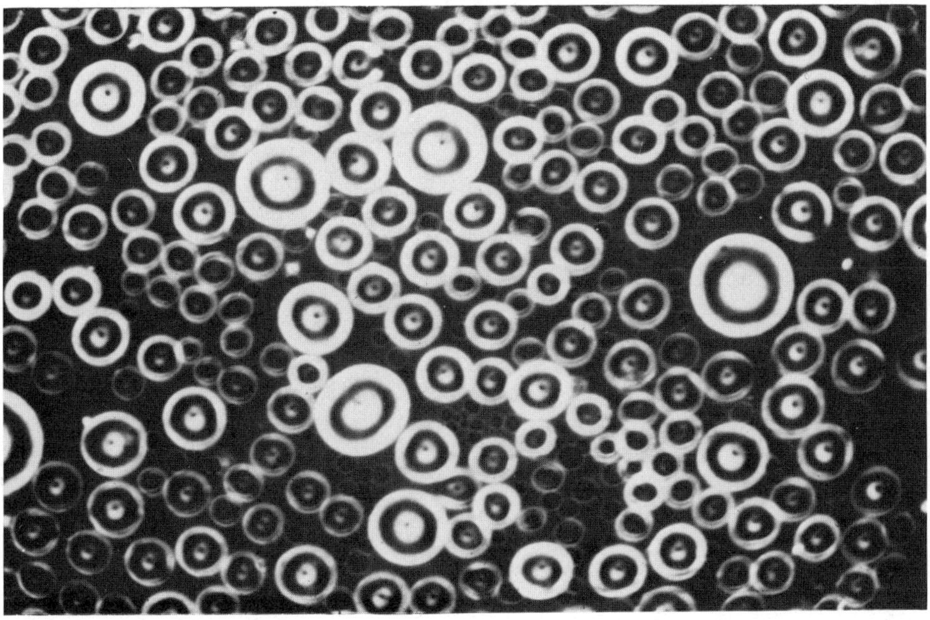

Fig. 8. Spherulites formed near the transition temperatures between nematic liquid crystals and the isotropic liquid. The spherulites are in a dynamic state; the small spherulites coalesce with larger ones and then divide.

Liquid crystals that possess one- and two-dimensional order will self-organize and orient. This "self-orientation" exhibited by spherulites provides a model then for how molecules form oriented structures that are of biological importance (Brady, 1973, 1974; Filas, 1977; King, 1959; Robinson, 1966).

The aggregation of chain molecules is important in the chemistry of liquid crystalline structures, in the formation of micelles, and as a possible mechanism by which the structural formation of cellular membranes took place.

Optical Activity

Another question that one must consider is optical activity. The "primordial soup" from which life is believed to have arisen would have to contain nearly equal numbers of laevo (L) and dextro (D) asymmetric molecules. The fact that only L optical isomers occur in the amino acids in polypeptides formed of living organisms is still open to question. This has challenged biochemists since Pasteur's early experiments with crystals of tartaric acid and sodium ammonium tartrate isomers (racemic forms). These experiments led Pasteur (1878) to an important observation that asymmetric molecules are always the products of living processes. This appeared to Pasteur to be one of the fundamental differences between the chemistry of life and that of inanimate matter. Pasteur discovered that living organisms selected only the L-amino acids in constructing their proteins. He postulated that this peculiar asymmetry might be the manifestation of asymmetric forces in the environment acting at the time of molecular synthesis on the evolving cells. The optical activity of organic compounds on the primitive Earth could have been accelerated by an asymmetric catalyst (Harada, 1971). Experimental studies on the heat polymerization of simple organic molecules, such as polypeptides and ribose, have found that either the pure L or the pure D optical form is more stable than the racemic mixture. The seemingly asymmetrical distribution of these optically active substances in living organisms may be a result of only one pure optical isomer surviving at a very early stage in evolution. Wald (1964, 1970) expressed this as natural selection on the molecular level.

Liquid crystalline systems may be helpful in looking at the asymmetry of molecules that have biological importance. A molecule that has an asymmetric atom can convert a nematic liquid crystal to a cholesteric liquid crystal. The resulting structure will rotate polarized light. A mechanically twisted nematic liquid crystal cell can be prepared in the laboratory by rubbing two glass surfaces in the same direction and then placing them so that the direction of rubbing of one is perpendicular to the direction of rubbing of the other. A nematic liquid crystal placed between these two planes will rotate polarized light through 90°.

That a similar situation in the evolutionary process may have occurred is worth speculation, for if organic chiral compounds were synthesized, a nematic liquid crystal was already in the environment. If a solid surface (inorganic or organic) were exposed to environmental forces, it would develop charges on the surface. The charges on the surface would interact with the polar groups in the liquid crystal molecules, thus orienting the liquid crystal. The surface charges could come from ions in the environment, from directed abrasion on the surfaces, from the action of water, and from wind moving in a given direction. These various forces acting on the surface will align the charges and, as a result, will orient the liquid crystal molecules so that their long axes are parallel. If another surface is similarly acted on, and these two surfaces make contact so that one is perpendicular to the other, a twisted nematic liquid crystal can be formed. The resulting twisted nematic liquid crystal can be used as a solvent for carrying on organic reactions that result in the preparation of D or L or a racemic mixture.

Thus, as a result of natural forces, an optically active compound could have been synthesized in a twisted nematic liquid crystal (to give the D or L form, depending on whether the twist of the liquid crystal solvent has a left-handed or right-handed twist). Such an optically active molecule would then have served as an intermediate for the synthesis of other cholesteric liquid crystals. The fact that the L-amino acids were isolated from meteorites suggests that such a process may have taken place in the early synthesis of polypeptides and, hence, in the assembly of proteins.

The Cell Membrane

The existence of the cell is dependent on the cell membrane. The cell membrane separates the internal environment from the external environment and provides a large surface area for bringing molecules together for interaction and energy transfer. The question then is how did the hydrocarbons in the environment self-order to form the cell membrane that was able to encapsulate other molecules from the environment for metabolic processes.

To partially answer this question we can examine the properties of lipids for the lipid bilayer is the basic structural unit of cell membranes (Danielli and Davson, 1935). The major molecules in the lipid bilayers are phospholipids (Fig. 9), and the molecular packing of these lipids dictates the skeletal structure of the membrane.

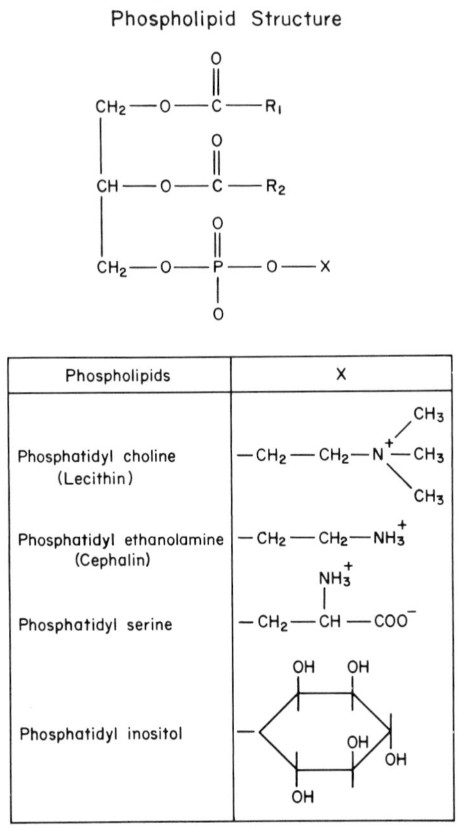

Fig. 9. Phospholipid structures.

Phospholipids are polar lipids; they possess strongly polar or charged groups (Figs. 10a-d).

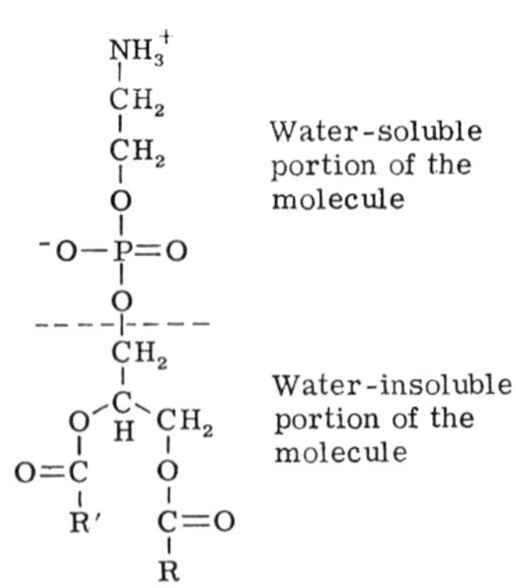

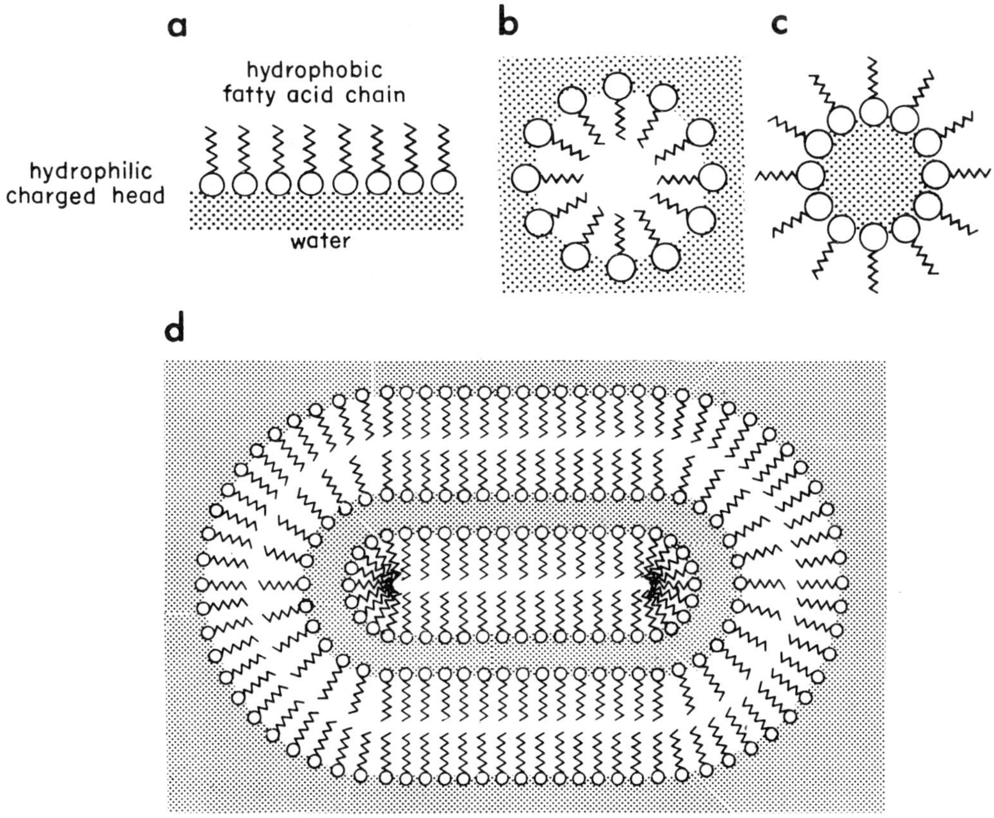

Fig. 10. (a) A monolayer of phospholipid molecules in water.
(b) The charged phosphate faces outward if the liquid is polar.
(c) They face inward if it is nonpolar. It can also exist as a combination of (b) and (c) as indicated in (d).
The phospholipids are symbolized by a circle representing the charged hydrophilic end, and the zig-zag line represents the hydrophobic fatty acid chain.

In water the charged phosphates face outward (Fig. 10a,b), and if the medium is nonpolar, they face inward (Fig. 10c). In other words, the molecules may be oriented toward water or other polar molecules, and their nonpolar groups are oriented away from the polar environment. Most phospholipids disperse molecularly in water to only a small extent, and, if large quantities are introduced into the aqueous medium, aggregates are formed, micelles. Phospholipids swell in water and form many spherical bodies composed of concentric layers, lamellae, with water trapped between them. If the spheres are surrounded by a single phospholipid bilayer, they are referred to as liposomes. Lecithin dispersed in water will form concentric bilayers, lamellae (Fig. 11), and are observed as myelin structures in cells. These lipid molecules will self-assemble into ordered replicating structures, cellular membranes. These molecules need no further specifications for alignment into an ordered structure. For example, lecithin in physiological saline will self-align into a precisely ordered replicating structural membrane, as observed in Fig. 12. All of these phospholipid mono and bilayer structures are liquid crystalline systems (Chapman, 1979).

Phospholipids depend very markedly on temperature: on heating they undergo an endothermic transition at a temperature well below the ture melting point, which would be at about 200°C. At this transition temperature, a change of state occurs from the crystalline or gel to the liquid crystalline state. This change is associated with increased conformational freedom for the lipid fatty acid chains. The transition temperature rises with increasing length of the fatty acid chain and lowers with increasing unsaturation in the the chain (Lee, 1975). Below the transition temperature, in the gel (crystalline) phase, the phospholipids adopt a bilayer structure in which the fatty acid chains are packed in ordered hexagonal arrays

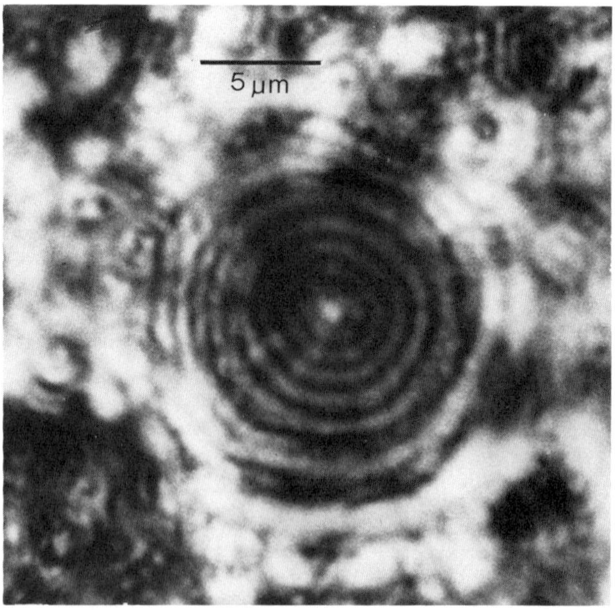

Fig. 11. Lecithin (phosphatidylcholine) dispersed in water.

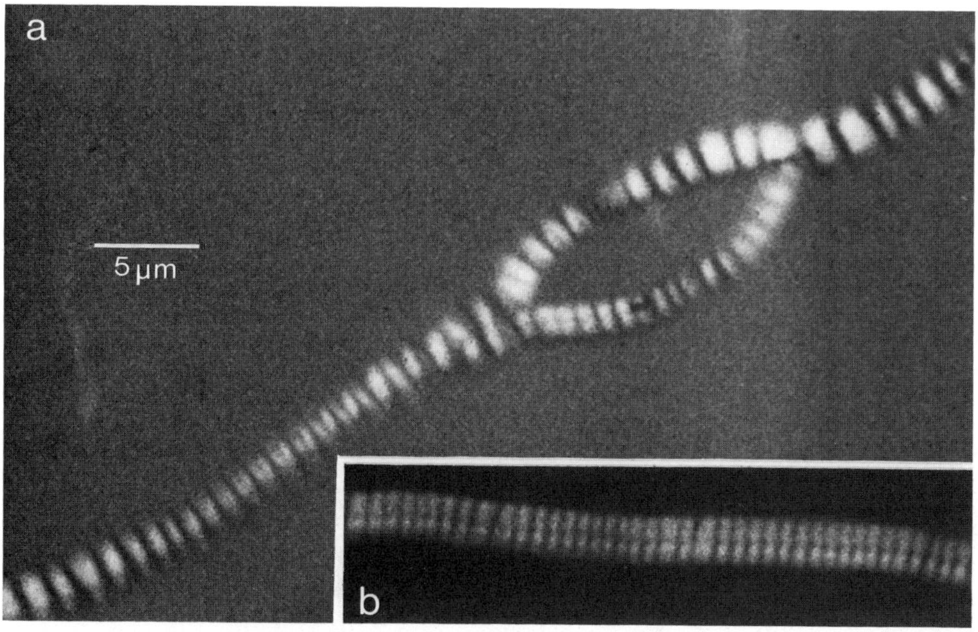

Fig. 12. Membrane formed from lecithin in physiological saline, using polarized light, a quarter-wavelength filter, and photograph taken by phase contrast microscopy (original in color).

where motion of the fatty acid chains is highly anisotropic and restricted. At the transition temperature, there is a 50% increase in the surface area occupied by the lipid, and appreciable motion becomes possible about the C-C bonds of the fatty acid. The motion about the C-C bonds in lipid bilayers has been studied by ^{13}C NMR, and ^{13}C spectra show resonances for many of the fatty acid chain carbons. It was found that there is a very marked motional gradient within the phospholipid molecule in the liquid crystalline phase. The motion about the C-C bonds with the fatty acid ^{13}C NMR data provided evidence suggestive of an axial rotation of the whole lipid molecule in the plane of the bilayer. The degrees of freedom possible for the phospholipid chains are schematized in Fig. 13.

One can say that a membrane containing phospholipids with little unsaturation is less fluid than one with much unsaturation. The control of fluidity of the components of cell membranes may be related to the diffusional characteristics of molecules and ions passing in and out of the membrane. The state of the phospholipid in a membrane, in a gel or in a liquid crystalline state, can be expected to have a marked effect on the function of the membrane.

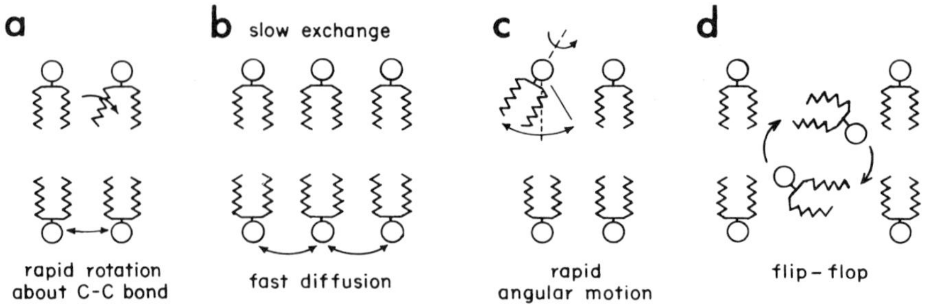

Fig. 13. Schematic showing mobility of the phospholipid hydrocarbon chain.
(a) Rapid rotation about C-C bond in the hydrocarbon.
(b) Lateral diffusion in the plane of the membrane.
(c) Angular motion of the phospholipid molecule.
(d) Flip-flop of the phospholipid molecules across the bilayer.

Thus, small molecules will be able to move relatively easily through a membrane in which the phospholipids are in a liquid crystalline state.

Proteins are molecularly associated with the lipid bilayer membrane for the lipid bilayer provides for the spatial relationship of proteins with their associated receptor molecules. Vanderkooi and Green (1971) suggested that some membrane proteins might not be electrostatic, but bimodal, that is, their protein molecules would possess both polar and nonpolar groups like the phospholipids. Proteins in a globular configuration would fit directly into a lipid bilayer, their hydrocarbon chains and their polar groups bonded to lipid heads. Singer and Nicholson (1972) proposed a model in which the protein can float in the lipid because lipids and proteins are mobile when they are in a fluid (or melted) state. Also, both fluid and solid regions may be present in the same membrane. Oldfield (1973) indicates that it is possible for membranes to contain some of their lipids in a crystalline state. These structural models for cellular membranes have been reviewed by Bretscher (1973), Capaldi et al. (1973), Eisenburg and McLaughlin (1976), and Meyers and Burger (1977).

Photochemical Activity

A liquid crystal cell system can be assembled to perform some of the photometabolic processes that we associate with photosynthesis. Surfactants are liquid crystals and when dispersed in water will form concentric lamellar structures like that of the phospholipid bilayers (Figs. 9-12). Of the non-ionic surfactants, digitonin (Fig. 2), a digitalis glyoside whose structure resembles cholesterol (Fig. 1) is of interest. Digitonin (1-2%) in water forms micelles which have a strong attraction for many complex molecules, particularly lipids and naturally occurring pigments (e.g., carotenoids, chlorophyll, heme, etc.). The role of the digitonin micelle is to react with one of the substrates while simultaneously attracting the other substrate to the same vicinity. This parallels the behavior of an enzyme in bringing the reactants together. The digitonin micelles can be used then for biological assemblies (membranes and protein tertiary structures) since the interactions responsible for micelle stability are similar to those which stabilize bio-aggregates. These supramolecular assemblies compartmentalize reacting molecules and have a pronounced catalytic effect on energy and electron transfer reactions by virtue of the potential gradients at the interface.

Chlorophyll molecules will orient in liquid crystals (Journeaux and Viovy, 1978) and a chlorophyll-lipid-protein can be structured into a liquid crystalline system (Wolken, 1967; Ke and Vernon, 1971). The chlorophyll-lipid-protein in the digitonin micelles is referred to as chloroplastin. Chloroplastin is birefringent when observed through crossed polarizers; hence, there is an alignment of the molecules in the micelle solution. Chloroplastin is photoactive for it can photoreduce a dye, evolve oxygen, and, in the presence of the right cofactors, perform some of the primary steps of photosynthesis, turning inorganic phosphate to organic phosphate (ATP) outside the living cell (Wolken, 1966, 1967, 1975). Serebrovskaya (1971) also observed similar photochemical activity with chlorophyll in surfactants-water and phosphatide-water systems.

If a drop of chloroplastin is evaporated rapidly from a surface, lamellae are formed. When the chloroplastin lamellar structure is scanned with the microspectrophotometer at the major absorption peak for chlorophyll (675mm), chlorophyll is found to be concentrated and oriented within the lamellae and not in the interspaces (Wolken, 1975; Brown and Wolken, 1979). This illustrates a process in which an amphiphile in water generates a series of structures, and by removing water a reverse in the order of mesophase formation takes place to an ordered lamellar structure. These experiments indicate a change from the micellar to lamellar liquid crystalline structure. The surfactant (digitonin) can be removed by various chemical and physical methods from the complex, and the chlorophyll-lipid-protein will remain oriented in the reconstituted membrane.

CONCLUDING REMARKS

How life originated on earth remains a msytery, but that life exists is a fact. Present thinking about the origin of life supports the view that life probably arose as a spontaneous event when the molecules in the environment found the right conditions to self-order and replicate into stable structures.

In the prebiological environment there was considerable photochemistry and chemistry resulting in the synthesis of many undetermined amino acids, protein-like molecules, lipids and phospholipids. These molecules were present on the primitive earth and on the oceanic surfaces. The physical and chemical forces in the environment shaped these molecules into particular molecular geometries that self-aligned and replicated into macromolecular assemblies. Their stability was dictated by their chemical constituents.

It is of great interest in pursuit of the origin of life to search for the mechanisms of how molecules self-order, replicate, and assemble into macromolecular structures that gave rise to protocells and hence to living cells. This search continues to be explored from experimental and theoretical considerations. Various hypotheses have been put forward by Calvin, 1969; Cairns-Smith 1982; Crick, 1981; Eigen and Schuster, 1979; Monod, 1971; Morowitz, 1967, 1981; and Prigogine, 1980; as well as by many others that have been cited.

Liquid crystalline systems as models have not been thoroughly investigated in regards to the question of biogenesis. If liquid crystal compounds originated in the primitive environment, they would have influenced the formation of micelles, coacervates, and protocells. The molecules of living cells are liquid crystalline in their chemical, mechanical, and physical properties. Liquid crystal molecules self-align and replicate into various structures that possess structural organization as well as mobility. These are the very properties and characteristics of a cell.

Therefore, a liquid crystal protocell was conceptualized. The protocell model would consist of a core of nematic liquid crystals surrounded by a smectic liquid crystal (Fig. 5). Nematic liquid crystals are self-ordering, and the long axes of their molecules are essentially parallel. As they organize, they generate a "layer" of molecules which serves as a substrate for chemical reactions to take place. The surface energy is not constant but varies as a function of the molecular alignment with the surface.

The smectic liquid crystal would form a layer, membrane, around the nematic liquid crystal and would be able to transfer ions across it. If a smectic membrane contacts a region which lowers its surface energy, it will expand its area of contact. As a result, the smectic membrane will engulf material from the environment that tends to lower its surface energy.

The shape of the liquid crystalline droplets that are uniformly aligned will vary from an oblate to a prolate spheriod, depending on the properties of the liquid in which they are immersed. The liquid crystalline protocell would respond to energy changes (e.g., light, temperature) and to changes in the chemical environment.

Proteinoids are thermal polycondensed amino acid polymers and in water give rise to microspheres, protocells. These protocells may well be liquid crystalline systems for they exhibit properties of self-assembly, growth and replication. Like the liquid crystalline protocell model, they possess a cell membrane that surrounds a molecular core. Such a system can trap molecules from the environment and exchange ions across the membrane. As a result, a variety of biochemical reactions can take place, depending on the chemical environment in which they are immersed. These are the very properties of liquid crystalline systems.

Experimental protocell models are suggestive of how molecules self-organize into a living cell and have influenced our thinking in attempts to create protobionts. As expressed by Needham (1968) ". . . to abandon our quest at this stage would be the height of folly."

REFERENCES

Bernal, J. D., 1933, Trans. Faraday Soc., 29:1082.
Bernal, J. D., 1951, "The Physical Basis of Life," Routledge & Kegan Paul, London.
Bernal, J. D., 1967, "The Origin of Life," World Publishing Company, Cleveland, Ohio.
Blum, H. F., 1968, "Time's Arrow and Evolution," 3rd edn., Princeton University Press, Princeton, New Jersey.
Brady, G. W., 1973, J. Chem. Phys., 58:3542.
Brady, G. W., 1974, Acc. Chem. Res., 7:174.
Bretscher, M. S., 1973, Science, 181:622.
Brown, G. H., 1977, J. Coll. Inter. Sci., 58:534.
Brown, G. H., and Mishra, R., 1971, J. Agr. Food Chem., 19:645.
Brown, G. H., and Wolken, J. J., 1979, "Liquid Crystals and Biological Structures," Academic Press, New York.
Bungenberg de Jong, H. G., 1936, "La Coacervation," Hermann, Paris.
Cairns-Smith, A. G., 1971, "The Life Puzzle," Oliver & Boyd, Edinburgh.
Cairns-Smith, A. G., 1982, "Genetic Takeover and the Mineral Origins of Life," Cambridge University Press, Cambridge.
Calvin, M., 1969, "Chemical Evolution," Oxford University Press, London and New York.
Calvin, M., 1975, Amer. Sci., 63:169.
Capaldi, R. A., Kamap, H., and Hunter, D. R., 1973, Biochem. Biophys. Res. Commun., 55:655.
Chapman, D., 1979, in: "Liquid Crystals," F. D. Saeva, ed., pp. 305-334, Marcel Dekker, New York.

Crick, F., 1981, "Life Itself: Its Origins and Nature," Simon and Schuster, New York.
Danielli, J. F., and Davson, H., 1935, J. Cell Comp. Physiol., 5:495.
Darwin, F., ed., 1892, "The Autobiography of Charles Darwin and Selected Letters," Appletin, New York (republished by Dover, New York, 1958, p. 220.
Dauvillier, A., 1965, "The Photochemical Origin of Life," Academic Press, New York.
Dupré, D. B., and Samulski, E. T., 1979, in: "Liquid Crystals: The Fourth State of Matter ," F. D. Saeva, ed., pp. 203-247, Marcel Dekker, New York.
Eigen, M., and Schuster, P., 1979, "The Hypercycle: A Principle of Natural Self-Organization," Springer-Verlag, New York.
Eisenburg, M., and McLaughlin, S., 1976, Bioscience, 26:436.
Fergason, J. L., and Brown, G. H., 1968, J. Am. Oil Chem. Soc., 45:120.
Filas, R. J., 1977, J. Phys., 39:49.
Fox, S. W., 1965a, Nature, 205:328.
Fox. S. W., ed., 1965b, "The Origins of Prebiological Systems," Academic Press, New York.
Fox, S. W., 1973, Naturwiss., 60:359.
Fox, S. W., 1980a, Naturwiss., 67:378.
Fox, S. W., 1980b, Comp. Biochem. Physiol., 67:423.
Fox, S. W., and Dose, K., 1972, "Molecular Evolution and the Origin of Life," Freeman, San Francisco, and in revised edition (1977) Marcel Dekker, New Yori.
Fox, S. W., Harada, K., Woods, K. R., and Windsor, C. R., 1963, Arch. Biochem. Biophys., 102:439.
Gladyshev, G. P., and Khasanov, M. M., 1981, J. Theor. Biol., 90:191.
Haldane, J. B. S., 1929, Rationalist Annual, 3 (Reprinted in Bernal, 1967, pp. 242-249).
Haldane, J. B. S., 1954, "The Biochemistry of Genetics," Macmillan, New York.
Haldane, J. B. S., 1966, "The Causes of Evolution," Cornell University Press, Ithaca, New York.
Harada, K., 1971, in: "Chemical Evolution and Origins of Life," R. Buvet, and C. Ponnamperuma, eds., pp. 71-79, American Elsevier, New York.
Journeaux, R., and Viovy, R., 1978, Photochem. Photobiol., 28:243.
Ke, B., and Vernon, L., 1971, in: "Photochromism," G. H. Brown, ed., p. 687, Wiley (Interscience), New York.
Kenyon, D. H., and Steinman, G., 1969, "Biochemical Predestination," McGraw-Hill, New York.
Keosian, J., 1964, "The Origin of Life," Van Nostrand-Reinhold, New York.
King, L. J., 1969, Bioscience, 19:505.
Leduc, S., 1911, "Mechanisms of Life," London.
Lee, A. G., 1975, Endeavour, 34:67.
Lehmann, O., 1904, "Fussige Kristalle, sowie Plästizitat von Kristallen im Allgemeinin, Moleculare Umlagerungen und Aggregatzumstandsänderugen," Englemann, Leipzig.

Lehmann, O., 1922, *in*: "Handbuch der biologischen Arbeitsmethoden," Physik-Chem. Methoden, Untersuchung des Verhaltens gelöster Stoffe; E. Arberhalden, ed., AB. III, Teil A2, pp. 123-352, Urban and Schwarzenberg, Munich.

Litster, J. D., and Birgeneau, R. J., 1982, *Physics Today*, 35:1.
Meyers, D. I., and Burger, M. M. 1977, *Chemistry*, 50:36.
Miller, S. L., 1953, *Science*, 117:528.
Miller, S. L., 1955, *J. Amer. Chem. Soc.*, 77:2351.
Miller, S. L., 1957, *Biochim. Biophys. Acta.*, 23:480.
Monod, J., 1971, "Chance and Necessity," Alfred A. Knopf, New York.
Morowitz, H. J., 1967, *Progr. Theor. Biol.*, 1:38.
Morowitz, H. J., 1981, *BioSystems*, 14:41.
Needham, J., 1968, "Order and Life," MIT Press, Cambridge, Massachusetts.
Oldfield, E., 1973, *Science*, 180:982.
Oparin, A. I., 1938, "The Origin of Life," S. Morgulis, transl. Macmillan, New York (2nd edn., Dover, New York, 1953).
Oparin, A. I., 1968, "Genesis and Evolutionary Development of Life," Academic Press, New York.
Pasteur, L., 1878, in: "Collected Works of Pasteur (Oeuvres de Pasteur)," by Vallery-Radot (1922-1939), Vols. I and II, Masson, Paris.
Pauling, L., 1960, "The Nature of the Chemical Bond," 3rd edn., Cornell University Press, Ithaca, New York.
Ponnamperuma, C., 1965, *in*: "The Origins of Prebiological Systems," S. W. Fox, ec., pp. 221-242, Academic Press, New York.
Ponnamperuma, C., 1972, "The Origins of Life," Dutton, New York.
Prigogine, I., 1980, "From Being to Becoming," W. H. Freeman, San Francisco.
Rao, M., Odom, G. D., and Oro, J., 1980, *J. Mol. Evol.*, 15:317.
Reinitzer, F. O., 1888, *Monastsch. Chem.*, 9:421.
Rinne, F., 1933, *Trans. Faraday Soc.*, 29:1016.
Robinson, C., 1956, *Trans. Faraday Soc.*, 52:571.
Robinson, C., 1958, *Trans. Faraday Soc.*, 54:29.
Robinson, C., 1961, *Tetrahedron*, 13:219.
Robinson, C., 1966, *Mol. Cryst. Liq. Cryst.*, 1:467.
Serebrovskaya, I., 1971, *in*: "Chemical Evolution and the Origin of Life," R. Buvet, and C. Ponnamperuma, eds., pp. 297-306, American Elsevier, New York.
Singer, S. J., and Nicholson, G. L., 1972, *Science*, 175:720.
Vanderkooi, G., and Green, D. E., 1971, *Bioscience*, 21:409.
Wald, G., 1952, *in*: "Modern Trends in Physiology and Biochemistry," E. S. G. Barron, ed., pp. 337-376, Academic Press, New York.
Wald, G., 1964, *Proc. Nat. Acad. Sci. U.S.*, 52:595.
Wald, G., 1970, *in*: "Miami Winter Symposia," pp. 1-32, North-Holland, Amsterdam.
Welch, G. R., Somogyi, B., and Damjanavich, S., 1982, *Prog. Biophys. Molec.* Biol., 39:109.
Wolken, J. J., 1966, *J. Am. Oil. Chem. Soc.*, 43:271.

Wolken, J. J., 1967, "Euglena: An Organism for Biochemical and Biophysical Studies," 2nd rev. edn., Appleton-Century-Crofts, New York.
Wolken, J. J., 1975, "Photoprocesses, Protoreceptors and Evolution," Academic Press, New York.

A COMPARISON OF PROTEINOID AND ALDOCYANOIN MICROSYSTEMS AS MODELS OF THE PRIMORDIAL PROTOCELL

D. H. Kenyon

Department of Biological Sciences
San Francisco State University
San Francisco, California 94132

INTRODUCTION

Perhaps more than any other investigator, Professor Sidney Fox has helped build the kind of solid empirical foundation upon which any meaningful discussion of the origin of life must be based. His work has had a significant impact on my own research on the origin of life. The proteinoid microsystem theory is one of the most important current models of protobiogenesis and I have followed its development over the years with interest and admiration. Especially impressive is the scope of this view, ranging as it does from the primordial synthesis of amino acids from simple gases to the beginnings of nucleic acid-directed polypeptide synthesis in microscopic units of bacterial dimensions (Fox, 1978; Fox and Dose, 1977).

Dr. Fox has repeatedly stressed the need to discipline one's speculations in this difficult and tantalizing field by actual laboratory data. I have appreciated this view, since all too often speculation on the origin of life has run far ahead of established facts. The result has been that the number of published hypotheses, some frankly untestable, others only partially testable, far exceeds the number of plausible models, or models worthy of serious consideration.

Needless to say, Dr. Fox's general outlook on the origin of life is quite compatible with the thesis of Biochemical Predestination (Kenyon and Steinman, 1969). Central to Fox's thought is the concept of spontaneous self-organization of organic matter; i.e., the tendency of carbonaceous matter in all stages of its development to move with relatively high probability to more complex configurations closer to the living state (Fox, 1978, 1980). The key example here is the demonstrated ability of free amino acids, under certain experimental conditions, to condense into polypeptides which exhibit some degree of sequence ordering (Fox, 1980, 1981; Fox and Dose, 1977). This process takes place in the complete absence of nucleic acids and under conditions that conceivably might have occurred on the primitive Earth. As tantalizing as these results are, the crucial question of course is whether the non-randomness in proteinoids has anything to do with biological information (Yockey, 1977, 1981). But a detailed discussion of this problem is beyond the scope of this paper.

What I will do in this essay in commemoration of Professor Fox's work is confine the discussion to one major aspect of the origin-of-life problem, the presumed formation of "protocell" units on the primitive Earth. It is in this area of the field that his contributions are perhaps the most distinctive. I shall be raising many criticisms of both my own and Dr. Fox's work, but all in the spirit of searching for a view of origins compatible with what we know in the various disciplines of science.

For the sake of discussion let us assume that there are only two alternative model systems for the primordial protocell, and that we wish to decide which of them is the more plausible. Although cogent reasons for not considering most of the other proposed "protocell" model systems may be given, I will not discuss those reasons here. My purpose is simply to compare Fox's proteinoid microsystems with one other system; i.e., aldocyanoin microspheres (Kenyon and Nissenbaum, 1976), in order to evaluate critically the strengths and weaknesses of each of these model systems. In this way I hope to clafiry some of the conceptual problems concerned with "protocell" model systems and to provide some suggestions for future research.

PROTEINOID MICROSYSTEMS

Preparation and Properties

The general features of Fox's proteinoid microsystems are well known to students of the origin of life, so I will not present a detailed description here. It will be sufficient simply to review some of the salient facts to provide a framework for the discussion to follow (Fox, 1978; Fox and Dose, 1977).

The starting mixture for the thermal preparation of proteinoids generally consists of pure protein-forming amino acids with an excess of the L- over the D-isomers (Fox and Harada, 1960; Fox and Yuyama, 1964). Other types of organic compounds considered likely to have been present on the surface of the primitive Earth; e.g., sugars, aldehydes, amines, purines, pyrimidines, porphines, fatty acids, are excluded from the reaction mixture. Typically a dry mixture of amino acids, containing a large molar excess of acidic or basic amino acid is heated under N_2 at 160-180°C for a period of several hours. (Proteinoids also form in glycine-rich amino acid mixtures and at temperatures below 100°C.) Upon cooling of the reaction mixture water is added to the glassy product causing the precipitation of proteinoids of fairly high molecular weights (3,000-9,000). In addition to α-peptide bonds proteinoids contain γ-peptide bonds and other unnatural linkages as well as branching chains.

Among the properties of these polypeptides that might be related to the origin of life are the following: (1) Non-randomness with respect to amino acid composition (compared to the composition of the initial reaction mixture), the frequency of occurrence of certain amino acids at the N- and C-termini of the polymers, and the frequency of occurrence of particular short sequences of amino acids within the polymers (Fox, 1978). (2) Weak catalytic activities for several classes of reaction including, for example, amination and light-induced decarboxylation. (3) The ability spontaneously to form uniform microscopic spheres of bacterial dimensions.

When either purified (dialyzed) or crude (undialyzed) samples of proteinoid are heated in aqueous media and allowed to cool, numerous microscopic spheres of uniform diameter spontaneously form. The spheres are of bacterial size (diameter 1-2 μ) and they often occur in doublets or larger clusters with "junctions" between the units (Fig. 1A). Electron microscopy reveals a finely granular ultrastructure with a complex, layered boundary structure. The crude preparations generally yield microspheres with greater complexity of ultrastructure than the purified preparations (Fox, 1976a). The sphere boundaries are semipermeable and the units exhibit growth by accretion and small protuberances resembling buds at the light microscope level. Weak catalytic activities of the kind detected in proteinoid solutions have been detected in the spheres. Of special interest is the reported ability of proteinoid microspheres to promote peptide and oligonucleotide synthesis (Fox et al., 1974; Jungck and Fox, 1973; Nakashima and Fox, 1980). Evidence has been reported by Fox (1974a) suggesting that these latter reactions may involve "codonic" interactions between amino acids and polynucleotides.

Conceptual Problems

While proteinoid microsystems do in fact exhibit a remarkable range of properties of possible relevance to the origin of life, there are some problems which suggest the need for further analysis and constructive criticism. For example, there is the question of the composition of the initial reaction mixtures for proteinoid synthesis. Typically these mixtures consist solely of amino acids. But if the results of the Miller-Urey primitive atmosphere simulation experiments are any guide, surely many other substances

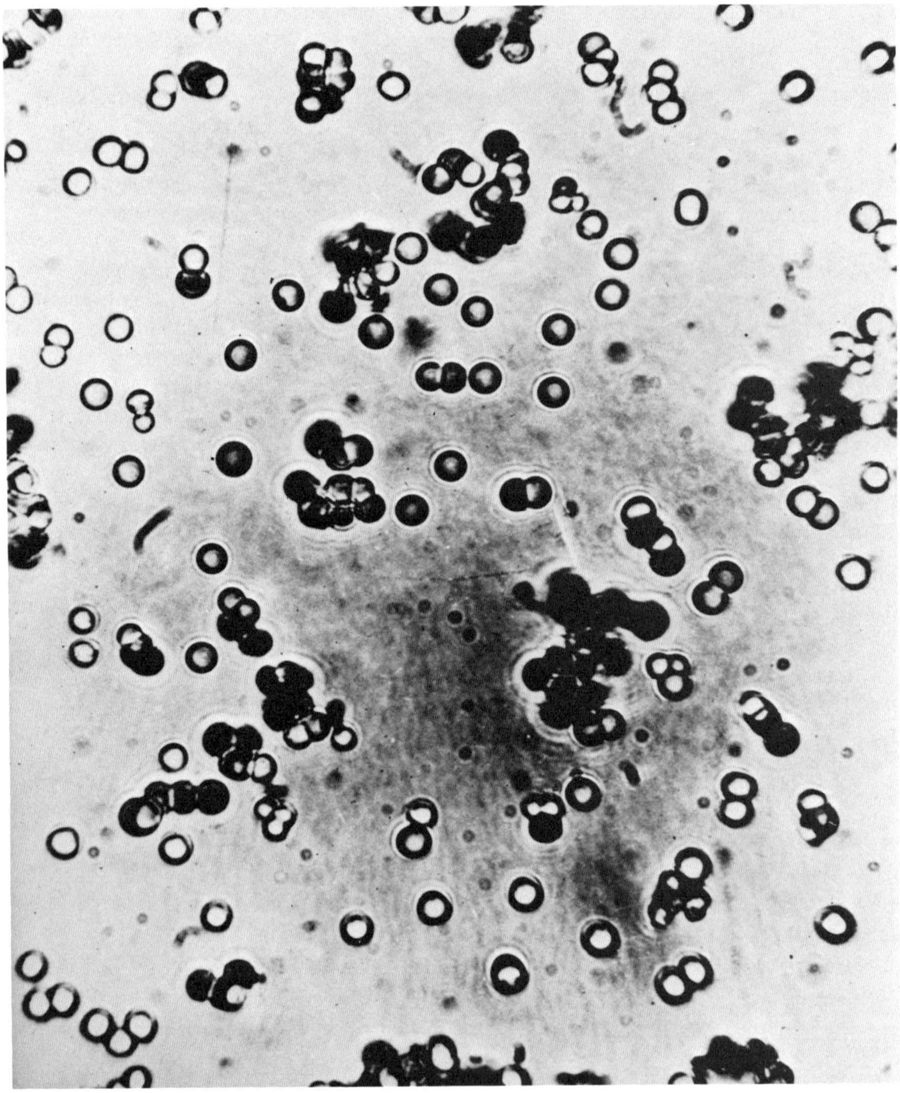

Fig. 1. (A) Proteinoid microspheres. Diameter approximately 2 μ. From Fox (1965b).

besides amino acids must have been present on the Earth's primitive surface (Fox and Dose, 1977; Kenyon and Steinman, 1969; Miller and Orgel, 1974). It would seem that a more realistic simulation of primitive conditions would require several different classes of compounds (i.e., not just amino acids) in the initial reaction

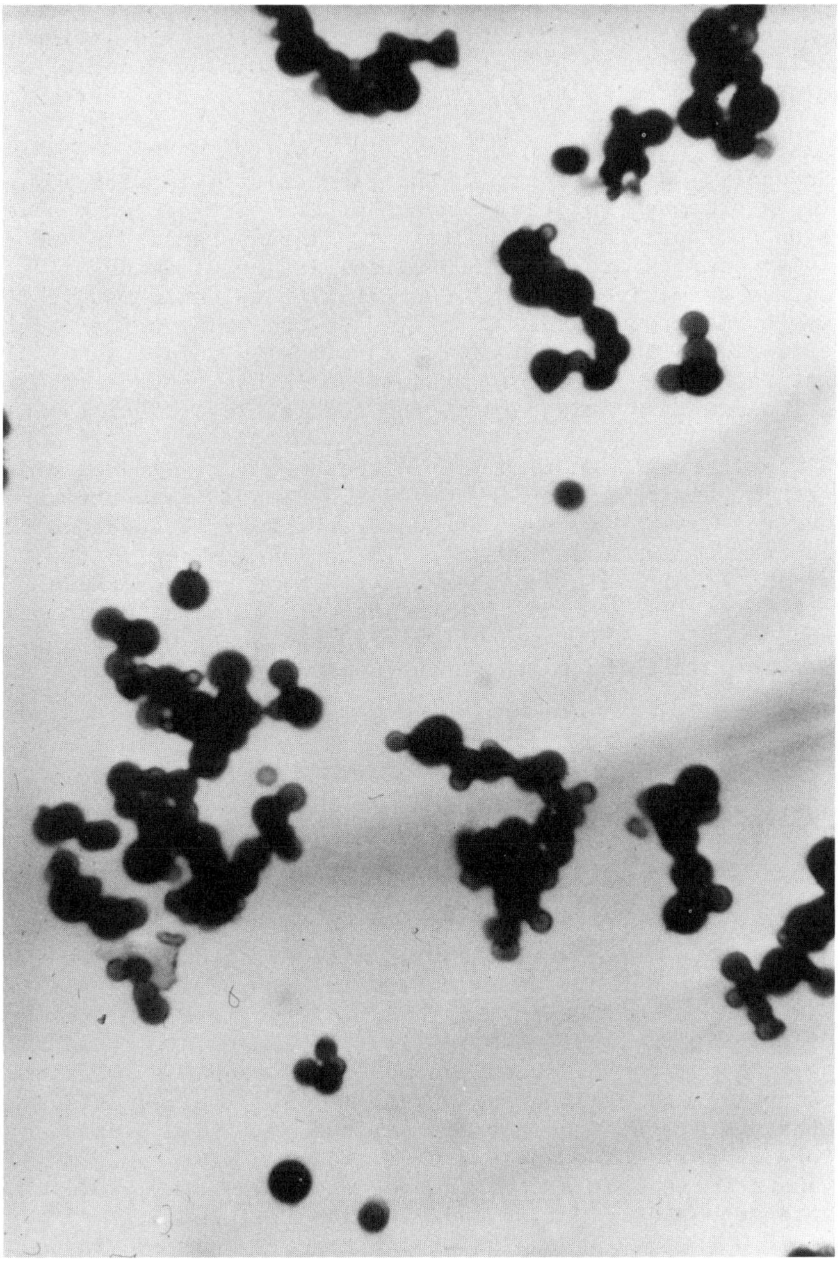

Fig. 1. (B) Aldocyanoin microspheres. Diameter of larger spheres approximately 2 μ. Kenyon, unpublished.

mixtures. But if other compounds such as aldehydes and sugars were present in significant amounts, would not some of these react with amino acids preventing to some degree the synthesis of polypeptides? One would expect the Maillard reaction (the condensation of sugars and amines) to occur with high probability in such mixtures (Ellis, 1959).

Secondly, one may question the geochemical plausibility of the conditions proposed for the sequence of events from amino acid synthesis to the appearance of proteinoid microsystems. In one scenario high temperature (900-1100°C) initiation of amino acid synthesis from primitive gases in contact with volcanic material is followed by low-temperature formation of the amino acids and their polymerization under dry conditions at 160-180°C followed rapidly by dissolution of the proteinoid product and subsequent microphase separation due to a rainstorm at the locale (Fox, 1964, 1974b).

Since we really know so little about the detailed nature of the Earth's primitive surface, we should keep in view the major alternative possibility that the entire historical sequence of microsphere-forming reactions might have taken place in the relatively cool primitive hydrosphere in the bulk aqueous phase. It was with this view in mind that an attempt was made to find an alternative model for protocell development that would overcome some of the objections to the proteinoid microsystem.

ALDOCYANOIN MICROSYSTEMS

Introduction

Since one possible difficulty with the proteinoid microsystem concerned the purity or simplicity of the initial reactant mixture (i.e., pure amino acids), a system was sought in which microspheres might form in the presence of a rich variety of substances thought to have been present on the Earth's primitive surface. Aldehydes and cyanides are characteristic intermediates in the Miller-Urey type of simulation experiment (Miller, 1957) and are known to generate a rich variety of organic compounds in aqueous solutions of the appropriate pH (Miller and Orgel, 1974). The resulting solution simulates the presumed composition of the primitive hydrosphere. Therefore, formaldehyde and hydrogen cyanide (plus some ammonium thiocyanate to include sulfur in the reaction system) were chosen as initial reactants (Kenyon and Nissenbaum, 1976). If microspheres of bacterial dimensions spontaneously appeared in such a system, they would have done so in spite of the presence of many potentially interfering substances. Thus the objection in the case of the proteinoid microsystems that the initial reactant mixture was artificially pure, would have been answered.

There were several indications that such a system would form microscopic structures. Ammonium cyanide in concentrated alkaline (pH 9.2) solution produces an insoluble macromolecular substance in the form of microscopic granules when heated for several hours under reflux (Labadie et al., 1967a). Ammonium thiocyanate (3 g) mixed with 7 ml of 37% aqueous formaldehyde generates a variety of complex microscopic structures (Herrera, 1940). Dilute solutions of these compounds subjected to ultraviolet radiation also produce microscopic particles (Smith et al., 1968).

Preparation of Microspheres

The typical reaction mixture is an aqueous solution of the following composition: NaCN (0.2 M), NH_4Cl (0.2 M), HCHO (0.2 M), and NH_4SCN (0.02 M) (Kenyon and Nissenbaum, 1976). The optimum pH for microsphere formation is 9.3 (the pK_a value of HCN). Solutions are allowed to stand at room temperature in glass-stoppered flasks for 2-4 weeks. During this time the solutions turn from colorless and limpid to dark brown and turbid and the water-air interface is covered with sheet-like flakes of organic matter. Microscopic examination of the resulting suspensions reveals the presence of numerous spherical particles of mean diameter approximately 2 µ (with occasional larger spheres present as shown in Fig. 1(B). This sphere-forming, insoluble material was named aldocyanoin (aldo for "aldehyde," and cyan for "cyanide").

General Properties

Except for their amber color aldocyanoin microspheres bear a remarkable resemblance at the light microscope level to Fox's proteinoid microsystems. Photomicrographs of the two types of microsphere are shown in Fig. 1 for comparison. The aldocyanoin microspheres are fairly uniform in size, possess "collar junctions," "bud-like" appendages of various sizes, occur in doublets, triplets and larger clusters including chains or "filament-like" arrays. Under both the light microscope and the transmission electron microscope aldocyanoin spheres exhibit no discernible internal structure.

The microspheres, like their proteinoid counterparts, are mechanically stable; i.e., they can be centrifuged and resuspended without losing their integrity. They are stable in aqueous buffers over a pH range from 2 to 11. The spheres do not swell or shrink in NaCl solutions ranging from 0-10% w/v. Except for an insoluble outer layer the aldocyanoin particles dissolve in 0.1 M NaOH, 0.1 N HCl, and 88% aqueous formic acid. Thus, while there is some differentiation between surface material and material inside the particles, there does not appear to be a semipermeable boundary of the type reported for proteinoid microspheres (Kenyon, unpublished).

Requirement for Molecular Oxygen

It has generally been assumed by students of the origin of life that the Earth's early atmosphere was a reduced one virtually devoid of molecular oxygen. On the other hand, some rather convincing evidence and argumentation have been recently presented which support the view that the Earth's atmosphere contained significant amounts of molecular oxygen from the earliest times (Clemmey and Badham, 1982; Dimroth and Kimberley, 1976). Of course if the latter view prevails, serious problems for the possibility of any significant chemical evolution on the primitive Earth will be raised. At any rate it is of interest to inquire whether any given protocell model system can form spontaneously in the absence of O_2.

In order to determine whether molecular oxygen is required for the synthesis of aldocyanoin microspheres the following experiments were performed (Kenyon, unpublished): Samples of freshly prepared reaction mixture of the standard composition (see above) were frozen in liquid nitrogen and evacuated on a vacuum line to less than 6 μ Hg pressure. Helium was admitted into the reaction tubes at a pressure of 720 mm Hg. The tubes were reevacuated to less than 6 μ Hg and He readmitted to 700 mm Hg. The tubes were then sealed, removed from the vacuum line and allowed to thaw and stand at room temperature for 3 weeks. Control tubes were prepared in an identical manner except that air was readmitted to the tubes after initial evacuation. After two weeks the control tubes were dark amber and turbid and contained numerous aldocyanoin microspheres. The tubes containing He in place of air were faint amber, clear and contained no microspheres. The experiments were repeated with identical results. An additional experiment in which N_2 replaced the He yielded the same result. Thus, molecular oxygen is required for formation of aldocyanoin microspheres.

The above result was confirmed by Pollock and Heiderer (1979). These investigators did find, however, that macromolecular material was formed in this system under anaerobic conditions. Further studies are required to determine the minimum threshold oxygen partial pressure for aldocyanoin microparticle formation.

Acid Hydrolysis

Proteinoids are formed from free amino acids and amino acids are liberated from the polymer upon acid hydrolysis. Although it contains no detectable amino acid residues the macromolecular material ("azulmic acid") forming Labadie's NH_4CN-microspherules does give rise to small amounts of amino acids upon acid hydrolysis, so at least some sort of amino acid precursor forms part of the "polymeric" structure (Labadie et al., 1968).

Are free amino acids detectable in the supernatant fluid when aldocyanoin microspheres are subjected to acid hydrolysis? In order to answer this question purified samples of aldocyanoin (1-2 mg) were hydrolyzed in 6 N HCl (2-2.5 ml) in sealed glass ampoules at 110-120°C for 24 hr. The hydrolyzates were dried in a vacuum dessicator over Drierite and NaOH pellets and stored in a freezer prior to analysis. Portions of the samples were examined for amino acids by the automatic amino acid analyzer and GC-MS (using the N-TFA-2-butyl esters of the amino acids). The qualitative results are given in Table 1. Generally similar results were obtained by Pollock and Heiderer (1979), who also detected iminodiacetic acid in the hydrolyzates.

Carbon Isotope Fractionation

Isotope fractionation in the same direction as occurs in living cells (i.e., preference for ^{12}C over ^{13}C) has been shown to occur in aldocyanoin synthesis, adding another example to the list of non-biological reactions exhibiting enrichment of ^{12}C in the product compared to the reactants. Triplicate samples of the NaCN, NH_4SCN, and HCHO used for aldocyanoin synthesis, and the aldocyanoin product were oxidized in a microcombustion vacuum system (Kenyon, unpublished) and the resultant CO_2 analyzed for carbon isotope content by low-resolution mass spectrometry (Studley, 1975). The results are shown in Table 2. These results suggested that carbon atoms derived from cyanide were isotope fractionated in the synthesis while those of formaldehyde were not. To shed further light on this question the stoichiometry of aldocyanoin synthesis was studied.

Two identical reaction mixtures were prepared (NaCN 0.2 M, NH_4Cl 0.2 M, NH_4SCN 0.02 M, HCl added at 0.1 N). Fifty microcuries of $H^{14}CHO$ were added to one solution and 50 μc of $Na^{14}CN$ to the other. The solutions were allowed to stand at room temperature in glass-stoppered Erlenmeyer flasks for 2 weeks, after which the aldocyanoin product was dialyzed and lyophilized. Weighed samples of each aldocyanoin preparation were counted on a liquid scintillation spectrometer to determine the ratio of specific activities of the two radioactive preparations. The following result was obtained:

$$\frac{\text{Aldocyanoin } (H^{14}CHO)}{\text{Aldocyanoin } (Na^{14}CN)} = 1.08$$

Therefore, carbon atoms from the two sources (HCHO and CN^-) are incorporated into the aldocyanoin microspheres in an approximately 1:1 ratio. We conclude that ^{12}C is preferentially incorporated over ^{13}C from among carbons contributed by CN^-, but that no such fractionation occurs for carbon atoms from HCHO.

These data may have some bearing on the question of biogenicity criteria applicable to organic matter in Archean sediments (Eichmann

Table 1. Amino acids detected in acid hydrolyzate of aldocyanoin*

Amino acid analyzer	GC	Confirmed by MS
Glycine	Glycine	+
Aspartate	Aspartate	+
Alanine	Alanine	
Serine		
Threonine		
Glutamate	Glutamate ?	
	β-Alanine	
	Isoleucine ?	

*Samples of aldocyanoin were hydrolyzed in 6N HCl in sealed glass ampoules at 110-120°C for 24 hr. Amino acids in the hydrolyzate were converted to the N-TFA-2-butyl esters for analysis by GC-MS.

Table 2. Carbon isotope fractionation in aldocyanoin synthesis*

Substance	$\delta^{13}C$
NaCN	-27.13 ± 0.02
NH_4SCN	-49.91 ± 0.14
HCHO	-40.55 ± 0.11
Aldocyanoin	-40.60 ± 0.15

*Triplicate samples of the reactants and product were oxidized in a microcombustion vacuum system and the resultant CO_2 analyzed for carbon isotope content by low-resolution mass spectrometry.

and Schidlowski, 1975; Oehler et al., 1972). That is, demonstration of a given negative $\delta^{13}C$ value for a sample of Archean organic matter need not support a biogenic origin for that material (Anders et al., 1974; Lancet and Anders, 1970).

COMPARISON OF THE TWO MODELS

The Proteinoid Scenario

Accumulation of amino acids. We now have two quite different alternative models of the historical protocell which may be carefully compared in order to exhibit the strengths and weaknesses of each model in sharp relief. An approximate overall flowchart for the proteinoid microsystem model of the origin of life is shown in Fig. 2.

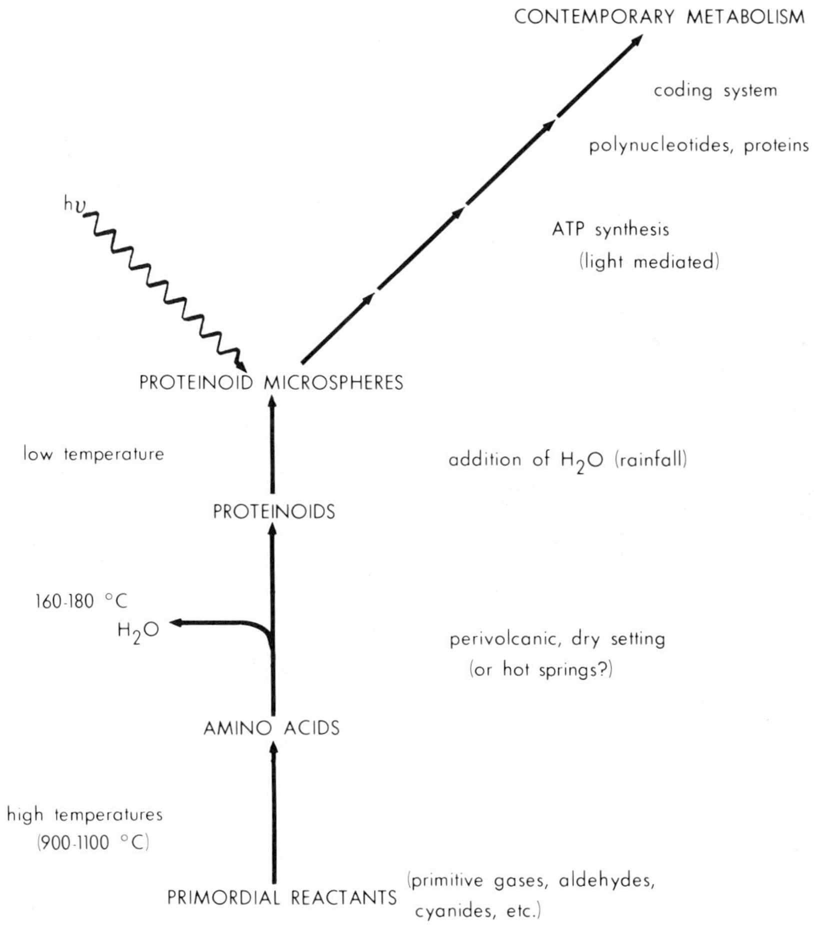

Fig. 2. The proteinoid model of the origin of life. After Fox (1978).

If the Earth's early atmosphere contained significant amounts of O_2 it is difficult to see how any appreciable quantities of amino acids could have accumulated on the primitive Earth (Abelson, 1957; Miller and Orgel, 1974). Aside from the oxygen question there are other cogent reasons for doubting that any significant quantities of amino acids, or indeed of any other organic compound of biologic interest could have built up on the primitive Earth. It will suffice here just to mention these two: (1) the thermodynamic equilibrium amounts of compounds such as amino acids and sugars are so minute as to suggest that no more than trace amounts of these substances could have accumulated over long periods of time on the prebiotic Earth (Hull, 1960), and (2) the same energy sources which promote the laboratory synthesis of simple biologic compounds (and their precursors) would be expected to destroy these same compounds in nature more effectively than they promoted their synthesis (Abelson, 1966; Hulett, 1969; Hull, 1960). However, since amino acids are characteristic and significant products in the various primitive Earth simulation experiments, let us assume for the sake of discussion that significant quantities of amino acids were available on the Earth's early surface.

Geochemical plausibility. The proteinoid model has been sharply criticized as implausibly requiring a highly improbable combination of rare circumstances (Miller and Orgel, 1974; Sagan, 1965). Especially problematical is the likelihood that the thermal energy required for proteinoid synthesis would also destroy the proteinoid polymer. Professor Fox has replied that the proposed volcanic scenario when integrated over the globe and over the prebiotic phase of Earth history was not nearly as improbable as his critics claim (Fox, 1965a). Moreover, proteinoids have been synthesized at much lower temperatures ($T < 100°C$) in dry or hypohydrous reaction mixtures with or without polyphosphoric acid (Fox, 1976b; Harada and Fox, 1965; Rohlfing, 1976).

In my view the perivolcanic setting is no worse than any of the other proposed settings for the origin of life.*

The Maillard reaction. If amino acids were present on the primitive Earth then other classes of organic compounds were also present, including aldehydes and sugars (Kenyon and Steinman, 1969).

*There now appear to be five major settings proposed for the origin of life: (1) the perivolcanic zone (Fox and Dose, 1977), (2) the hydrosphere generally (Oparin, 1957); the primitive seashore (Bernal, 1949), (3) submarine thermal vents (Corliss et al., 1981), (4) a microdroplet phase in the primitive atmosphere (Woese, 1979), and (5) an extraterrestrial setting (Arrhenius, 1908; Crick and Orgel, 1973; Hoyle and Wickramasinghe, 1979).

But because of the great range of conditions over which the Maillard reaction (the condensation of amines with sugars) readily takes place (Ellis, 1959), it seems very likely that sugars would cross-react with free amino acids with a greater probability than the self-condensation of amino acids, even in the dry molten state (see Fig. 3).

The condensation of amino acids and sugars has been demonstrated to occur at temperatures from -20 to at least 110°C, and from acidic to slightly alkaline pH's (Ellis, 1959; Nissenbaum et al., 1975). The reaction proceeds over a wide range of reactant concentrations, over a wide range of moisture contents including a nearly dry state, and under anaerobic as well as aerobic conditions. Thus it appears that if any of these reactions took place on the primitive Earth the Maillard reaction would have effectively competed with and overwhelmed the proteinoid-forming reaction. The result would have been the formation of the insoluble, amber "polymeric" material known as melanoidin (Kenyon and Nissenbaum, 1976; Nissenbaum et al., 1975). This substance forms microscopic spheres which closely resemble the aldocyanoin and proteinoid microspheres at the light microscope level. It has been argued that melanoidin formation was a significant process in the primitive hydrosphere which acted to remove much of the dissolved organic matter from the primitive seas thus inhibiting the formation of an "organic soup" (Nissenbaum et al., 1975).

The proteinoid model assumes that proteinoid microspheres were formed in substantial numbers <u>in spite of</u> the presence of highly probable competing cross-reactions. However, experimental support for this contention is presently lacking. In addition, it is conceivable that proteinoids themselves (assuming they could have been formed on the primitive Earth) might have been inhibited from forming microspheres by other substances present in solution with them.

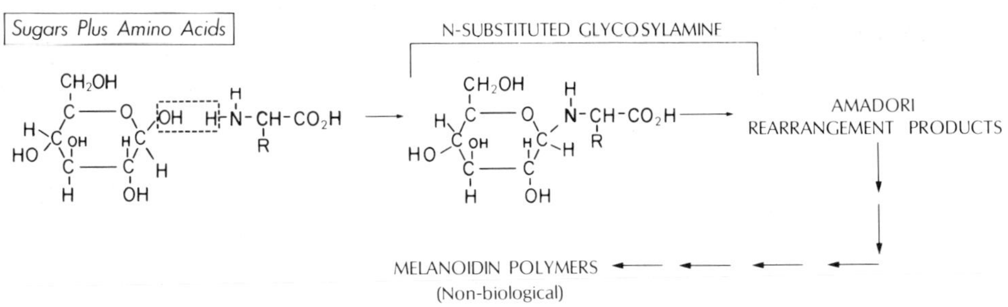

Fig. 3. The Maillard reaction followed by melanoidin synthesis. After Ellis (1959).

The Maillard reaction would also have posed a serious problem at a later stage of the proteinoid model, namely, that stage at which polynucleotides presumably formed. Free amino acids would have reacted with the ribose and deoxyribose molecules with a greater probability than the incorporation of these sugars into polynucleotides. That is, the cross-reaction between sugars and amino acids (initiating melanoidin synthesis) would have effectively blocked polynucleotide formation since the enzymes guiding nucleic acid synthesis today could not have been present then.

Non-randomness. The specific ways in which proteinoids are non-random arrangements of amino acids are potentially significant in the context of the origin of life since the ordering process takes place without the participation of nucleic acids. However, some degree of sequence ordering would be expected simply because of the different R-groups of the reacting amino acids; i.e., steric and electronic factors would affect the rates at which amino acids are added to the growing polymers (Fox, 1980b). It is possible that the only relationship the non-randomness of proteinoids bears to biologic information is a negative one. That is, the greater is the non-randomness of amino acid sequences in proteinoids, the lower is their information content (Yockey, 1981).

The Aldocyanoin Scenario

General features. Let us now look at the flowchart for the aldocyanoin microsphere model (Fig. 4). "Protocell" units are imagined to have arisen before the appearance of the contemporary classes of biological macromolecules. The building material for these "protocell" units is the non-biological, "polymeric" substance aldocyanoin which is derived from aldehydes and cyanides reacting in aqueous solution at room temperature in the presence of some molecular oxygen. Again, I shall assume that the partial pressure of O_2 was not so high as to have prevented any substantial accumulation of organic matter on the Earth's early surface.

The "sorting out" of the contemporary classes of biopolymer is imagined to have taken place within the aldocyanoin microspheres. Based on their probable general similarity to azulmic acids (Labadie et al., 1968) and melanoidin (Nissenbaum et al., 1975), it is likely that aldocyanoin macromolecules contain a variety of heterocyclic ring structures, as well as amino acid precursor moieties (Pollock and Heiderer, 1979). In the aldocyanoin model the entire chain of events from the primordial appearance of aldehydes and cyanides to the emergence of metabolism takes place at the relatively low temperature of the primitive hydrosphere.

Molecular oxygen. The requirement of molecular oxygen for aldocyanoin synthesis would appear to present a major difficulty. If O_2 was a significant component of the prebiotic atmosphere, no

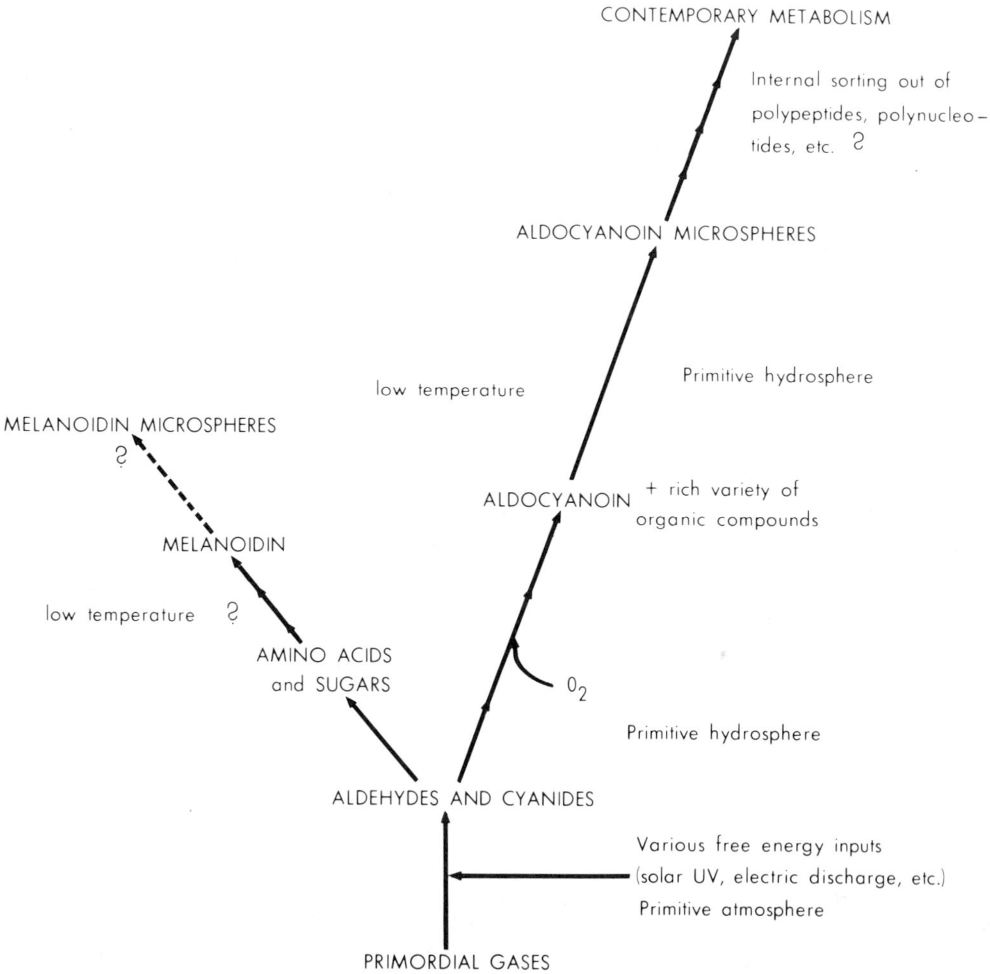

Fig. 4. The aldocyanoin model of the origin of life.

substantial chemical evolution could have taken place. If molecular oxygen was absent, no aldocyanoin microspheres could have formed. It is of course possible that at some low level of oxygen partial pressure both significant chemical evolutionary accumulation of simple organic substances and aldocyanoin microsphere formation might have occurred, but information is lacking on this point. At any rate, given the presence of amino acids, molecular oxygen is of course not required for the laboratory production of proteinoids.

Concentration of reactants. Another potential problem for the aldocyanoin model involves the concentrations of the initial reactants. In the laboratory studies the concentration of formaldehyde and hydrogen cyanide was typically 0.2 *M*, a value which would appear to be much too high for the primitive hydrosphere (Miller and Orgel, 1974). Experiments at substantially lower concentrations are yet to be done. We know that aldocyanoin microspheres do not form (within the four-week standard reaction time) if the molar ratio of aldehyde to cyanide is less than about 0.5 or greater than about 2.0. So it would appear that the concentrations of aldehydes and cyanides would have had to have been roughly comparable (and sufficiently high) in order for aldocyanoin microsphere formation to have been an important process on the primitive Earth.

Future research. Perhaps the greatest difficulty facing the aldocyanoin model is the lack of clear indications about the directions future work should take. One could perhaps search for weak catalytic activities of the kind reported for the NH_4CN-microspherules (Labadie et al., 1967b) and the proteinoid microspheres (Fox, 1978). The initial reactant mixtures could be varied in order more precisely to delimit the conditions under which microsphere formation takes place. Scanning electron micrographs could be obtained, and more detailed studies of the molecular structure of aldocyanoin (i.e., partial degradation plus characterization of the products) could be undertaken. But the real stumbling block is the complex, non-biological nature of the aldocyanoin macromolecules. How could such a material have spontaneously given rise to living matter with its clearly defined classes of biological homopolymers? How could polynucleotides and polypeptides have "sorted themselves out" from such a semi-random, high molecular weight organic matrix? It is difficult, if not impossible, to visualize the relevant experiments.

Problems for Both Models

Neither model sheds any light on the difficult problem of optical isomer preferences (i.e., L-amino acids and D-sugars) in living matter. Needless to say, the origin-of-life problem cannot be said to have been "solved" without first resolving this great question. The published experiments overwhelmingly support the view that any amino acids and other organic compounds that were synthesized on the primitive Earth were precisely racemic (Dickerson, 1978; Folsome, 1979). Moreover, none of the attempts so far undertaken to find a lawful basis for the origin of the optical isomer preferences of living matter has been successful (Bonner, 1972; Bonner et al., 1980). In view of these facts one may qurstion why most of the published experiments on proteinoid synthesis employ significant excesses of L- over D-amino acids (Fox and Harada, 1960; Fox and Yuyama, 1964). There are of course reasons of convenience and the possible experimental fact that proteinoid derived from an excess of L-amino acids consistently gives better microsphere preparations (i.e., size, yield, rate of formation) than does proteinoid derived from racemic mixtures of amino acids.

But none of these reasons would appear to justify the continued routine use of large excesses of L-amino acids in the starting mixtures. Similar criticism may be applied to other "protocell" model systems; e.g., the "marisomes" of Yanagawa et al. (1980), who use pure L-amino acids in their initial solution.

Neither model provides any significant insight into the origin of the phospholipid membrane of contemporary living cells, although some discussion of this point has appeared in the literature of the proteinoid model (Bahn and Fox, 1981; Fox and Dose, 1977; Fox et al., 1969). It should be stressed that in both models the ultrastructural and biochemical gap between the proposed "protocell" system and the simplest living cells is enormous, and not likely to be bridged by years of future work.

A Curious Dilemma

We are faced with the following dilemma. On the one hand we have a model which is based on some questionable assumptions (e.g., the availability on the primitive Earth of reactant mixtures containing only amino acids, and appropriate temperature and moisture regimes in proper, rapid time sequence), but which nevertheless generates an enormous quantity of laboratory work. On the other hand we have another model which is based on more accurate simulation of presumed primitive conditions; i.e., low temperatures and an aqueous phase containing a variety of organic substances. Moreover, the major initial reactants are characteristic intermediates in the primitive atmosphere simulation experiments. This model, however, for reasons described above, is not nearly as fruitful as the first model.* If fruitfulness were the sole criterion of a good theory we would easily decide that the proteinoid theory is the better view.

At this point the cynic might tentatively conclude that (1) the proteinoid model is very fruitful and largely correct in its assertions about the self-organizing properties of amino acids and proteinoids, but that it is not about the origin of life, and (2) the aldocyanoin model, although it is not notably fruitful, may be correct concerning the origin of life, but we have no way of finding out! I would put the question somewhat differently. If the remarkable morphogenicity and chemical properties of proteinoids

*It might be argued that C. E. Folsome's production of organic microstructures in a spark-discharge apparatus represents an even more accurate simulation of primitive Earth conditions (Folsome, 1979). But here again, due to the intractable nature of the structure-forming "polymer," the problem is one of defining the direction future research should take.

demonstrated by Professor Fox and his colleagues are not related to the historical origin of life, then what is their significance? To what extent is microsphere-forming ability unique to proteinoids?

PROTEIN MICROSPHERES

Introduction

It is clear from the comparison of proteinoid and aldocyanoin microsystems that many of the morphologic features of the former (spherical shape and size, uniformity of size, junctions, doublets, clusters, chains, "buds") are not unique to that system. Thus, proteinoids (derived from pure amino acids), aldocyanoin (derived from aldehydes and cyanides), and melanoidin (derived from amino acids and sugars) all yield broadly similar microscopic structures. Depending on one's frame of mind these facts may be either encouraging or discouraging in the search for the origin of life.

One might argue that with three (or more) alternative models for the origin of the "protocell" it is all the more likely that some spontaneous process generated the first cells on Earth. Perhaps all three (as well as others) contributed to the origin of life. On the other hand, if one is seeking a unique explanation of protocellular origins the fact that several different macromolecular materials generate microspheres of similar morphology may raise serious doubts about the validity of any given model. Of course one can hope that future research will eliminate all but one model.

Let us for the sake of argument assume that neither the proteinoid nor the aldocyanoin microsystem is (was) relevant to the spontaneous origin of the first living cells on the Earth. What then is the significance of the remarkable morphogenicity of proteinoids and aldocyanoin? A partial answer may be provided by the fact that many contemporary proteins under a variety of simple conditions also spontaneously form microspheres in initially clear aqueous solutions.

Formation

Just as proteinoid microspheres form in cooling solutions of proteinoid, so do microspheres (with varying degrees of complex internal morphology) form in cooled solutions of soluble liver proteins (Anderson, 1959). Nelson (1958a) has shown that spherical microparticles form in sterile media containing red cell lysates, liver extract proteins, and soluble proteins from spleen, kidney, testis and heart tissue. Microspheres formed from hemoglobin and DNA at pH 5 -5.4 exhibit osmotic properties (Nelson, 1958b). Cinátl (1967, 1969) has shown that serum proteins form microstructures under various experimental conditions. It is likely that such processes taking place in vivo help generate much of the cell's ultrasturcture (Anderson, 1959).

Superficial Resemblance to Bacteria and Mycoplasma

It is precisely their resemblance to bacteria and mycoplasma (at the light microscope level) that makes the protein microstructures of vital interest in pathology and tissue culture work (Cinátl, 1967; Smith and Kenyon, 1972). In fact elaborate criteria have been devised to distinguish these structures from bacteria and mycoplasma (Hijmans et al., 1969). "It should be emphasized that any medium containing body fluids, whole tissues, or tissue extracts may give rise to particles and fibers in vitro which can easily be confused with bacterial forms" (Nelson, 1958a, p. 766). In some cases enzyme action may be required to initiate microparticle formation (Nelson, 1958a), which of course is not the case for aldocyanoin or proteinoid. One would expect to find in the literature the suggestion that such processes might have contributed to the origin of the first life, or to hypothetical neobiogenetic events (Cinátl, 1967; Smith and Kenyon, 1972).

Protein and Proteinoid Microsystems

Might it be the case that the morphogenicity and the weak catalytic activities of proteinoids are simply lower level analogs or reflections of the ultrastructure-generating and biochemical properties of true proteins upon which so much of present cellular function depends? In other words, one could ask why it would be surprising to discover that proteinoids exhibit morphogenetic properties since these after all, as Professor Fox and others have shown, broadly resemble true proteins at least in molecular weight, overall composition of subunits, and their polypeptide nature.

Since proteinoids are composed of branching chains and generally have molecular weights at the small end of the spectrum of values for proteins (Fox and Dose, 1977), strictly linear polypeptides of high molecular weight (e.g., > 50,000) are not required for some degree of morphogenicity. Moreover, the aldocyanoin and melanoidin studies show that the sphere-forming polymer need not even be a polypeptide, but may be some sort of a mixed polymer (i.e., derived from aldehydes and cyanides, or from amino acids and sugars). But not every insoluble organic polymer forms microspheres of bacterial size. Azulmic acid, for example, which is derived from ammonium cyanide only (Labadie et al., 1967a; Kenyon, unpublished), forms very small (< 1 μ) oblong granules instead.

The amino acid sequences of microsphere-forming proteins vary widely from system to system, and in a given proteinoid preparation many different sequences presumably are present. Therefore, a specific primary structure cannot be required for microsphere formation. Melanoidin and aldocyanoin "polymers" do not have primary structures in the sense of definite linear sequences of subunits, and yet they form microspheres. It is doubtful, therefore, that any significant link exists between the "sequence ordering" exhibited by proteinoids and their morphogenicity. Moreover, this writer cannot see any connection between this sequence ordering (which is imprecise and results largely from steric factors) and biologically useful information (Yockey, 1977, 1981), but this point is beyond the scope of this paper. The forces which generate microspheres of bacterial size from all of these organic macromolecules are probably physico-chemical forces of aggregation similar to those studied by colloid and polymer chemists. "We now understand the mechanism of the organization of the simulated protocells as one based on the powerful binding ability of the polyampholytic molecules. . . . We believe that it is when the solubility of proteinoid in water is exceeded that the molecules in the phase-separated material bind to each other to form cell-like structures" (Fox, 1978, p. 48).

SUMMARY AND CONCLUSION

Strengths and Weaknesses of the Two Models

From the above discussion it is apparent that the proteinoid and aldocyanoin models of the primeval protocell have both strong and weak points. A partial summary of the major strengths and weaknesses is given below:

Proteinoid Microsystems (Strengths)

1. α-Amino acids are characteristic products of primitive atmosphere simulation experiments.
2. Proteinoids form readily from free amino acids.
3. Proteinoids readily generate numerous microsystems of uniform diameter in the range of bacterial sizes.
4. The microstructures exhibit granular ultrastructure and a semipermeable boundary.
5. A (perivolcanic) geochemical scenario for Nos. 2 and 3 is at least conceivable.
6. The spheres are mechanically stable, and pH stable over the range 3-9 (depending on the conditions of synthesis of the proteinoid) (Fox and Dose, 1977; Snyder and Fox, 1975).
7. Junctions, "budding," growth by accretion, weak catalytic activities, semipermeable boundaries, and the ability to promote some degree of peptide and oligonucleotide synthesis, can be seen as possibly relevant to the origin of protocells.
8. These microsystems represent one phase of an overall theory of origins starting with primitive gases.
9. The model presents some possibility of investigating the

origin of the genetic code (Fox, 1974a, 1981).

10. The model has been very fruitful in generating laboratory research. There are clear lines of future investigation.

Proteinoid Microsystems (Weaknesses)

1. The starting mixture for proteinoid synthesis consists of pure amino acids. Other substances, such as sugars, also likely to have been present on the primitive Earth according to current ideas, are excluded. Thus potentially interfering cross-reactions (e.g., amino acids with sugars as in the Maillard reaction) are avoided.
2. Proteinoids are generally synthesized from mixtures containing an excess of L-amino acids, whereas simulation experiments routinely produce precisely racemic amino acids. Moreover, the model does not account for the origin of optical isomer preferences.
3. The perivolcanic geochemical scenario is questionable. An improbable combination of temperature, moisture, and timing conditions appears to be required.
4. Microsphere formation and the morphological properties of microspheres such as "budding," junctions, and clusters, are not unique to proteinoids, but also occur in melanoidin, aldocyanoin, and globular proteins.
5. The ultrastructure of the spheres bears no significant resemblance to the ultrastructure of the simplest living cells.
6. The catalytic activities so far demonstrated are minor and not (in most cases) significantly related to biochemistry.
7. The model does not adequately account for the origin of polynucleotides, polysaccharides, lipids, and the phospholipid plasma membrane.
8. Self-ordering of amino acids in proteinoid synthesis is unrelated to morphogenicity and to biologic information, but is an expected consequence of differing steric and other factors among the reacting amino acids.

Aldocyanoin Microspheres (Strengths)

1. The microspheres are synthesized in solutions initially containing simple reagents likely to have been present on the primitive Earth (according to current ideas). These reagents quickly generate a rich chemical mixture simulating the primitive seas. The spheres form in spite of the chemical complexity of the mixture.
2. The microspheres form in an aqueous system at room temperature over the pH range 7-11, which is consistent with the presumed value of 8.1 for the primitive oceans (Sillén, 1967).
3. The spheres are of bacterial size and relatively uniform in diameter. They exhibit doublets, junctions, clusters, "filaments," and "bud-like" appendages.

4. The spheres have mechanical stability and a wide range of pH stability (2-11).
5. Aldocyanoin yields amino acids upon hydrolysis and may contain heterocyclic rings.
6. In aldocyanoin synthesis carbon isotope fractionation occurs in the same direction as in living cells.
7. Aldocyanoin is probably broadly similar to the insoluble pigmented organic macromolecules which dominate the chemistry of the primitive atmosphere simulation experiments.

Aldocyanoin Microspheres (Weaknesses)

1. The initial concentration of reagents in the usual experimental protocol (i.e., 0.2 M) is unrealistically high for the primitive hydrosphere.
2. Microsphere formation requires O_2, but this may or may not be seen as a deficiency of the model depending on one's view of the composition of the primitive atmosphere.
3. Aldocyanoin is not a biological substance; it is difficult to visualize its transformation into biopolymers.
4. The boundary of the microspheres does not exhibit osmotic properties (in NaCl solutions ranging up to 10% w/v).
5. The spheres are uniformly clear, i.e., without any apparent ultrastructure throughout.
6. The possibility of catalytic activities has not yet been investigated.
7. The model does not account for the origin of the phospholipid-protein plasma membrane, optical isomer preferences, or the genetic coding system.
8. The lines future research should take are not clear. The model is not nearly as fruitful, in this sense, as the proteinoid model. It is not possible to study the origin of the genetic code, for example, since neither polypeptides nor polynucleotides are present.

Conclusion

The proteinoid microsystem model of the primordial protocell has without doubt been very fruitful experimentally, but it has some conceptual flaws. The aldocyanoin model on the other hand apparently overcomes some of the conceptual flaws, but it is not nearly as fruitful because of the intractable nature of the sphere-forming substance. One might be tempted to conclude (paradoxically) that the more realistic the simulation, especially perhaps with respect to the chemical complexity of the reaction mixtures, the less clear is it that the system is moving in the direction of the living state. In fact the dominant trend of the organic chemistry in such systems, including the Miller-Urey electric discharge experiments, seems to point away from the living state. That is, over 60% and possibly much more of the carbon in reaction products is in non-biological macromolecular material (Folsome, 1979; Folsome et al., 1975; Fraser and Folsome, 1974). At least it is not clear which experiments should be done next to determine how aldocyanoin microsystems might have moved closer to the living state. This problem is of course far less acute for investigators studying artificially (and often inappropriately) simplified model systems.

At this point we may ask whether the ability of proteinoids to form microspheres under certain experimental conditions by itself implies that such a process was involved in the origin of the first life. I would argue that the morphogenicity of proteinoids is not by itself sufficient to demonstrate that these polymers played an important role in the origin of life (any more than arguing that morphogenicity alone demonstrates that aldocyanoin microspheres were the historical protocells). Such a conclusion must be defended along other lines as well, such as overall geochemical, organic chemical, and thermodynamic plausibility of the proposed events. Also, the mere existence of a relatively simple, non-enzymatic way to synthesize some biologic compound does not imply that that reaction was involved in the origin of life, even if the substance is a polypeptide or a polynucleotide. The possibility cannot be ruled out that proteinoid morphogenicity is merely a simplified expression of protein morphogenicity, and is unrelated to the origin of life.

We of course do not know whether in the light of future research the proteinoid or aldocyanoin microsystem, or some other model system, will turn out to be the best approximation of the historical protocell. Indeed we must even hold open the possibility, at least in our most general reflections on this great subject, that it may prove impossible to find any naturalistic explanation whatsoever for the origin of living cells (Wilder-Smith, 1970). We certainly cannot rule out this option on _a priori_ (i.e., logical) grounds, or on the basis of any empirical data currently in hand, but only on grounds of _methodological necessity_, as I have argued elsewhere (Kenyon and Steinman, 1969, p. 31).

In spite of the criticisms raised in this paper (and they were all offered in the "constructionistic" spirit I know Professor Fox appreciates), I believe that Sidney Fox's experimental investigations of possible mechanisms for the origin of life on Earth will long be regarded by many scientists throughout the world as the single most impressive body of work on the problem.

ACKNOWLEDGMENT

This work was supported in part by the National Aeronautics and Space Administration (Grant No. NSG-7234).

REFERENCES

Abelson, P. H., 1957, Annals N.Y. Acad. Sci., 69:276.
Abelson, P. H., 1966, Proc. Nat. Acad. Sci. U.S., 55:1365.
Anders, E., Hayatsu, R., and Studier, M. H., 1974, in: "Cosmochemical Evolution and the Origins of Life," J. Oró, S. L. Miller, C. Ponnamperuma, and R. S. Young, eds., pp. 57-67, D. Reidel, Dordrecht, Holland.
Anderson, N. G., 1959, Exp. Cell Res., 16:42.
Arrhenius, S., 1908, "Worlds in the Making: The Evolution of the Universe," Harper, New York.
Bahn, P. R., and Fox, S. W., 1981, Bio-Systems, 14:3.
Bonner, W. A., 1972, in: "Exobiology," C. Ponnamperuma, ed., pp. 170-234, North Holland, Amsterdam.
Bonner, W. A., Blair, N. E., and Lemmon, R. E., 1980, in: "Limits of Life: Proceedings of the Fourth College Park Colloquium on Chemical Evolution," C. Ponnamperuma, and L. Margulis, eds., pp. 183-194, D. Reidel, Dordrecht, Holland.
Bernal, J. D., 1949, "The Physical Basis of Life," Routledge and Kegan Paul, London.
Cinátl, J., 1967, Exp. Cell Res., 47:123.
Cinátl, J., 1969, J. Theor. Biol., 23:1.
Clemmey, H., and Badham, N., 1982, Geology, 10:141.
Corliss, J. B., Baross, J. A., and Hoffman, S. E., 1981, Oceanologica Acta, No. SP:59.
Crick, F. H. C., 1981, "Life Itself: Its Origin and Nature," Simon and Schuster, New York.
Crick, F. H. C., and Orgel, L. E., 1973, Icarus, 19:341.
Dickerson, R. E., 1978, Sci. Amer., 239:70.
Dimroth, E., and Kimberley, M. M., 1976, Canad. J. Earth Sci., 13:1161.
Eichmann, R., and Schidlowski, M., 1975, Geochim. Geophys. Acta, 39:585.
Ellis, G. P., 1959, in: "Advances in Carbohydrate Chemistry," M. L. Wolfram, ed., Vol. XIV, pp. 63-134, Academic Press, New York.
Folsome, C. E., 1979, "The Origin of Life: A Warm Little Pond," W. H. Freeman, San Francisco.
Folsome, C. E., Allen, R. D., and Ichinose, N., 1975, Precamb. Res., 2:263.
Fox, S. W., 1964, Nature, 201:336.
Fox, S. W., 1965a, in: "The Origins of Prebiological Systrms and of Their Molecular Matrices," S. W. Fox, ed., p. 197, Academic Press, New York.
Fox, S. W., 1965b, in: "The Origins of Prebiological Systems and of Their Molecular Matrices," S. W. Fox, ed., pp. 361-373, Academic Press, New York.
Fox, S. W., 1974a, Mol. Cell. Biochem., 3:129.
Fox, S. W., 1974b, Amer. Biol. Teacher, 36:161.
Fox, S. W., 1976a, Origins Life, 7:49.

Fox, S. W., 1976b, J. Mol. Evol., 8:301.
Fox, S. W., 1978, in: "The Nature of Life," W. H. Heidcamp, ed., pp. 23-92, University Park Press, Baltimore.
Fox, S. W., 1980a, The Sciences, Jan. 1980:18.
Fox, S. W., 1980b, Naturwiss., 67:378.
Fox, S. W., 1981, Int. J. Quantum Chem. QBS, 8:441.
Fox, S. W., and Dose, K., 1977, "Molecular Evolution and the Origin of Life," 2nd edn., Marcel Dekker, New York.
Fox, S. W., and Harada, K., 1960, J. Amer. Chem. Soc., 82:3745.
Fox, S. W., and Yuyama, S., 1964, Comp. Biochem. Physiol., 11:317.
Fox, S. W., McCauley, R. J., Montgomery, P. O'B., Fukushima, T., Harada, K., and Windsor, C. R., 1969, in: "Physical Principles of Biological Membranes," F. Snell, J. Wolken, G. J. Iverson, and J. Lam, eds., pp. 417-430, Gordon and Breach, New York.
Fox, S. W., Jungck, J. R., and Nakashima, T., 1974, Origins Life, 5:227.
Fraser, C. L., and Folsome, C. E., 1975, Origins Life, 6:429.
Harada, K., and Fox, S. W., 1965, in: "The Origins of Prebiological Systems and of Their Molecular Matrices," S. W. Fox, ed., pp. 289-297, Academic Press, New York.
Herrera, A. L., 1940, Bull. Lab. Plasmogenie, Mex., 2:3.
Herrera, A. L., 1942, Science, 96:14.
Hijmans, W., Van Boven, C. P. A., and Claesner, H. A. L., 1969, in: "The Mycoplasmatales and the L-Phase of Bacteria," L. Hayflick, ed., pp. 67-143, Appleton Century Crofts, New York.
Hoyle, F., and Wickramasinghe, C., 1979, "Diseases from Space," J. M. Dent, London.
Hoyle, F., and Wickramasinghe, C., 1981, "Evolution from Space," J. M. Dent, London.
Hulett, H. R., 1969, J. Theor. Biol., 24:56.
Hull, D., 1960, Nature, 186:693.
Jungck, J. R., and Fox, S. W., 1973, Naturwiss., 60:425.
Kenyon, D. H., and Nissenbaum, A., 1976, J. Mol. Evol., 7:245.
Kenyon, D. H., and Steinman, G., 1969, "Biochemical Predestination," McGraw-Hill, New York.
Labadie, M., Cohere, G., and Brechenmacher, C., 1967a, Comp. Rend. Soc. Biol., 161:1689.
Labadie, M., Jensen, R., and Neuzil, E., 1967b, Bull. Soc. Chim. Biol., 49:673.
Labadie, M., Jensen, R., and Neuzil, E., 1968, Biochim. Biophys. Acta, 165:525.
Lancet, M. S., and Anders, E., 1970, Science, 170:980.
Miller, S. L., 1957, Biochim. Biophys. Acta, 23:480.
Miller, S. L., and Orgel, L. E., 1974, "The Origins of Life on the Earth," Prentice-Hall, Englewood Cliffs, New Jersey.
Nakashima, T., and Fox, S. W., 1980, J. Mol. Evol., 15:161.
Nelson, E. L., 1958a, J. Expt. Med., 107:755.
Nelson, E. L., 1958b, J. Expt. Med., 107:769.

Oehler, D. Z., Schopf, J. W., and Evenvolden, K. A., 1972, Science, 175:1246.
Oparin, A. I., 1957, "The Origin of Life on the Earth," Academic Press, New York.
Nissenbaum, A., Kenyon, D. H., and Oró, J., 1975, J. Mol. Evol., 6:253.
Pollock, G. E., and Heiderer, R., 1979, J. Mol. Evol., 13:253.
Rohlfing, D. L., 1976, Science, 193:68.
Sagan, C., 1965, in: "The Origins of Prebiological Systems and of Their Molecular Matrices," S. W. Fox, ed., pp. 374-375, Academic Press, New York.
Sillén, L. G., 1967, Science, 156:1189.
Smith, A. E., and Kenyon, D. H., 1972, Perspectives Biol. Med., 15:529.
Smith, A. E., Silver, J. J., and Steinman, G., 1968, Experientia, 24:36.
Snyder, W. D., and Fox, S. W., 1975, BioSystems, 7:222.
Studley, S., 1975, personal communication.
Wilder-Smith, A. E., 1970, "The Creation of Life: A Cybernetic Approach to Evolution," Harold Shaw, Wheaton.
Woese, C. B., 1979, J. Mol. Evol., 13:95.
Yanagawa, H., Kobayashi, Y., and Egami, F., 1980, J. Biochem., 87:855.
Yockey, H., 1977, J. Theor. Biol., 67:377.
Yockey, H., 1981, J. Theor. Biol., 91:13.

THE INFLUENCE OF SOME PHYSICO-CHEMICAL FACTORS ON SELF-ORGANIZATION IN PREBIOTIC SYSTEMS

Mella Paecht-Horowitz

The Seagram Centre for Soil and Water Sciences
The Hebrew University of Jerusalem
Rehovot, Israel

INTRODUCTION

Since Oparin (1924) came forth with his model for prebiotic synthesis, namely that simple organic substances were formed in an abiogenic form, and that the evolution of these substances has to lead to the formation of protein-like compounds and then to colloid systems capable of gradually improving their internal organization by the operation of natural selection, the interest in this sort of reactions grew steadily. Many hypotheses were presented, many experiments were planned and carried out; the picture became fuller and fuller, though it is still far from being complete. The subjects which evolved, and seem to be important for the understanding of the formation of prebiotic systems, are so far: Synthesis of organic molecules out of the gas mixture of the then existing earth atmosphere, polymerization of these molecules to biopolymers*, self-organization of these biopolymers to cell-like particles in which certain prebiotic reactions could have taken place and finally the development of these particles into self-reproducing units, a process which is considered possible by living organisms only.

*Some authors (Mathews and Moser, 1966; Simionescu and Denes, 1981) believe that polymerization preceded the formation of monomers, namely that HCN yields on polymerization directly polypeptides. Others (Ferris, 1979) claim that the substance formed by irradiation of HCN and yielding upon hydrolysis amino acids is not a polyamino acid but diaminomaleonitrile which condenses into HCN oligomers. The latter, upon hydrolysis, yield, among other substances, also amino acids.

Many of the reactions would have been reversible, especially those of self-organization, but only those yielding products fitted for further development in the evolution of life would survive, while the other products would decompose and finally perish. The reactions leading to the formation of self-reproducing units would follow a time arrow and would have been irreversible. Each of the reactions leading to this final stage of chemical evolution and the beginning of life depend on a lot of physical and chemical factors. Some of them shall be discussed in this paper.

CONCENTRATION OF MOLECULES NEEDED FOR PREBIOTIC SYNTHESIS

One of the first problems to present itself when approaching the study of prebiotic evolution was how the various molecules needed for prebiotic chemical reactions could have been concentrated to a level sufficiently high so as to be able to interact. One possibility of "natural" concentration is by the means of coacervates. Bungenberg de Yong (1937) showed that by mixing solutions of gelatin and gum arabic, phase separation takes place and droplets containing both polymers precipitate. He called these droplets coacervates. Evreinova et al. (1962) showed that when gelatin and gum arabic interact in a solution containing amino acids, the coacervates which precipitate carry inside them amino acids in a concentration 10-100 fold greater than their concentration in the solution. Oparin et al. (1964) showed that the phenomenon of coacervates is not something restricted to gelatin and gum arabic, but that any two polymers could produce coacervates, which might be able to concentrate inside them monomers vital for the formation of life, thus enabling the monomers to interact with each other.

As polymers able to form coacervates would also need to be at a certain level of concentration in order to interact, the problem arises again by what "natural" means could concentration have taken place. Bernal (1951) suggested that clay particles on the shores of seas and rivers might have adsorbed and thus concentrated the varous organic monomers. He even went further and proposed that life itself might have originated there. Since then, quite a number of papers have appeared probing into this possibility, and measurements of adsorption on clays have been carried out, in context with origins of life and otherwise. Siesskind (1960) and Greenland et al. (1965) have reported on the adsorption of amino acids on montmorillonite. We (Paecht-Horowitz, 1978) have measured the adsorption of amino acid adenylates and also of hexametaphosphate on montmorillonite. Eirich (1981) and Hsu (1977) found by X-ray measurements that polypeptides are intercalated between parallel montmorillonite platelets and that the interplatelet distances increase with the amount adsorbed while showing distinct steps at multiples of 1/2 of the surface occupation density of a full flat monolayer. The spacing $\theta = 1/2$ corresponds to a single monolayer of oligomers between 2 platelets and the distance of $\theta = 1$ corresponds to 2 monolayers per 2 platelets. It is interesting to note that at $\theta = 1/2$ the lightly dried clay-peptide sandwiches are very poorly redispersible in water, while at $\theta = 1$ they are readily so. Risphon et al. (1982) studied the adsorption of ATP and ADP on montmorillonite and kaolinite as a function of pH and of the ionic composition of the system. They found that Mg^{2+}- and Zn^{2+}-montmorillonite adsorb the two nucleotides to a greater extent than does Na^{+}-montmorillonite. As to kaolinite, the various ions have only a small effect. In all cases, adsorption is highest at pH 2-3, as was found also by the previously-mentioned authors with regard to amino acids and polypeptides. The adsorption of ATP and ADP in most cases reached a minimum at pH 6-7, and then increases again, but to a lesser extent than at the low pH values.

In an attempt to repeat the experiments on lysine adsorption of Greenland et al. (1965), we found that the results obtained depend on the technique used (M. Paecht-Horowitz and A. Banin, in preparation). Our general technique was to agitate suspensions of Na^{+}-montmorillonite with various amounts of lysine for 24 hours, then to centrifuge the mixtures and to establish the amount of lysine in the supernatant. The difference between the amount of lysine added initially and that found in the supernatant should indicate the amount of lysine adsorbed. In fact, what is usually obtained is an apparent adsorption, much higher than the actual one as is shown subsequently. We found that the results obtained depend on the ratio between the volume of the supernatant to that of the precipitated suspension, which in turn depends on the rate of centrifugation. Fig. 1 shows the results obtained in one of these experiments. When the rate of centrifugation is very low, the volume of supernatant is very small and that of the suspension very large. In this case, the amount of lysine having disappeared from the supernatant, as related to the amount added initially, is small, but still higher than can be explained by a mere mechanism of cation exchange. With the increase of the centrifugation rate, the volume of supernatant increases, yet for quite a big range the amount of lysine apparently adsorbed remains constant. When the volume of suspension decreases to about 100-200 ml/litre, the amount of apparently adsorbed lysine increases sharply and decreases again only when ultracentrifugation is applied.

When one examines the X-ray picture of montmorillonite on which lysine has been adsorbed (Fig. 2), one sees that the basal spacing between the clay layers increases until 0.2 m moles lysine/g clay are adsorbed, and does not change any more whatever the adsorption. (Some rearrangement takes place at an adsorption of about 0.4-0.5 m moles/g clay, but the increase in basal spacing is much too small to allow for another layer of lysine to be adsorbed.) The displacement of Na^{+} ions measured simultaneously with lysine adsorption indicates an exchange of 0.3 m moles/g clay only. This value is

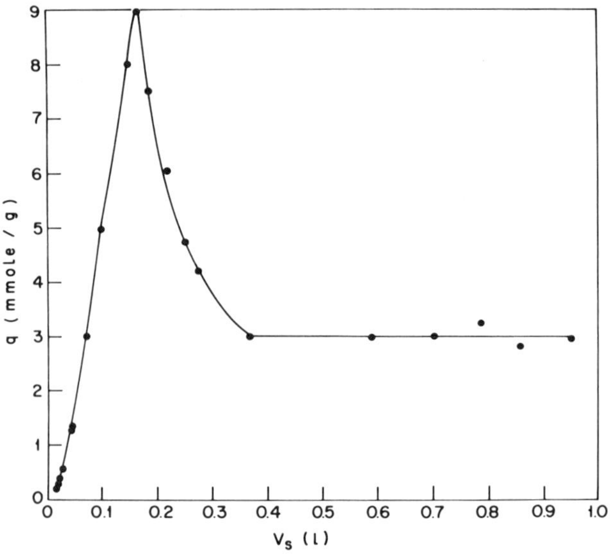

Fig. 1. Apparent adsorption of lysine on Na^+-montmorillonite. Initial concentration of lysine: 60 m moles/litre. Concentration of Na^+-montmorillonite: 1g/litre. q = Difference of initial concentration of lysine and that found in supernatant. V_s(l) - Volume of suspension after centrifugation (from 1 litre initial volume).

therefore the true adsorption, based on cation exchange. The much larger apparently adsorbed amount is just lysine dissolved in the water occluded between the layers of the clay, in a manner similar to what takes place in coacervates. The shape of the adsorption figures are very similar, not only for various concentrations of lysine, but also for other amino acids, as well as for Ca^{2+}-montmorillonite and Na^+-kaolinite. There are differences in the amount of adsorbed and occluded amino acids, depending on their initial concentration, as well as on their type and configuration. The volume of suspension where maximal occlusion occurs shifts to lower values with Ca^{2+}-montmorillonite and Na^+-kaolinite, but the shape of the curve is always the same.

The same phenomena obtained at various centrifugation rates are also observed when the mixture is allowed to stand and separate by gravitation. For the same ratios of supernatant/suspension, the same results are obtained as with centrifugation. Such a phenomenon, of precipitation merely by gravitation, could easily have occurred in nature and could have provided an ideal means of concentrating molecules, and this by substances and under conditions which almost certainly existed in the prebiotic era.

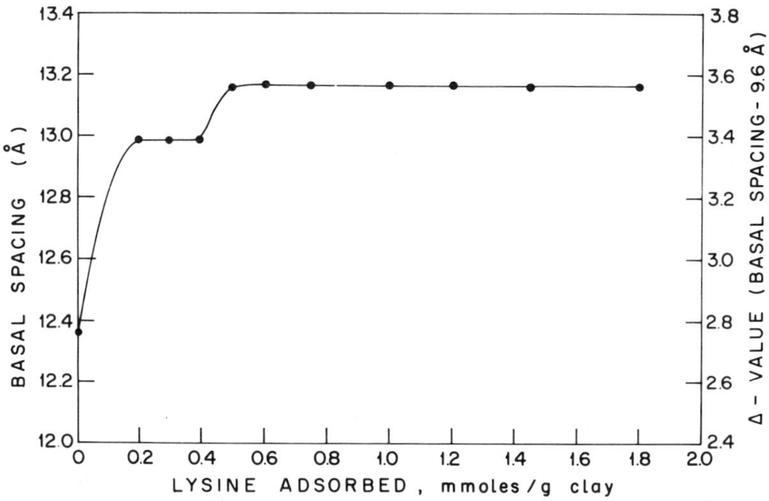

Fig. 2. Change of basal spacing of Na^+-montmorillonite with adsorption of lysine at pH 5.8. X-ray measurements were carried out on clay preparations dried at 50°C. (9.6 Å - the width of the montmorillonite platelet without any water.)

By gravitation alone one does not reach the values of suspension volumes where occlusion disappears again; as mentioned before, this takes place only when ultracentrifugation is applied. Hence, in nature, the precipitation of the clay suspension would stop at a volume where maximal occlusion occurs, and therefore the highest concentrations of molecules are obtained.

SPECIFIC INTERACTIONS OF OPTICAL ISOMERS

One of the typical criteria of living systems is that they use only one type of isomer (L amino acids and D sugars). This feature might have evolved in later steps, as part of the biological evolution, but researchers get more and more interested in the problem of chirality and whether there could have been a way in the prebiotic era to distinguish and separate one isomer from the other.

Jackson (1971) reported selective adsorption of optical isomers of amino acids up to 4% on kaolinite. He found also that L-aspartic acid polymerized in the presence of kaolinite up to 20%, while D-aspartic acid polymerized at the same time only up to 7%. Bondy and Harrington (1979) found that L-leucine, L-aspartate, and D-glucose are adsorbed on bentonite 2-10 times more than are their non-biological enantiomers.

Friebele et al. (1981) criticize the analytical technique of the above-mentioned authors. Jackson measured differences in ultraviolet light absorption of supernatant solutions of mixtures of clay with D- and L-phenylalanine. Bondy and Harrington measured the radioactivity of the bentonite suspension, after interacting with the labeled compound. Before measuring, they precipitated, washed, and reprecipitated the clay, but neither Jackson nor Bondy and Harrington isolated the adsorbed substance and determined it directly.

Friebele et al. believe that indirect measurements of adsorption introduce too great an error. They think that, in the study of preferential adsorption of isomers by clays, the D and L enantiomers have to be measured in a direct manner. They studied adsorption of amino acids on Na^{+}-montmorillonite and their method was, after adsorption, to centrifuge and wash the clay, and then to extract the adsorbed amino acid with HCl and to determine the ratio between the two isomers adsorbed. In this way, they found a preferential adsorption of L-amino acids of 0.5-2%. From the point of view of origins of life, it is important to notice that Na^{+}-montmorillonite has no known chirality, but that some assymetry might result from the arrangement in the structural layers of the clay when Al^{3+} in the octahydral lattice is replaced by Mg^{2+}, as is the case with montmorillonite up to about 18%.

Kim et al. (1981) showed that certain small molecules, mainly vitamin B_2, but also adenosine and AMP (3 and 5), are able to interact selectively with enantiomers of helicenes. The latter are ortho-condensed polyaromatic hydrocarbons whose aromatic rings overlap and lead thus to a chiral molecular shape.

On the whole, there are not many promising experiments to show how nature could have distinguished between one isomer and the other. Yet nature had much more time at its disposal than has one individual researcher, and the amount of isomer necessary for further evolution could have accumulated little by little.

PRECELLULAR FORMATION

In his paper "The Origin and Nature of Protolife," Fox (1978) states that the emphasis should not be on the question: nucleic acid first--protein first, but on <u>cells first</u>. Fox et al. (1959) have shown that by heating mixtures of amino acids containing large amounts of aspartic and glutamic acid proteinoids are obtained which, upon boiling in sea water, yield microspheres which provide a boundary similar to a cellular membrane (Fig. 3). Such a boundary could offer the same advantages as a present-day cell membrane, like

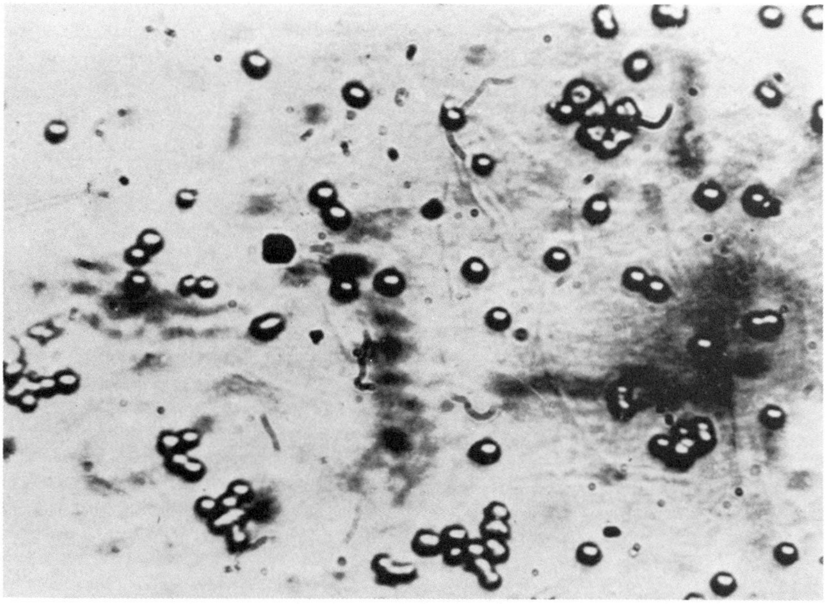

Fig. 3. Microspheres, obtained by boiling proteinoids with sea water (Fox et al., 1959).

organization of chemical reactions, protection of reactants, etc. These microspheres swell or shrink in hypotonic or hypertonic solutions respectively, they are able to bind other polymers selectively, are affected by pH, salt concentration, pressure, and temperature, and are dynamic in nature. The expression "dynamic" cannot be stressed enough in this context, as this is the basis of life. Life is a dynamic process, where each cell is in continuous interchange with its environment; should this exchange stop, life itself would stop.

Although protocell formation might be considered an advanced part in prebiotic evolution, it is, from a physical point of view, rather easy to explain. Bénard (1900, 1901) has shown many years ago that when a homogeneous liquid is heated slowly at the bottom of its container, certain geometrical structures are obtained at a certain critical temperature. These structures, at first ring-like, become honeycomb-like on further heating. The range of critical temperatures is very narrow, and with further increase in temperature the liquid becomes homogeneous again (Fig. 4). Segel (1966) has presented the theoretical basis of this phenomenon. The reason for these structure formations is the reciprocal interaction of two types of flows, heat flow in one direction and convective flow in the other one. The same phenomena, and for the same reason, can also be observed in crystallization processes. It seems

Fig. 4. Structure formation in a fluid heated from below. (a) The beginning of heating with formation of circular structures.

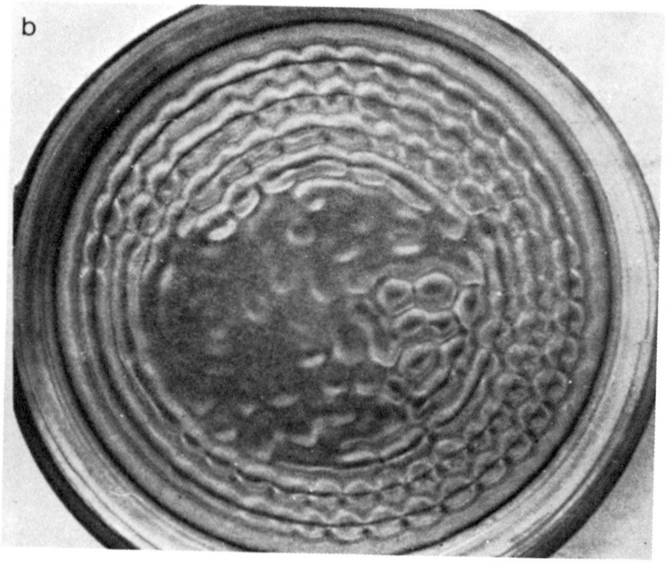

Fig. 4. (b) Continuation of heating and the beginning of cell formation.

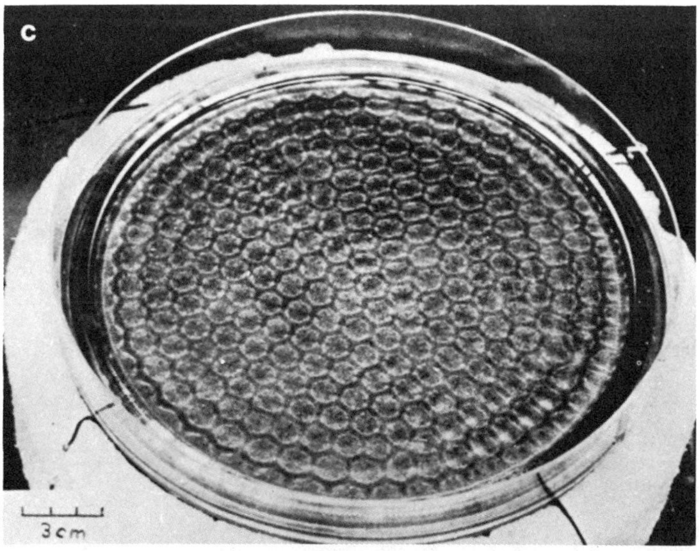

Fig. 4. (c) Further heating and formation of regular hexagonal cells (Segel, 1966).

that this, and the increasing salt concentrations, is the mechanism behind the proteinoid microspheres obtained by Fox. A further indication that this is a phenomenon depending on heat flow, rather than on temperature increase, can be seen in the work of Young (1965) who obtained microparticles by saturating solutions of amino acids at 25°C and chilling them to 0°C, and by the work of Simionescu et al. (1981) who obtained microspheres very similar (Fig. 5) to those obtained by Fox, but by cooling the reaction mixtures. Here again, temperature differences occur which produce two types of flow; hence, if indeed these flows are responsible for the structure formation, it is not surprising that the same sort of microspheres are obtained by heating as well as by cooling.

Microspheres were obtained also by Yanagawa et al. (1980) when they kept glycine with either acid, basic, or aromatic amino acids, under a nitrogen atmosphere in modified sea medium, enriched with essential transition elements, at 105°C for 8 weeks. They called these microspheres marigranules and showed them to be susceptible to pronase and elastase, which proves that peptide bonds were present. The authors consider the essential factor in the formation of these marigranules the high concentrations of salts. Later on, Yanagawa and Egami (1981) have shown that the essential factor is not just any salt, but that marigranules need for their formation Cu^{2+} ions, which could be replaced to a certain extent by Zn^{2+} ions, but by no other divalent or any other ion.

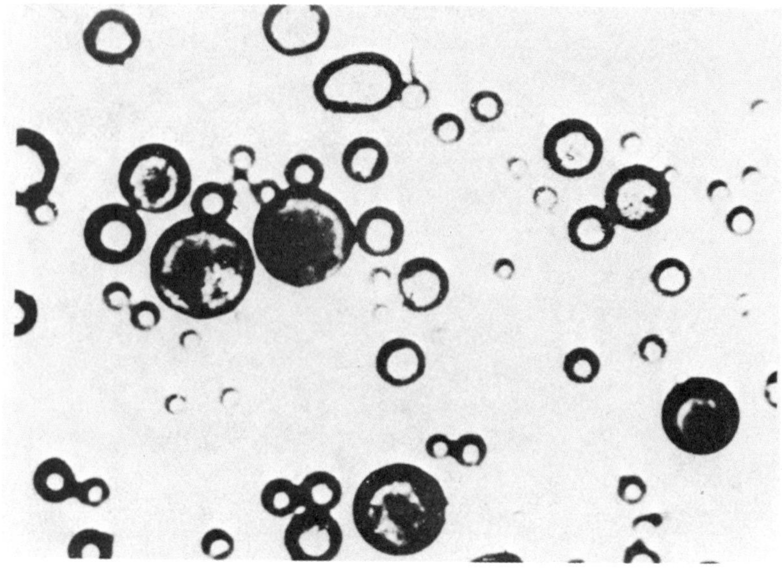

Fig. 5. Photomicrograph of self-assembled protobiopolymers obtained by cooling the reaction mixture (Simionescu et al., 1981).

THE INFLUENCE OF VARIOUS IONS

Many biological reactions are strongly influenced by various ions. The latter certainly existed in the prebiotic era. Hence, it is only natural that scientists should have looked into the possibility of ions influencing chemical reactions which might have been part of chemical evolution.

We have shown (1978a) that various cations influence the distribution and yields of molecular weights of polyamino acids obtained in the presence of montmorillonites. As a general trend, we have found that for the same amino acid, higher degrees of polymerization are obtained when the cation in the octahydral lattice of the clay is divalent (like in hectorite) rather than trivalent (like in montmorillonite or in nontronite). With exchangeable cations, the order is reversed. In this case, highest degrees of polymerization are obtained with monovalent cations, lowest with divalent ones and intermediate values are obtained with clay salts of trivalent cations. We have not found in our reactions any specificity of certain cations except for their valency.

Yet there are reactions where a specific cation is absolutely necessary, even if the reaction is a non-enzymatic one. Mullins and Lacey (1981) have shown that for the non enzymatic activation of amino acids by ATP, a divalent ion catalyst is absolutely necessary. At 23°C, Be^{2+} gives the highest yields of activation. At 50°C, the yields of activation are much higher than at 23°C, and at this temperature there is no difference among the results obtained with Be^{2+}, Mg^{2+}, and Zn^{2+}.

As mentioned in the previous section, Yanagawa and Egami (1981) have shown that, for the formation of marigranules, Cu^{2+} ions are essential. They can be replaced to a small extent by Zn^{2+} ions, but by no other divalent ion.

An inhibitory effect of Fe^{3+} has been pointed out by Ferris et al. (1981). These authors have reported that while HCN condenses in aqueous solutions to oligomers which yield purines, pyrimidines, and amino acids when hydrolyzed, this oligomerization is inhibited by montmorillonite clays containing Fe^{3+}. This inhibitory effect is due to the oxidation by Fe^{3+} of diaminomaleonitrile, the intermediate of HCN oligomerization. The oxidation of diaminomaleonitrile yields diaminosuccinonitrile, which itself is rapidly hydrolyzed to HCN and oxalic acid derivatives. When Fe^{3+} in the montmorillonite is reduced, diaminomaleonitrile is not oxidized, and the clay has no inhibitory effects on oligomerization of HCM.

CATALYTIC PROPERTIES OF CLAYS

As the choice of substances able to enhance evolutionary processes in the prebiotic era was not very great, and clays were most probably one of them, we thought it worthwhile to consider the possibilities of reactions in which clays might have taken part. We have seen previously that they could act as concentrating agents, as well as specific adsorbents. We will now look into their ability to catalyze certain reactions.

We have observed (Paecht-Horowitz et al., 1970) that when amino acid adenylates are added to a suspension of Na^+-montmorillonite at slightly alkaline pH ($\sim$7.8), polypeptides of a degree of polymerization up to about 50 can be reached, yet not in a continuous spectrum but in a discrete one. This is in contrast to the reaction carried out under the same conditions, but in the absence of montmorillonite, when a continuous spectrum of degrees of polymerization is obtained, stopping mostly at degrees of polymerization 5-6 (Lewinsohn et al., 1967) but eventually showing traces up to 10-12 (Paecht-Horowitz and Katchalsky, 1967). Relatively high polymerization of alanine starting from alanine adenylate in the presence of montmorillonite has been observed also by Calvin (1974). He reported to have detected also the simultaneous polymerization of adenylic acid.

Changing the surface of the clay might affect strongly its catalytic properties. Coyne et al. (1981) have shown that when kaolin is preannealed and then mixed with $CaCO_3$, or preirradiated with X-rays, oligomerization of glycine is enhanced. The authors believe that the reason for this higher catalytic activity is the formation of "defects" of the clay crystal, defects produced either in nature or by the various treatments of the clay. These defects can serve as traps for migrating electrons and would thus be placed for storage of energy (it was observed that they emit light), and this stored energy might very well be related to catalytic activity.

We (F. R. Eirich and M. Paecht-Horowitz, in preparation) have observed another phenomenon, related also to the changes of the clay surface. When montmorillonite is pretreated with a polyamino acid and alanine adenylate is added to this pretreated montmorillonite, the resultant average degrees of polymerization of alanine pass through a maximum when the polyamino acid added before polymerization is continuously increased (Fig. 6): With the first additions of polyamino acid, the degree of polymerization decreases, then increases sharply and reaches a maximum at additions of about $10 \cdot 10^{-5}$ moles/g clay. With further addition of polyamino acid to the clay, the average degree of polymerization obtained from alanine adenylate drops again and falls to values lower than those obtained when no polyamino acid is added.

When we measure the increase of optical density of the clay suspension with the addition of polyamino acid (Fig. 7), we see that at small amounts of polyamino acid the optical density changes very little, although adsorption takes place. This would mean that the

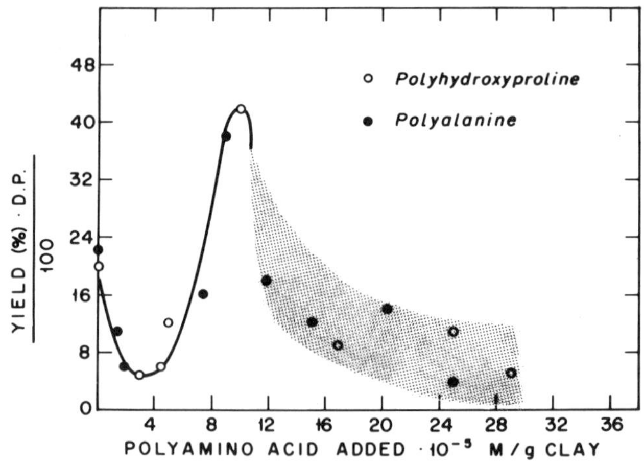

Fig. 6. Average degrees of polymerization (D.P.) obtained from alanine adenylate in the presence of Na^+-montmorillonite suspensions in water (200-500 mg/litre) when the clay was pretreated with various amounts of polyalanine or polyhydroxyproline.

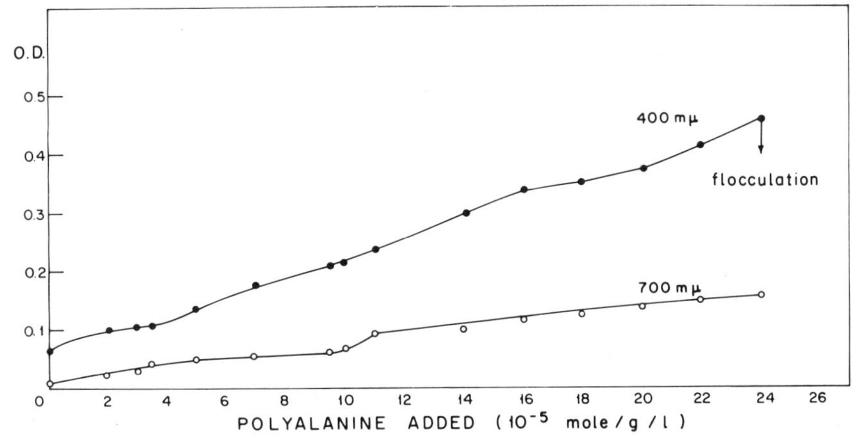

Fig. 7. Dependence of optical density (O.D.) on the addition of polyalanine (degree of polymerization: ∿60) to Na^+-montmorillonite.

molecules adsorbed are bound to the edges of the clay and cause only a very small increase of the size of the clay particle. But we have shown previously (Paecht-Horowitz, 1977) that free edges and their adjacent sites on the interfacial layers are imperative for the polymerization of alanine adenylate on montmorillonite. Hence, it is not surprising that with small amounts of polyamino acid added to the clay, polymerization of alanine adenylate decreases. With further adsorption of polyalanine the peptide is not only pushed into the interfacial layers, which fact is reflected by wider basal spaces on the X-rays diffractograms, but produces also a high degree of order in the clay particles. This order reaches its maximum with the addition of $10 \cdot 10^{-5}$ moles polyamino acid/g clay (Fig. 8) and would be the optimal proportion for effective polymerization. Eirich (1981) tried to relate the amount of adsorbed polyamino acid to the geometry of the stacks of montmorillonite which would on its part influence the polymerization of the amino acid adenylate. He found that $10 \cdot 10^{-5}$ moles polymer/g clay corresponds to two layers of polymer between the platelets of the clay. This would correspond to a much greater expansion of the clay stacks than if only one monolayer of polypeptide were encased between those platelets. Such an expansion, in addition to regular stacking, would cause polycondensation to be much more effective and would be responsible for the maximum in the degree of polymerization obtained with the addition of $10 \cdot 10^{-5}$ moles polypeptide/g clay.

We have seen previously that the oxidation state of the clay is responsible for the results of certain reactions (Ferris et al., 1981). The authors point out that this oxidation level must be

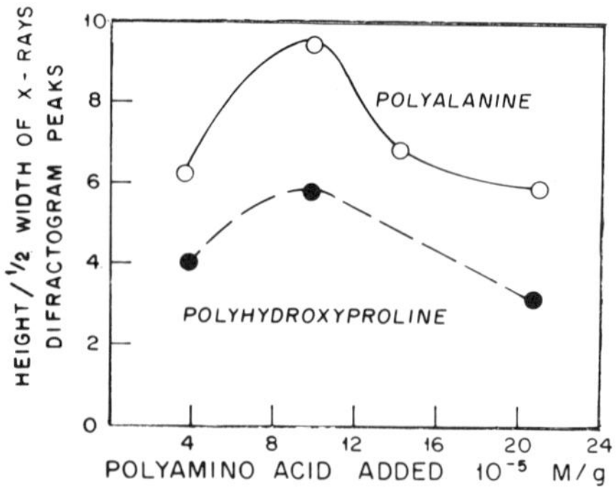

Fig. 8. The dependence of the ratio height/1/2 width of the peaks of the X-ray diffractograms on the addition of polyamino acid to the clay. This ratio reflects on the order in the clay particle. The sharper the peak--the higher the order.

considered as a variable in investigations of clays as potential catalysts of chemical events on the primitive Earth. They believe that at their formation, clays existed in their reduced form. With the increase of O_2 in the atmosphere, the oxidation state of the clay would change as well. They thus postulate that iron substituted montmorillonites would have had the function of electron transfer systems in primitive forms of life.

INFORMATION TRANSFER

The beginning and the mechanism of self-replication, which is the main characteristic of a living organism, are still unsolved problems. In order to approach them, scientists have looked for systems which might produce, under idential conditions, idential products. This would not be a self-replication system, but at least a system of information transfer which eventually might lead to a replication system.

Many years ago, S. W. Fox postulated that for the first primitive self-ordering of amino acids, even for biologically active proteins, no nucleic acids were necessary. His arguments for this point are summarized in his paper of 1978a.

Cairns-Smith (1971) suggested that the most primitive "genes" might have been mineral crystals which would have served as templates for the formation of polypeptides and polynucleic acids. Different imperfections in the crystal would have yielded different resulting products, of which the most suitable ones would have survived and evolved further.

In the last years, researchers have looked for the possibility of information transfer in the structure of the polypeptides themselves. Brack and Orgel (1975) have shown that when hydrophylic and hydrophobic residues alternate along polypeptide chains, the chains are forced to adopt the β-structure in salted water. Due to alternative repetition of the side-chains, one gets a polypeptide bilayer with a hydrophobic interior and a charged hydrophylic exterior. Brack and Spach (1979) have shown also that when samples of alternating poly (Leu-Lys) with increasing amounts of L-residues randomly distributed along the chain were prepared, the samples exhibited a mixture of coil and β-sheet conformation in salted water; the random coil contribution became dominant when the composition was close to the racemic, and the number of β-sheet nuclei increased the more the mixture departs from racemic. Brack and Spach (1981) have also shown that β-nuclei are much more resistent to hydrolysis and to heat than are helix structures. Thus, once β-structures, fromed by L-enantiomers, are formed, their chances for survival are much higher than those of α-helices formed by racemic mixtures. This would finally lead to an enrichment in L-residues. Baltscheffsky (1981) stresses also the possibility that the structure of peptides might be responsible for their function, survival, and further evolution. He has shown that ferredoxin apoproteins of the Clostridia-type tend to form β-structures. This formation is strongly enhanced in ethanol medium, and, according to the author, might be responsible for the binding of 2 iron atoms by the sulfurs of two cysteine residues. But what Baltscheffsky considers the most important feature of polypeptide structures is their ability to form 3-D structures by hydrogen-bounded α- and β-forms. He believes that these 3-D structural elements might have been significant "growing points" in the molecular evolution of polypeptides.

An interesting type of information transfer possibility was pointed out by Root-Bernstein (1982). According to his models, amino acid pairing is possible only on parallel β-ribbons and involves both the polypeptide backbones and the side chains. From all 210 possible amino acid pairs only 26 could be built to meet standard criteria for bonding. Of these, 14 were found to be genetically encoded when the codons are read as if they paired in a parallel manner. Thus, glycine and proline "templates" are absolutely specific, only glycine is small enough and has the proper stereochemistry to fit proline stereochemistry, and only the presence of a glycine paired to a proline affords a peptide sufficient flexibility to complement a proline bond in a parallel β-ribbon. Similarly, only arginine is long enough to interact across a β-ribbon with alanine.

Some experimental evidence for the above-mentioned theory is found in the paper by Poland and Sheraga (1967). These authors report interactions between 1:1 copolymers of glutamic acid and leucine. They observed a "triple transition" upon heating the copolymer. Between 0-40° the sample began to melt, and the helical content simultaneously dropped. Between 40-60° the helical content increased again as melting ceased, and decreased again as melting began again over 60°. Root-Bernstein (1982) believes that the changes in the content of α-helices in the above-mentioned experiment is a result of a transition from β-ribbon to a α-helix to random coil. Such a transition is not observed with other copolymers of glutamic acid or of leucine. Another example can be seen in the work of Reiser (1974). He reported on "supramolecular aggregates" formed by 1:1 copolymers of lysine and tyrosine. Again, such aggregates were not formed with any other copolymer of lysine or of tyrosine. Root-Bernstein (1982) finds indications of arginine and alanine pairing, by observing a reduction of osmotic pressure of the mixture. All these specific pairings indicate protein-to-protein information.

CONCLUSION

We see that in the study of the origins of life scientists look for reactions which might have occurred during the prebiotic era, reactions purely chemical which might have been influenced by the extant substances and situations. Although self-replication systems have not been discovered so far, the study as a whole has greatly advanced. We know by now how cells could have been formed, how polypeptides could have been synthesized, what the influence of various ions was, and how clays and minerals could have catalyzed certain reactions. Some experiments and some theories have been advanced to indicate information transfer and prebiological "genetic codes," based on the configuration of amino acids. The jump to self-replication has not yet been achieved, but it seems that we are not very far from it.

REFERENCES

Balscheffsky, H., 1981, Proc. of the 3rd ISSOL Meeting on Origin of Life, Y. Wolman, ed., p. 481, D. Reidel, Dordrecht, Holland.

Bénard, H., 1900, Rev. Gen. Sci. Pures Appl., 11:1309.

Bénard, H., 1901, Ana. Chim. Phys., 23:62.

Bernal, J. D., 1951, "The Physical Basis of Life," Routledge and Kegan Paul, London.

Bondy, S., and Harrington, M., 1979, Science, 203:1243.

Brack, A., and Orgel, L. E., 1975, Nature, 256:383.

Brack, A., and Spach, G., 1979, J. Mol. Evol., 13:35.

Brack, A., and Spach, G., 1981, Proc. of the 3rd ISSOL Meeting on Origin of Life, Y. Wolman, ed., p. 487, D. Reidel, Dordrecht, Holland.

Bungerberg de Yong, H., 1937, Koll. Zeitschr., 80:221.

Cairns-Smith, A. G., 1971, "The Life Puzzle," p. 118, Oliver and Boyd, London.

Calvin, M., 1974, Angew. Chem. Intern. Ed., 121.

Coyne, L. M., Lawless, J., Lehav, N., Sutton, S., and Sweeney, M., 1981, Proc. of the 3rd ISSOL Meeting on Origin of Life, Y. Wolman, ed., p. 115, D. Reidel, Dordrecht, Holland.

Eirich, F. R., 1981, Proc. of the 3rd ISSOL Meeting on Origin of Life, Y. Wolman, ed., p. 285, D. Reidel, Dordrecht, Holland.

Evreinova, T. N., Pagosova, P., Tsukawara, T., and Lapinova, T., 1962, Nauch. Dokl, Vyssh, Shk., 1:159.

Ferris, J. P., 1979, Science, 203:1135.

Ferris, J. P., Alwis, K. W., Edelson, E. H., Mount, N., and Hagan, Jr., W. J., 1981, Proc. of the 3rd ISSOL Meeting on Origin of Life, Y. Wolman, ed., p. 125, D. Reidel, Dordrecht, Holland.

Fox, S. W., 1978, "The Nature of Life, XIIIth Nobel Conference," H. Heidcamp, ed., p. 23, University Park Press, Baltimore.

Fox, S. W., 1978a, "The Nature of Life, XIIIth Nobel Conference," H. Heidcamp, ed., p. 28, University Park Press, Baltimore.

Fox, S. W., Harada, K., and Kendrick, J., 1959, Science, 129:1221.

Friebele, E., Shimoyama, A., and Ponnamperuma, C., 1981, Proc. of the 3rd ISSOL Meeting on Origin of Life, Y. Wolman, ed., p. 263, D. Reidel, Dordrecht, Holland.

Greenland, D. J., Laby, R. H., and Quirk, J. P., 1965, Proc. Faraday Soc., 61:2013.

Hsu, Sh. H., 1977, Ph.D. Dissertation, Polytechnic Inst., New York.

Jackson, T. A., 1971, Experientia, 27:242.

Kim, Y. H., Tishbee, A., and Gil-Av., E., 1981, Proc. of the 3rd ISSOL Meeting on Origin of Life, Y. Wolman, ed., p. 331, D. Reidel, Dordrecht, Holland.

Lewinsohn, R., Paecht-Horowitz, M., and Katchalsky, A., 1967, Biochim. Biophys. Acta, 140:24.

Mathews, C. N., and Moser, R. E., 1966, Proc. Nat. Acad. Sci. U.S., 56:1087.

Mullins, Jr., D. W., and Lacey, Jr., J. C., 1981, Proc. of the 3rd ISSOL Meeting on Origin of Life, Y. Wolman, ed., p. 209, D. Reidel, Dordrecht, Holland.

Oparin, A. I., 1924, "Proiskhozdeniye zhizmi" ("The Origin of Life"), Izdatel'stov "Moskovskiy Rabochly," Moscow.

Oparin, A. I., Serebrovskaya, K., Vasilyeva, N., and Balayevskaya, T., 1964, Dok. Acad. Nauk, SSSR, 154:471.

Paecht-Horowitz, M., 1977, BioSystems, 9:93.

Paecht-Horowitz, M., 1978, Proc. of the 2nd ISSOL Meeting on Origin of Life, H. Noda, ed., p. 289, Center for Academic Publications Japan, Tokyo.

Paecht-Horowitz, M., 1978a, J. Mol. Evol., 11:101.

Paecht-Horowitz, M., Berger, J. D., and Katchalsky, A., 1970, Nature, 228:638.

Paecht-Horowitz, M., and Katchalsky, A., 1967, Biochim. Biophys. Acta, 140:4.

Poland, D., and Sheraga, H. A., 1967, "Poly-α-amino Acids," G. Fasman, ed., p. 481, Marcel Dekker, New York.

Reiser, H. J.-P, 1974, "Peptides, Polypeptides, and Proteins," E. R. Blout et al., ed., p. 621, Wiley-Interscience, New York.
Rishpon, J., O'Hara, P. J., Lahav, N., and Lawless, J. G., 1982, J. Mol. Evol., 18:179.
Root-Bernstein, R. S., 1982, J. Theor. Biol., 94:885.
Segel, L. A., 1966, "Non-Equilibrium Thermodynamics, Variational Techniques and Stability," R. J. Donnelly et al., ed., p. 165, University of Chicago Press, Chicago.
Siesskind, O., 1960, Comp. Rend., 250:2228.
Simionescu, C. I., and Denes, F., 1981, Proc. of the 3rd ISSOL Meeting on Origin of Life, Y. Wolman, ed., p. 255, D. Reidel, Dordrecht, Holland.
Simionescu, C. I., Denes, F., and Totlin, M., 1981, Proc. of the 3rd ISSOL Meeting on Origin of Life, Y. Wolman, ed., p. 263, D. Reidel, Dordrecht, Holland.
Yanagawa, H., Kobayashi, Y., and Egami, F., 1980, J. Biochem. 87:855.
Yanagawa, H., and Egami, F., 1981, Proc. of the 3rd ISSOL Meeting on Origin of Life, Y. Wolman, ed., p. 309, D. Reidel, Dordrecht, Holland.
Young, R. S., 1965, "The Origins of Prebiological Systems," S. W. Fox, ed., p. 347, Academic Press, New York.

ON PROTOCELLS, PREPROKARYOTES, AND EARLY PROKARYOTES

Herrick Baltscheffsky and Jerzy Jurka

Department of Biochemistry, Arrhenius Laboratory
University of Stockholm
S-106 Stockholm, Sweden

INTRODUCTION

It has been well documented that chemical reactions occurring on the surface of the prebiotic Earth could have given large amounts of monomeric organic compounds necessary for the origin of life. Less clear is how activated monomers obtained from such compounds could have formed larger molecules, particularly those heteropolymers made from nucleotides and amino acids, the nucleic acids and proteins, which are essential components in all contemporary living organisms.

In the present context the process of molecular coupling of specific components into living systems will be discussed. Informational and functional coupling between nucleic acid and protein would appear to be an essential property of such systems (Eigen, 1971). Significant evolutionary steps in the emergence of life from inanimate matter have long been sought in this area. Among the many different complex entities, which have been considered in connection with molecular evolution leading to the origin of life, particular attention will be given to protocells, preprokaryotes, and early prokaryotes. The emergence of the translation apparatus with its triplet code in biological template-directed protein synthesis is generally considered to be the result of a multi-step evolutionary process (Oparin, 1938; Orgel, 1968; Fox and Dose, 1977).

THREE EARLIER ALTERNATIVES

Proponents of protocellular models of the earliest cells have argued that the template-directed protein synthesis could have appeared as a result of protocellular evolution (Fox and Dose, 1977).

An alternative model has the origin of translation before the appearance of the earliest cell (Kuhn and Waser, 1981, 1982). In this model inorganic porous materials are assumed to provide the necessary compartments in which the earliest translation machinery could be assembled.

These models lead to the different alternatives: a protocell evolved a translation system and became the earliest cell, or, a translation machinery originated independently of protocellular organization, and the cell was organized afterwards as a result of translation. In a more classical sense, the first type of approach gives a primary role to proteins and the second to nucleic acids.

A third alternative, with a simultaneous, stepwise emergence of interacting nucleic acids and proteins, from interacting oligonucleotides and oligopeptides, can be visualized with the following scheme:

```
nucleotides → oligonucleotides            polynucleotides (nucleic
                                  ↖      ↙  acids)
                                  coupling          ↓↑
amino acids → oligopeptides      ↙        ↘polypeptides (proteins)
```

which is based in particular on detailed ideas and models given by Eigen and Schuster (1979) and Carter and Kraut (1974). Fragmented as our current knowledge still is about the actual steps involved in the emergence of that coupling between molecular information and function which may well have been realized with chains of oligonucleotides and oligopeptides only 10-15 links long, it may nevertheless serve as a point of departure for new experiments and concepts.

A NEW PROPOSAL

If we with a living system mean one which can realize a capacity for replication, accepting errors allowing evolution, and which can make a boundary between itself and its surroundings, then also the question of information for making a boundary must be brought into focus. This implies a coupling or simultaneity, at the macro-level, between the appearance of information, function, and space limitation.

Thus, we are extending the concept of simultaneity in the molecular evolution of life from information at the level of, essentially, nucleic acids and function at the level of, essentially, proteins, to include also space-limitation at the level of, essentially, membranes.

PRECELLULAR EVOLUTION

Clays, Polymers, and Protocells

The period of chemical evolution preceding the origin of life on the Earth is considered to have resulted in amino acids, sugars, purines, and pyrimidines, among other compounds (see Miller and Orgel, 1974; Schwartz, 1979). From such building blocks the heteropolymeric polypeptides and polynucleotides may have been formed by, for example, evaporation, freezing, or selective adsorption on clays. According to Weiss (1981) clays can adsorb amino acids and show pronounced selectivity to complementary purine-pyrimidine base pairs, in comparison to the non-complementary ones. In addition, they can form ferredoxin-like catalysts which are believed to be of importance for the evolution of the earliest metabolism. Other functions of clays could be protection from the photochemical degradation and providing of catalytically active surfaces.

Katchalsky's group (Paecht-Horowitz et al., 1970; Katchalsky, 1973) demonstrated that the montmorillonite clays can catalyze the polymerization of polypeptides from amino acid adenylates up to chain length of about 100.

The biologically important principle of replication is not a unique property of living systems. Again clays can be taken as an example. It has been shown that montmorillonites can disintegrate into single two-dimensional units under the influence of water and new units can crystallize on isolated matrix layers (Weiss, 1981). However, no answer can yet be given to the question of whether clay minerals really were a link to nucleic acid replication.

Evidence for heteropolymer formation under simulated prebiotic conditions, with both nucleotide (Lohrmann et al., 1980; Bridson and Orgel, 1980; Inoue and Orgel, 1981, 1982) and amino acid (Fox, 1975; Fox et al., 1982) building blocks have been obtained from numerous experiments in recent years. The template-directed condensation of activated nucleotide has been shown to give oligomers with a mean chain length as high as 15 and a predominance of 3', 5' linkages. Certain random copolymers gave complementary base incorporation, modelling in agreement with earlier speculations (Orgel, 1968) replication in enzyme-free systems (Inoue and Orgel, 1983). Remarkable features of the largely linear protein-like polymers, called proteinoids, which were obtained by simple heating of α-amino acids (Fox, 1975), are enzymic activities (Dose and Zaki, 1971) and formation, by aggregation in water, of microspheres, or protocells (Fox, 1975).

Although living organisms today are known to employ essentially polynucleotides for information and polypeptides for enzymic and other biological function, it would appear necessary to give increasing attention, in connection with early molecular evolution, to emerging reports on nucleotide heteropolymers with enzymic (Kruger et al., 1982) or enzyme model (Inoue and Orgel, 1983) properties, as well as on peptide heteropolymers with informational capacities (Fox et al., 1982).

The experiments with proteinoid containing protocells have led to the proposal that these microspheres represent a step in the molecular evolution toward the living cell (Fox, 1975, 1982). According to this view, protocells would provide the locale for emerging interactions between nucleotides and peptides giving rise to life. The left side of Fig. 1 shows this suggested sequence of events. Other models with some apparent similarities to protocells are coacervates (Oparin, 1938) and marigranules (Yanagawa and Egami, 1980).

PROTOCELLS, PREPROKARYOTES AND EARLY PROKARYOTES

In Fig. 1 is seen our representation of some proposed sequences of events leading to the origin of life on the earth. Although the figure can be regarded as self-explantory, and parts of it have been discussed above, we would like to comment on certain points:

1. Protocells, aggregates in inorganic compartments, or other structural entities may well have been of profound significance during the early phase of molecular evolution leading to living systems.

2. In our opinion, protein-like, polynucleotide-like and protocellular structures may well have emerged early in the evolutionary process. But a more essential series of events was the stepwise emergence of specific interactions between nucleotide and peptide oligomers and polymers which constituted that feedback system between information and function which led to the translation, transcription, and replication mechanisms of the early living cells--those with coded--for cellular membranes. According to this view, the simultaneity between emergence of information and function at the nucleic acid and protein levels extends to space limitation at the coded membrane level. The primitive coding patterns envisaged by Orgel (1972) and Brack and Orgel (1975) to possibly provide for early coded-for membranes may well be of significance in this connection.

3. We suggest that the term preprokaryote (Baltscheffsky and Baltscheffsky, 1981) be restricted to early life without DNA (Eigen and Winkler-Oswatitsch, 1981; Follmann, 1982) and that emergence of DNA be the basis for classification of emerging prokaryote species.

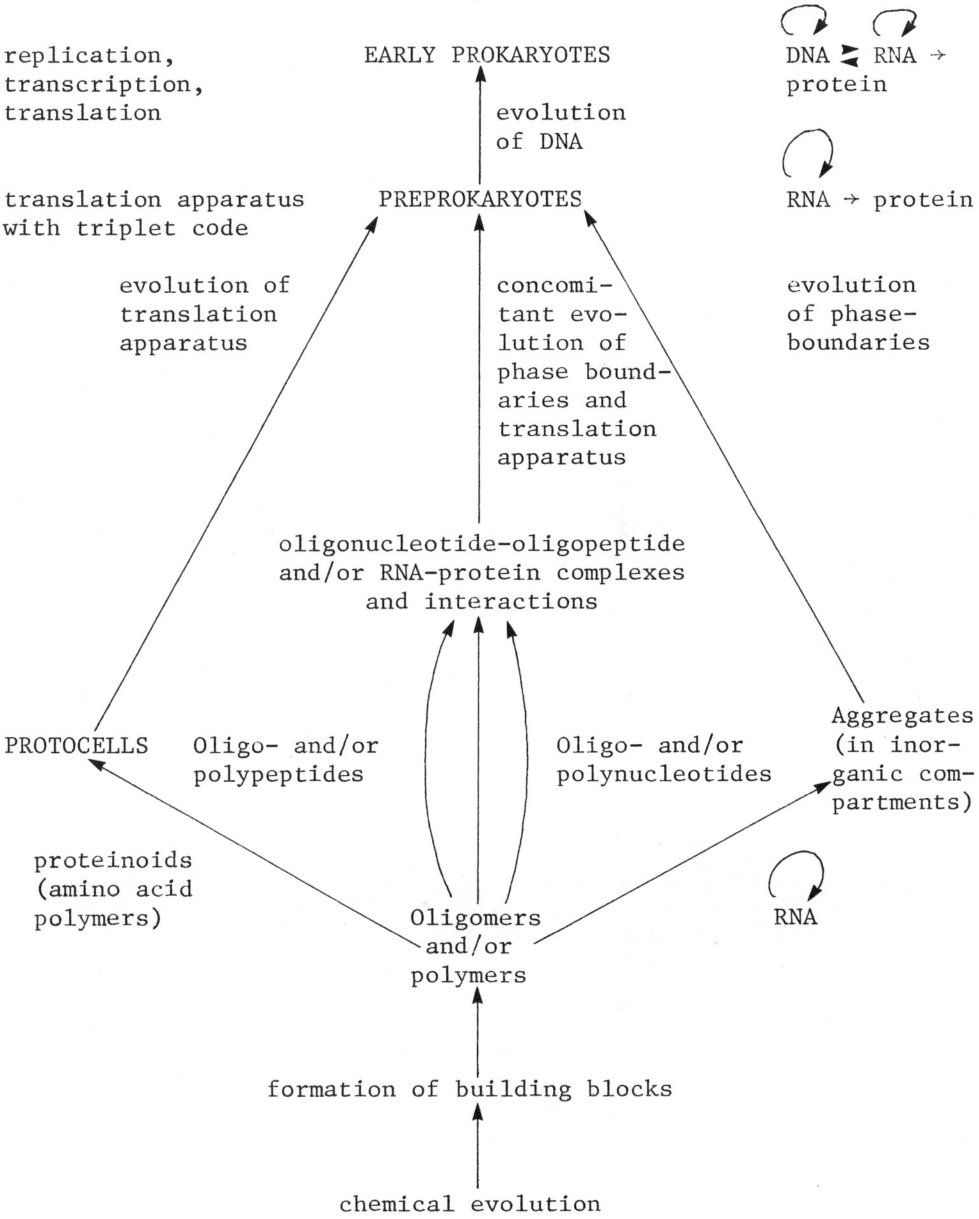

Fig. 1.

With the view that life in essence encompasses molecular information, function, and space-limitation, we endeavor to present a practical scheme (Fig. 2) for further discussion. On a scale from 0 to 100% life, life with DNA is put as 100%. Unspecified molecules may be put at the other extreme, 0% life. As 50% life we have chosen

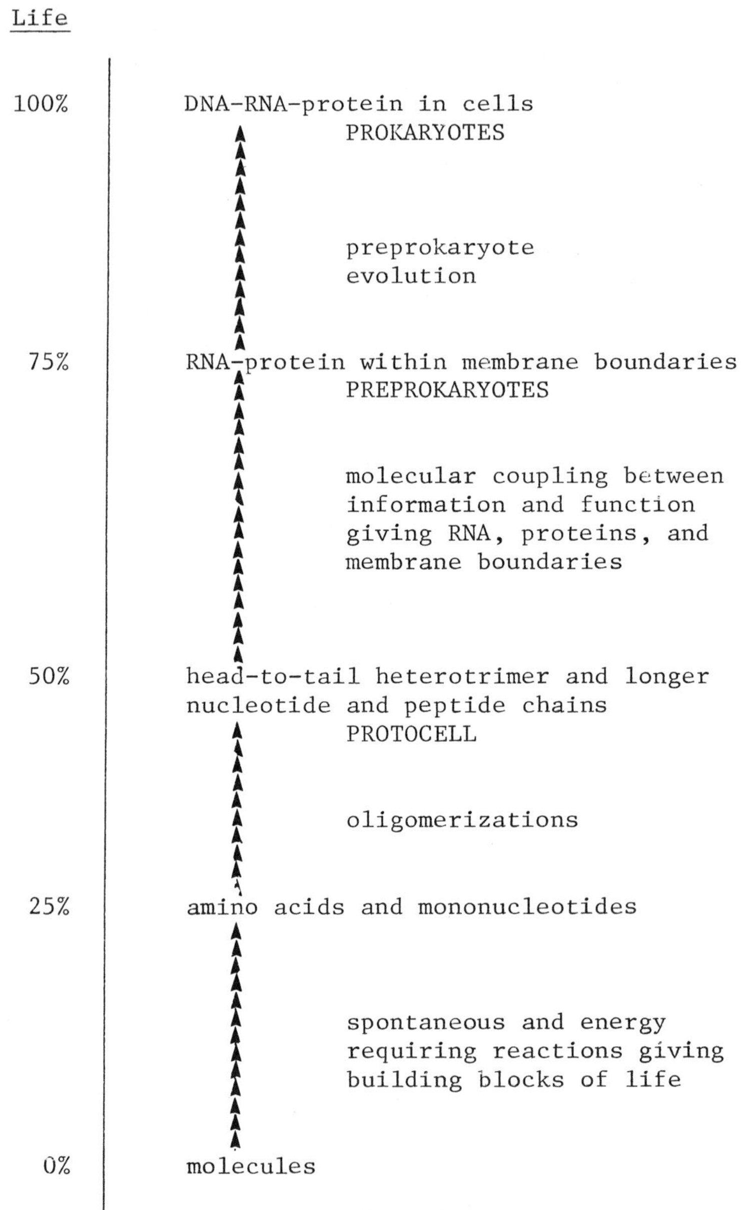

Fig. 2.

the emergence of head-to-tail heterotrimers and longer nucleotide and peptide chains, which could possibly constitute a well specified prerequisite for the start of the interaction between head-to-tail heteropolymer information and function, which on a most fundamental level, and with a continuous, stepwise evolution up to the present day constitutes the molecular essence of a living entity. Other, in certain ways related, schemes are those of Fox (1981) and Eigen and Schuster (1982). The divisions in our scheme between 0 and 100% should, of course, not be taken literally--in a sense they are really only reflecting the ease of saying "quarter," "half," and "three-quarters."

We hope for incisive comments, not least to our placement between 50 and 75% life of the molecular coupling between information and function which gave RNA, proteins, and membrane boundaries.

REFERENCES

Baltscheffsky, H., and Baltscheffsky, M., 1981, in: "Mitochondria and Microsomes," C. P. Lee, G. Schatz, and G. Dallner, eds., pp. 519-540, Addison-Wesley, Boston.
Brack, A., and Orgel, L. E., 1975, Nature, 256:383.
Bridson, P. K., and Orgel, L. E., 1980, J. Mol. Biol., 144:567.
Carter, Jr., C. S., and Krant, J., 1974, Proc. Nat. Acad. Sci. U.S., 71:283.
Dose, K., and Zaki, L., 1971, Z. Naturforsch. 26b:144.
Eigen, M., 1971, Quart. Revs. Biophys., 4:149.
Eigen, M., and Schuster, P., 1982, J. Mol. Evol., 19:47.
Eigen, M., and Winkler-Oswatitsch, R., 1981, Naturwiss., 68:282.
Follman, H., 1982, Naturwiss., 69:75.
Fox, S. W., 1975, BioSystems, 7:22.
Fox, S. W., 1981, Amer. Bio. Teacher, 43:127.
Fox, S. W., and Dose, K., 1977, "Molecular Evolution and the Origin of Life," 2nd edn., Marcel Dekker, New York.
Fox, S. W., Nakashima, T., Przybulski, A., and Syren, R. M., 1982, Int. J. Quantum Chem., QBS9:195.
Inoue, T., and Orgel, L. E., 1981, J. Amer. Chem. Soc., 103:7666.
Inoue, T., and Orgel, L. E., 1982, J. Mol. Biol., 162:201.
Inoue, T., and Orgel, L. E., 1983, Science, 219:859.
Katchalsky, A., 1973, Naturwiss., 60:215.
Kruger, K., Grabowski, P. J., Zang, A. J., Sands, J., Gottschling, D. E., and Cech, T. R., 1982, Cell, 31:147.
Kuhn, H., and Waser, J., 1981, Angev. Chem. (Engl. edn.), 20:500.
Kuhn, H., and Waser, J., 1982, Nature, 298:585.
Lohrmann, R., Bridson, P. K., and Orgel, L. E., 1980, Science, 208: 1464.
Miller, S. L., and Orgel, L. E., 1974, "The Origins of Life," Prentice Hall, Englewood Cliffs, New Jersey.
Oparin, A. I., 1938, "The Origin of Life," MacMillan, New York.
Orgel, L. E., 1968, J. Mol. Biol., 38:381.
Orgel, L. E., 1972, Isr. J. Chem., 10:287.
Paecht-Horowitz, M., Berger, J.. and Katchalsky, A., 1970, Nature, 228:636.

Schwartz, A. E., 1979, *in*: "Marine Organic Chemistry", E. K. Omersma, and R. Dawson, eds., pp. 7-30, Elsevier, Amsterdam.
Weiss, A., 1981, *Angev. Chem.* (Engl. edn.), 20:850.
Yanagawa, H., and Egami, F., 1980, *BioSystems*, 12:147.

PROTORIBOSOMES

Tadayoshi Nakashima

Institute for Molecular and Cellular Evolution
University of Miami
Coral Gables, Florida 33134

CONTEMPORARY RIBOSOMES

Ribosomes, ubiquitous nucleoprotein particles, are the only cell organelles found in all organisms. These supramolecular complexes, essential in the biosynthesis of proteins, translate genetic information into protein. The ribosomes of all organisms are constructed by the same architectural plan. Prokaryotic cells contain one chromosome and a large number of ribosomes. The cytoplasmic ribosomes of eukaryotes occur either in free form or bound to the surface of the endoplasmic reticulum. In eukaryotic cells, ribosomes are also found in the cell nucleus and in organelles such as mitochondria and chloroplasts. Mitochondrial and chloroplastic ribosomes from eukaryotic cells resemble those of bacteria in many respects. This fact is explained by a hypothesis of metabolic endosymbiosis (Lehninger, 1975).

Ribosome Structure

Prokaryotic (E. Coli) ribosomes have a sedimentation coefficient of 70S, a diameter of about 18 nm, and a particle weight of about 2.8 megadaltons. When not attached to mRNA they dissociate into 50S and 30S subunits. The small robosomal subunit contains one 16S RNA molecule and 21 different proteins. The large ribosomal subunit possesses two RNA molecules, namely a 23S component and a 5S component. The 50S subunit has 34 different proteins. Much more is known about ribosomal structure and function of E. coli than about all other ribosomes together. The ribosomes in the extramitochondrial portions of eukaryotic cells are larger and more complex than prokaryotic ribosomes. Such cytoplasmic ribosomes have a sedimentation coefficient of 73S to 80S, a dimeter about 21 nm, and a particle weight of about 4.0 megadaltons. They dissociate into 60S and 40S subunits. The small subunit contains one 18S RNA and 30 different proteins. The large subunit contains three RNA molecules; 5S, 5.8S and 28S. The 60S subunit has 50 different proteins.

Ribosomal RNA (rRNA) constitutes up to 65% of the mass of ribosomes. Each rRNA macromolecule contains four or more types of nucleotides arranged in specific sequences, and appear to serve as framework structures. In studies of E. coli ribosomes, the primary structures of the 5S RNA with 120 nucleotides (Brownlee et al., 1967), of the 16S with 1542 nucleotides (Brosius et al., 1978; Carbon et al., 1978), and of the 23S with 2904 nucleotides (Brosius et al., 1980) are known. Each of the three τRNA molecules has a specific three-dimensional conformation conferred by intrachain base pairing. In comparison of the primary and secondary structures of ribosomal RNAs from prokaryotic and eukaryotic species, the secondary structure of the RNA molecules has been strongly conserved throughout evolution, and the secondary structure is in general more conserved than the primary structure of rRNAs (Zwieb et al., 1981; Glotz et al., 1981). Functionally important sites of 16S rRNA appear to be located in unpaired regions and are phylogenetically highly conserved (Noller and Woese, 1981). Complementary interactions between regions of RNA are becoming increasingly implicated in various steps of the protein synthetic process. The effect of Mg^{2+} is exerted on the RNA moiety of the ribosomes, with little change in the conformation of the proteins, that is, Mg^{2+} ions stabilize the ribosome by stabilizing the RNA conformation (Spitnik-Elson et al., 1976).

Each ribosomal protein macromolecule has a specific information rich sequence of some 20 different amino acids. Ribosomal proteins are insoluble in water at pH 7.0 under normal conditions but can be extracted with concentrated salt solutions. Ribosomes contain acidic proteins in addition to the preponderance of basic proteins. Most of the ribosomal proteins are present in single copy. A large amount of amide nitrogen from the proteins (approximately 5 moles per 100 moles) suggests that many of the acidic residues are amidated (Wool et al., 1974). The primary structures of all the E. coli ribosomal proteins are known, and most of the 30S proteins are rich in α-helix and contain a small amount of β-sheet, whereas the proteins from the large subunit are more diverse in their α-helix content, almost all of the 50S proteins are relatively rich in β-sheet (Wittmann, 1982). There are 21 proteins, designated S1, S2, etc., in the 30S small subunit and 32 different proteins, L1, L2, etc., in the 50S large subunit. Ribosomal proteins from various prokaryotic and eukaryotic species have been compared by two-dimenional gel electrophoresis, immunological crossreactions, reconstitution, and amino acid sequence analysis. These studies show a strong homology among prokaryotic ribosomal proteins but only a weak homology between proteins from prokaryotic and eukaryotic ribosomes (Wittmann, 1982; Brimacombe et al., 1978). The A-proteins which are acidic ribosomal proteins corresponding to the proteins L7/L12 of the E. coli ribosomes can replace the corresponding proteins from yeast, reticulocytes, and rat liver ribosomes without loss of activity (Wool et al., 1974). There is a very weak homology between the amino acid sequences of A-protein from eubacteria and eukaryotes, whereas a strong homology has been found between cytoplasmic ribosomal proteins isolated from eukaryotic species (Matheson et al., 1980; Wittmann-Liebold et al., 1979). In spite of changing of ribosomal proteins during evolution, the function of the ribosome has remained unaltered.

The nucleic acid and protein components are held together by relatively weak noncovalent forces, such as hydrogen bonds, hydrophobic interactions, and van der Waals interactions. Proteins and RNA can exert a strong mutual influence on their respective conformations. The noncovalent association of macromolecules into supramolecular complex is highly ordered and specific.

Ribosome Function

Ribosmoes serve as the organizing center of the protein synthesizing apparatus. As such they interact with mRNA, aminoacyl- and peptidyl-tRNA and protein factors (initiation, elongation, and termination). The synthesis of proteins from activated amino acids takes place on the surface of ribosome. The 70S ribosome first dissociate into 50S and 30S subunits before peptide synthesis can begin, mRNA and tRNA are first bound to 30S subunit which then reassociate with the 50S subunit. The peptidyl transferase activity which is responsible for peptide bond formation is located on the 50S subunit. The ribosomes undergo a conformational change cycle in rapid movement with its growing peptide chain along the mRNA. Such fast cycling occurs by noncovalent dissociation and reassociation of subunit components or by noncovalent changes in the conformation of subunits. Colicin E 3-induced nucleolytic cleavage of the region near the 3' end of the 16S RNA inactivates ribosome functions, especially the ability of the ribosome to initiate on natural mRNA (Tai et al., 1974; Nomura et al., 1974b). Some of the ribosomal proteins participate directly in ribosome function. For example, protein S12 deficient particles have very low activity in chain initiation (Ozaki et al., 1969), and show a pronounced reduction in translational error frequency, in contrast, S11 deficient particles show a pronounced increase in the error frequency (Nomura et al., 1974a). The catalytic ribosomal functions, such as GTPase or peptidyl transferase involve a direct participation of some ribosomal proteins. GTP is required for both the binding of the aminoacyl-tRNA and for the translocation step. The hydrolysis of GTP provides the energy for conformational change that moves the ribosome to the next codon of the mRNA and shifts the peptidyl-tRNA from the A site to the P site. On ribosomes, the peptide bond is formed by nucleophilic attack of the amino group of the aminoacyl-tRNA on the esterified carboxyl carbon of the peptidyl-tRNA. This reaction is facilitated by peptidyl transferase which is one of the proteins in the 50S ribosome.

A Hypothesis for Ribosome Structure and Function

As has been mentioned earlier, the ribosomes of all organisms are constructed on the same architectural plan, i.e., all ribosomes consist of two subunits of unequal size, each unit has rRNA(s) and proteins at the ratio of 1.0-1.5. The small subunit has one rRNA and 21-30 proteins, the large subunit has 2-3 rRNA and 34-50 proteins. The secondary structure of the rRNA molecules from various organisms has been strongly conserved throughout evolution. There are no detectable similarities between ribosomal proteins of E. coli and those of eukaryotes (Wittmann, 1982). The ribosomal proteins are believed to play a specific role in the synthesis of polypeptides, either as an enzyme or as a helper in the overall process, however the specific function of only a few of the proteins is known (Lehninger, 1982). If ribosomal proteins do not have any direct functional role, it will be difficult to explain why ribosomes contain so many different proteins rather than many copies of a few proteins (Nomura, 1974a).

In comparing the molecular masses of ribosomal proteins from three organisms, E. coli (Wittmann, 1982), yeast (Saccharomyces cerevisiae) (Otaka et al., 1978, 1982), and rat liver (Wool et al., 1974; Wool, 1980), the author points out that the proteins from each subunit of the three organisms form a graded sequence of closely related molecular weights within a certain range. Figs. 1a and b show the pattern of sequentially increasing molecular weights of E. coli and rat liver ribosomal proteins. The consecutiveness of the protein masses of each subunit shows uninterrupted gradient steps. Each protein with its proper number is placed underneath each step, and proteins which have closer molecular weights share a step. Every small subunit of the three organisms has 20 steps. The ribosomal protein topography of E. coli has been extensively studied. Three-dimensional models of the 30S and 50S subunits of E. coli ribosomes with the location of the centers of antibody binding sites for the ribosomal proteins have been reported (Stöffler et al., 1977, 1980; Brimacombe et al., 1978). In the topography of the 50S subunit, all proteins have not been located. The consecutiveness of the E. coli ribosomal proteins which is shown in Fig. 1a has been traced on the protein topography of each subunit (Figs. 2a and b). In the figures, there seems to be some degree of ordering based on molecular weights as observed on the surface of each subunit, although some of these delineations are somewhat arbitrary. In the delineation of the small subunit, one mass from S16 and S18, two masses from S9 have been assigned according to their locations. That the ribosomal proteins would consist of graded sequences of sizes is not so surprising, but that there should also be some

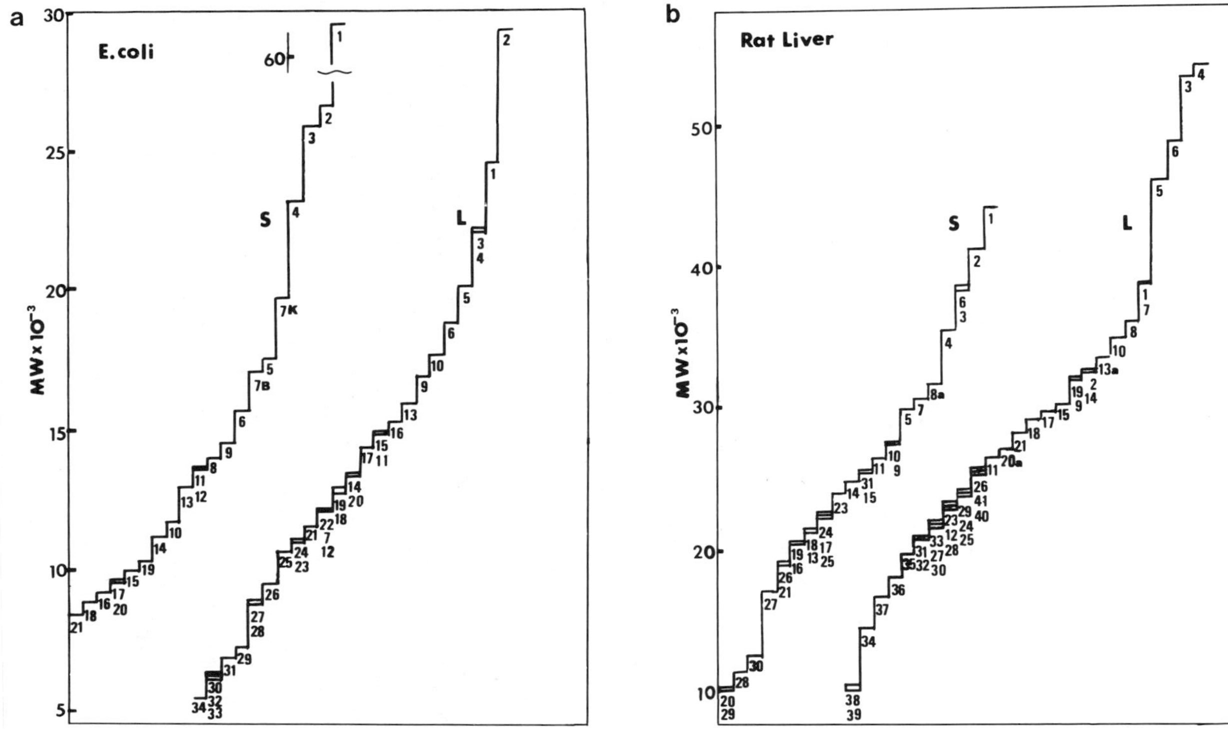

Fig. 1. Consecutive molecular weights of ribosomal proteins. (a) E. coli, and (b) Rat Liver. S = small subunit, L = large subunit. The data in Fig. 1a is from Wittmann (1982). The data in Fig. 1b is from Wool et al. (1974).

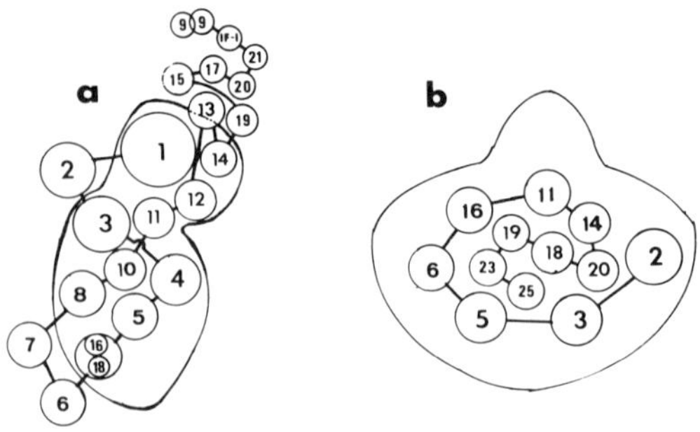

Fig. 2. Consecutive proteins on the E. coli ribosomal subunits. (a) left, small subunit (b) right, large subunit. IF-1 is inserted between S9 and S21 according to their molecular weights compared. The insertion is not based on the location of IF-1 on the 30S protein topography. The data derived from Brimacombe et al. (1978) and Stöffler et al. (1980).

ordering based on size on the ribosome seems unlikely just on a chance basis. There must be some functional basis for these facts. Such arrays of consecutive proteins are also observed in part of the assembly maps of E. coli ribosomal proteins (Nierhaus, 1980) and in the peptidyltransferase center proteins (Harris and Pestka, 1977) of E. coli 50S subunit. The consecutiveness of the ribosomal proteins might be the homology between ribosomal proteins from prokaryotes and eukaryotes, that is, this fact may have been conserved during evolution and may be a necessary structure for ribosome function. Comparisons which depend on the primary sequences of the ribosomal proteins would not necessarily detect the presence of homologous tertiary structures in different proteins (Kurland, 1977). It is conceivable that tertiary structures of ribosomal proteins may have been conserved throughout evolution. The array of the consecutive proteins may be oriented by the RNA and by their inherent force to spread over a large space of the RNA.

The author presents a hypothesis that the graded sizes of ribosomal proteins may act as gears which move ribosome subunits along a mRNA during peptide synthesis. Energy of GTP may be transferred to a protein, and the energy of the protein to the next and so on. The proteins may roll over sequentially by their specific force, for example, by their amphiphilicity (Eisenberg et al., 1982; Perutz et al., 1965) and/or electrostatic force. Water, Mg^{2+}, protein factors, aminoacyl-tRNA, peptidyl-tRNA and mRNA might be involved in this mechanochemical system. The difference of protein size may be required to provide synchronized different speeds for mRNA, tRNA and nascent peptides, as if a clock has various gears for the long, short, and second hands. The gradual sizes of proteins may also be necessary for transfer of energy.

PROTORIBOSOMES

According to the concept of chemical evolution (for example, Calvin, 1969; Fox and Dose, 1977) and the knowledge of contemporary ribosomes, protobiological models of ribosomes might be nucleoprotein microparticles which synthesize peptide coded for by nucleotides.

Ribonucleoprotein Microparticles

Many supramolecular systems assemble themselves from their subunits without any external instruction. When macromolecules are combined into supramolecular phase-separated systems, their interactions are brought to a mutual adjustment of intramolecular structure and function. Some examples of studies on a supramolecule, nucleoprotein microparticle are shown as follows. E. coli 16S RNA and 30S ribosomal proteins reconstitute active 30S subunits (Traub and Nomure, 1968). Twenty mM Mg^{2+} is required, optimum ionic strength is 0.37, and optimum temperature is 40-50°C. The reconstitution of active E. coli 50S subunits requires a two-step incubation procedure (Nierhaus and Dohme, 1974; Nierhaus, 1980). The interaction of single-stranded polyadenylic acid (Poly A) with polylysine has been studied by optical rotatory dispersion and ultraviolet spectroscopy (Davidson and Fasman, 1969) as follows. Polylysine forms a well-defined complex with poly A, altering the conformation of the latter. The primary association occurs with a 1:1 residue stoichiometry relatively little insolubility, the secondary aggregative interaction occurs near residue equivalence, and insolubility increases. With the complex formed at a ratio of 1.3 (lysil/adenyl) or less, reversibility is observed. At the ratio of 1.3, mixing at 3°C produces a predominantly primary complex, which is 45% temperature reversible, while mixing at 10°C results in formation of the secondary complex, which is totally temperature nonreversible. In the initial stage of the primary complex formation the positively charged polylysine molecules bind to a large chain of poly A until most of the negatively charged nucleotide sites have been covered by lysyl residues. As suitable lengths of poly A sequence become increasingly unavailable, additional polylysine chains cross-link the primary complex, resulting in an aggregated secondary complex. The intercomplex interaction might involve lysyl-lysyl, lysyl-phosphate, or lysyl-adenyl interchain association. In the electron micrographic observations, the particles composed of poly A and polylysine are globular, and the particles composed of poly A, poly U, and polylysine are fibrous shape (Oparin and Gladilin, 1980). The aggregates of rice dwarf virus RNA plus polylysine prepared in a 0.5 M salt solution are observed as vermiform clumps of almost equal size (about 500 Å x 2500 Å), and this value suggests that each clump consists of about 500 triple-stranded chains associated side by side (Matsuo et al., 1969).

Mixing of dilute solutions of soluble yeast RNA and lysine-rich proteinoid yields microparticles (Waehneldt and Fox, 1968). The mean diameter of microparticles is about 0.5 μm. The composition and yield of microparticles are pH-dependent. Strong acid or strong alkali or high salt concentration dissolves the microparticles. Near neutrality the proportion of RNA in these particles is close to constant (35-46%) over a wide range of composition of reacting mixtures. The relatively constant ratio of complexing RNA and lysine-rich proteinoid indicates a fixed stoichiometry. This behavior is consistent with the interpretation of an electrostatic interaction between the phosphate groups of the RNA and the amino-groups of lysine residues in the proteinoids. Both the amount of microparticles formed and the percentage of RNA incorporated increase with the lysine contents of proteinoid. Proteinoid containing both lysine and arginine interacts to form particles with each of the five homopolyribonucleotides (Yuki and Fox, 1969). When arginine is omitted, the proteinoid does not react with purine polynucleotides, when lysine is omitted the proteinoid does not interact with poly C. Omission of histidine has virtually no effect. The turbidity is expressed by optical density at 600 mμ. The role of Mg^{2+} within the interaction between polyribonucleotides and lysine-rich proteinoids has been studied (Lacey et al., 1979a) as follows. The formation of microparticles was monitored both by the turbidity and by the mass of precipitate formed. With the proteinoid containing 24.5 mol% lysine, the total amount of particles is decreased as the Mg^{2+} concentration is increased. The major effect on decreasing the amount of particles occurs with 10 mM Mg^{2+}, further increases in Mg^{2+} have less effect. Increasing concentration of Mg^{2+} tends to displace proteinoid from the complex with polyribonucleotide. In the study on the formation of microparticles composed of homopolyribonucleotides and several basic proteinoids, each rich in an individual amino acids, three of the four proteinoids yield results consistent with a matrix of anticodonicity (Lacey et al., 1979b).

The proteinoids are known generally to be of limited heterogeneity, due to the self-ordering effect of the amino acids reacting to yield proteinoids (Fox and Nakashima, 1967). The studies on thermal condensation of amino acids agree with the view that proteinoid might be the first informational macromolecules (Fox and Dose, 1977). Comparable data on the abiotic formation of self-ordered polynucleotides are not yet known. The formation of prebiotic ribonucleoprotein particles could have occurred as a result of the interaction between basic proteinoids and polyribonucleotides (Fox and Dose, 1977), and the latter could have been formed separately or formed by proteinoid (for example, Nakashima et al., 1970; Jungck and Fox, 1973). As an experiment for protoribosomes, microparticles composed of enzymically synthesized homopolyribonucleotide and lysine-rich proteinoid have been studied (Fox et al., 1974; Nakashima and Fox, 1981).

Peptide Synthesis on Model Protoribosomes

Microparticles composed of lysine-rich proteinoid and poly A cause the conversion of ATP and amino acids to various peptides. To prepare nucleoproteinoid microparticles in suspension, 0.05 M Tris buffer (pH 7.2) containing poly A is mixed with the buffer containing lysine-rich proteinoid. To this is added ATP and radioactive amino acid dissolved in the buffer. After incubation at 25°C for 3 days, the sample is chromatographed on paper. The radioactivity is counted on a radiochromatogram scanner. The experiments reveal that ATP is necessary to activate amino acids and to energize the snythesis. Without the microparticles no peptides result from amino acid. Oligopeptides of each of four amino acids (Gly, Pro, Lys, and Phe) in the suspensions of the microparticles are synthesized (Nakashima and Fox, 1981). In studies on the peptide formation by contemporary ribosomes and aminoacyl-tRNA in the absence of supernatant enzymes, diphenylalanine is the major product (Pestka, 1968). E. coli ribosomes in the absence of messenger polynucleotides are capable of synthesizing some polypeptides from aminoacyl-tRNAs as substrates (Nadezhda, 1982). These results are explained by the conception that the properties of the protoribosomes have subtracted functions from those of the contemporary ribosomes. The synthesis of peptide has been tested at pH 7.2 in a suspension of acidic + basic proteinoid particles (Nakashima and Fox, 1981). The activity of the complex particles is several times as large as that of the basic proteinoid solution alone. A measurable amount of peptide is not indicated for acidic proteinoid solution alone. It is conceivable that hypohydrous or surface conditions provided by an extrasolution particles are necessary to the peptides synthesized.

Lysine-rich proteinoids in aqueous solution catalyze the formation of peptides from free amino acid and ATP (Nakashima and Fox, 1980), Fig. 3 shows glycine peptides produced. This catalytic activity is not found in acidic proteinoids and in neutral proteinoids, even though they contain some basic amino acids. The pH optimum for the synthesis is about 11, but is appreciable below 8 and above 13. No peptide was formed in controls containing only unheated amino acids. The proteinoid polymer is thus essential. Peptide syntheses have been demonstrated by the mixing of eighteen amino acids, histidine and lysine-rich proteinoid, ATP and $MgCl_2$ at pH 11 (Fox and Nakashima, 1980). The mixture was incubated at 20°C for 5 days and fractionated on Sephadex G-25. Most of the oligopeptides appear to be in the dipeptide-tripeptide range or larger. Little or no free amino acids survive from the reaction. Virtually all types of amino acid appear to yield peptides. The concentration of reactants employed is high compared to traditional prebiotic simulation experiments. It is however characteristic of

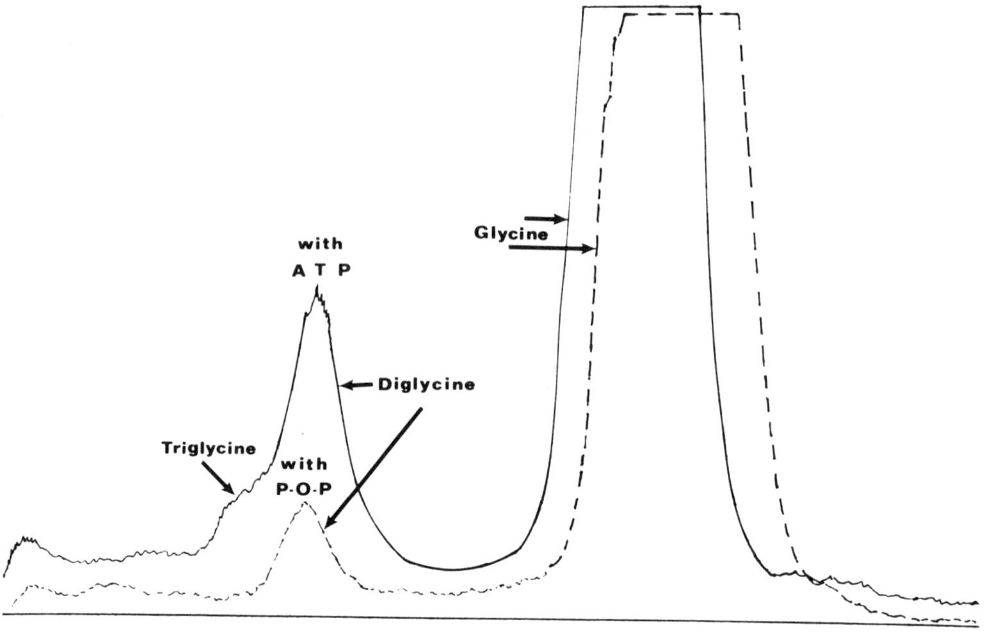

Fig. 3. Synthesis of gylcine peptides by lysine-rich proteinoid and ATP (or pyrophosphate).

biochemical reactions, such as peptide bond synthesis, to occur at the surface of ribosomes. Such conditions provide the catalytic agent in a concentrated condition. Results presented here are explicable in the data obtained in contemporary ribosomal experiments. An examination of the pH dependence of transpeptidation demonstrates that imidazole residue or N-terminal α-amino group is essential for catalysis, possibly an additional group of ε-amino- or phenolic hydroxyl- is necessary (Pestka, 1972a,b). An evidence for the involvement of a histidine residue in the functioning of the ribosomal peptidyltransferase has been shown by the chemical modification experiments of E. coli ribosomes with methyl p-nitrobenzenesulfonate, and this result suggests that the ribosomal peptidyltransferase active site geometry may be similar to the active site of proteolytic enzymes (Glick, 1980). Studies of the peptidyltransferase acceptor site of E. coli 70S ribosome show that the peptidyltransferase A site appears to be strikingly similar to the active sites of proteolytic enzymes (Bhuta et al., 1981).

Prebiotic simulation synthesis of peptides from amino acids in aqueous solution has been studied with the aid of ATP and free imidazole (Weber et al., 1977). The possibility for the origin of ATP has been discussed, and of the various ribonucleotide triphosphates, ATP was found to be most effective in synthesis of peptide bonds (Ryan and Fox, 1973). Both ribosomal and aribosomal syntheses of polypeptides in modern cells occur with the aid of ATP (Lee and Lipmann, 1974; Lipmann, 1974). Pyrophosphate can be used instead of ATP, with lesser yields resulting (Nakashima and Fox, 1980). The experiments of Fig. 3 compare equimolar amounts of these two energy-rich phosphates. The ratio of yield from ATP to that from pyrophosphate is 3.7. The observation that the high-energy phosphate is available from pyrophosphate is consistent with the view (Baltscheffsky, 1971) that pyrophosphate played an early role in metabolic evolution, to be largely replaced later by ATP.

The origin of the ribosomes can be conceptualized by the results that either acidic-basic proteinoid particles or polynucleotide-basic proteinoid particles catalyze the synthesis of peptides from free amino acids and ATP. Acidic-basic proteinoid particles may have been an original type of ribosome. The latter, the polynucleotide-basic proteinoid particles, may have been a rather evolved ribosome, that is, a prototype of contemporary ribosome. This kind of protoribosome model has provided data interpretable in the context of the genetic code mechanism (Nakashima and Fox, 1972, 1981).

Selective Condensation of Amino Acids by Model Protoribosomes

Microparticles composed of each of four enzymically synthesized homopolyribonucleotides and the same lysine-rich proteinoid have been found to influence the condensation of the AMP-anhydrides of each of four amino acids. The conditions of preparation of the particles and other variables of the experiments control the type of reaction. When a specified set of conditions was identified empirically, the incorporation favored the amino acids whose present-day codons are related to the homopolynucleotide in the particles. To study the charging of microparticles by adenylates, a measured aliquot of each of the individual radioactive adenylates was added to the microparticles, and after centrifugation the radioactivity of the precipitate was measured in a scintillation counter. The kind of result obtained with 0.25mg/ml of polynucleotide is shown in Fig. 4. All results proved to be codonic. When the concentration of polynucleotide was tripled to 0.75mg/ml, two of these results have the appearance of being codonic (Phe and Pro), another is strongly anticodonic (Gly), and the other (Lys) has no evident code-related quality.

When equimolar solutions of mixtures of glycine and phenylalanine with ATP are tested in suspensions of microparticles of lysine-rich proteinoid and each of the various polyribonucleotides, both homopeptides and heteropeptides are produced. Glycylphenylalanine or phenylalanylglycine is the principal product. In a suspension of poly U + or poly C + lysine-rich proteinoid microparticles phe-gly

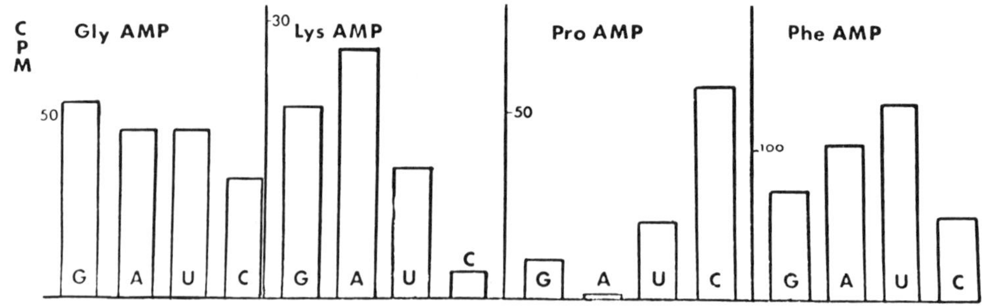

Fig. 4. Relative incorporation into nucleoproteinoid microparticles of radioactive amino acids in aminoacyl adenylates.

is formed predominantly, but in poly A + lysine-rich proteinoid, gly-phe mostly results (Nakashima and Fox, 1981). The affinity difference of purine and pyrimidine for amino acids, or the hydrophobicity of polynucleotides, may account for these priorities in peptide sequence. It is reasonable to assume that one kind of amino acid is bound by the polynucleotide while the other amino acid is reacted with the bound one. In exhibiting the ability to synthesize peptide bonds and to influence amino acid sequences, the nucleoproteinoid microparticles have been confirmed to be suitable models for protoribosomes.

Proteinoid as Protoribosomal Proteins

When amino acids are heated in dry conditon, they form copolyanhydroaminoacids (Fox and Dose, 1977). When all of the amino acids common to protein are simultaneously condensed by heat, the polymer (proteinoid) is found to contain some proportion of each of the eighteen amino acids and to possess molecular weights of many thousands.

An acidic proteinoid has been fractionated, and the major fractions have been characterized by several methods (Fox and Nakashima, 1967). The acidic proteinoid was treated with liquid ammonia to convert all imide and lactam residues to amides in order to minimize variation in the product. The product thus obtained was almost neutral in reaction. The crude proteinoid amide was separated into six fractions by DEAE-cellulose column chromatography. The three major fractions were individually examined for heterogeneity by Sephadex exclusion and Bio-Gel exclusion, by high-voltage electrophoresis, and by ultracentrifugal analysis. The total results and related inferences are interpreted to signify a high degree of compositional uniformity throughout the original polymer. Partial hydrolyzate of the three major fractions yielded fingerprints which are similar and which contain 39-41 peaks in chromatograms. Small compositional differences between three of the fractions and the crude polymer were found in amino acid analysis. The total proteinoid, according to the analyses, clearly does not approach randomness in its degree of heterogeneity. Neither is it homogeneous. Amino acids compositions of lysine-rich proteinoid (Fox et al., 1962) and neutral proteinoid (Fox and Waehneldt, 1968) have been studied. Fractionations of acidic, neutral, and basic proteinoids on Bio-Gel indicate that acidic proteinoid has mol. wt. < 4000, neutral proteinoid is > 4000, and basic proteinoid is > 6000 (Fox and Waehneldt, 1968). A review of linkages in proteinoids indicates α- and β-linkages for aspartic acid, α- and γ-linkages for glutamic acid, α- and ε-linkages for lysine, and α-linkages for other amino acids (Fox and Suzuki, 1976). The proportion of α- and ε-linkages of lysine residues in a thermal polymer of lysine and alanine is 35:65 (Harada and Fox, 1965).

A test for randomness in tyrosine peptides isolated from a thermal polyamino acid has been demonstrated (Nakashima et al., 1977). A thermal polymer of glutamic acid, glycine and tyrosine was fractionated on Sephadex and on paper. Amino acid compositions of some of the tyrosine containing peptides were determined. Fig. 5 shows a chromatogram of subfractions of tyrosine containing fractions 2, 3, and 4. The amino acid composition (Glu : Gly : Tyr) of each

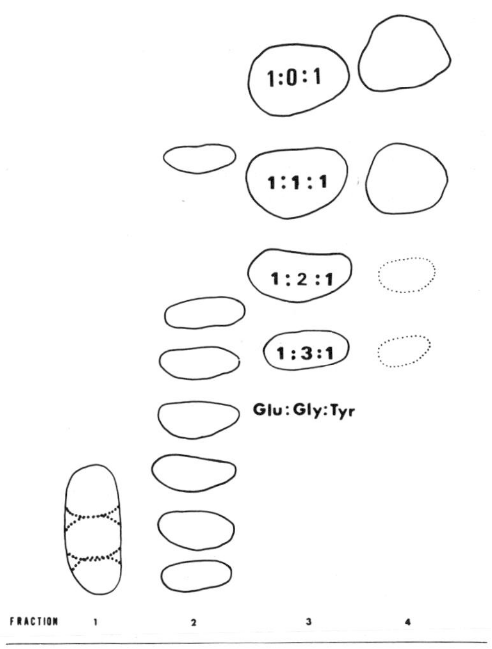

Fig. 5. A chromatogram of subfractions of fractions 2, 3, and 4. Amino acid compositions are indicated in numbers.

subfraction of fraction 3 is indicated in numbers inside a spot. In sequence studies of the tripeptides, pyroglutamyltyrosylglycine and pyroglutamylglycyltyrosine were identified (Nakashima et al., 1968, 1977; Hartmann et al., 1981). Results shown here confirm the limited heterogeneity of the thermal polyamino acid.

The acidic proteinoidamide has three major fractions, of which the molecular weights by the sedimentation analysis are 4070, 5160, and 5800. The molecular masses of these fractions are of consecutiveness. It is reasonable to assume that the basic proteinoid is a mixture of fractions having limited consecutive masses since the basic proteinoid is an analog (Fox and Waehneldt, 1968) of acidic and neutral proteinoids. The results of fractionation of tritonic copolymer show the consecutive glycine-containing peptides. The evolution of protein (or of proteinoid) would not be possible without some diversity (Fox, 1953). The limited heterogeneity of ribosomal proteins and proteinoids suggests that the proteinoid could have served to form primitive ribosome-like proteins. Homologous tertiary structures between ribosomal proteins would not necessarily relate to their primary sequences (Kurland, 1977). In this context, the remarks (Oparin et al., 1980; Andini et al., 1975) for the proteinoids differing from proteins in the type of bonds and primary structure may not be impediments to the proteinoid theory.

CONCLUSIONS AND PERSPECTIVES

The protoribosome model composed of polyribonucleotide and lysine-rich proteinoid demonstrated the peptide synthesis activity with some selectivity in incorporating amino acid. Within the concept of chemical evolution, the data reviewed here provide support for the idea that thermal proteinoid could have participated in the formation of primitive ribosome-like proteins. The tertiary structures and consecutive molecular weights of ribosomal proteins may have been conserved during evolution. The author presents a hypothesis that the narrow sequential grading of sizes of ribosomal proteins may be an essential requirement for ribosome function. Experiments on the protoribosome models which have incomplete activity will remain conceptual, as long as the function of modern ribosomes is not completely understood. In this context, models of more advanced protoribosomes would be required. Plausible prebiological synthesis of polyribonucleotide is one of the most difficult unsolved problems. The protoribosome research cannot be considered apart from the studies of the origins of aminoacly-tRNA, of mRNA and of the genetic code. It is useful for further studies to employ the mixture of proteinoids (modern ribosomes have acidic and basic proteins), the mixtures of different nucleoproteinoid particles (simulation of subunits), salts (Mg^{2+}, K^+ or NH_4^+), contemporary ribosomal components. The advance of studies on the supramolecules especially on the nucleoprotein microparticles is mostly desirable.

The study of protobiological models which consist of relatively simpler components will facilitate the understanding of the relationship between chemical structure and biological function at a level of complexity, and vice versa.

ACKNOWLEDGMENTS

I thank the National Aeronautics and Space Administration for Grant No. NGR-10-007-008.

REFERENCES

Andini, S., Benedetti, E., Ferrara, L., Paolillo, L., and Temussi, P. A., 1975, Origins Life, 6:147.

Baltscheffsky, H., 1971, in: "Chemical Evolution and the Origin of Life," R. Buvet and C. Ponnamperuma, eds., pp. 466-474, North-Holland, Amsterdam.

Bhuta, A., Quiggle, K., Ott, T., Ringer, D. P., and Chládek, S., 1981, Biochemistry, 20:8.

Brimacombe, R., Stöffler, G., and Wittmann, H. G., 1978, Ann. Rev. Biochem., 47:217.

Brosius, J., Palmer, M. L., Kennedy, P. J., and Noller, H. F., 1978, Proc. Nat. Acad. Sci. U.S., 75:4801.

Brosius, J., Dull, T. J., and Noller, H. F., 1980, Proc. Acad. Sci. U.S., 77:201.

Brownlee, G. G., Sanger, F., and Barrell, B. G., 1967, Nature, 215: 735.

Calvin, M., 1969, "Chemical Evolution," Oxford University Press, New York.

Carbon, P., Ehresmann, C., Ehresmann, B., and Ebel, J. P., 1978, FEBS Lett., 94:152.

Davidson, B., and Fasman, G. D., 1969, Biochemistry, 8:4116.

Eisenberg, D., Weiss, R. M., and Terwilliger, T. C., 1982, Nature, 299:371.

Fox, S. W., 1953, Amer. Nat., 87:253.

Fox, S. W., Harada, K., and Rohlfing, D. L., 1962, in: "Polyamino Acids, Polypeptides and Proteins," M. A. Stahmann, ed., pp. 47-53, The University of Wisconsin Press, Madison.

Fox, S. W., and Nakashima, T., 1967, Biochim. Biophys. Acta, 140: 155.

Fox, S. W., and Waehneldt, T. V., 1968, Biochim. Biophys. Acta, 160:246.

Fox, S. W., Jungck, J. R., and Nakashima, T., 1974, Origins Life, 5:227.

Fox, S. W., and Suzuki, F., 1976, BioSystems, 8:40.

Fox, S. W., and Dose, K., 1977, "Molecular Evolution and the Origin of Life," 2nd edn., Marcel Dekker, New York.

Fox, S. W., and Nakashima, T., 1980, BioSystems, 12:155.

Glick, B. R., 1980, Can. J. Biochem., 58:1345.
Glotz, C., Zwieb, D., Brimacombe, R., Edwards, K., and Kössel, H., 1981, Nucl. Acids Res., 9:3287.
Harada, K., and Fox, S. W., 1965, Arch. Biochem. Biophys., 109:49.
Harris, B. J., and Pestka, S., 1977, in: "Molecular Mechanisms of Protein Biosynthesis," H. Weissbach, and S. Pestka, eds., pp. 413-442, Academic Press, New York.
Hartmann, J., Christel Brand, M., and Dose, K., 1981, BioSystems, 13:141.
Jungck, J. R., and Fox, S. W., 1973, Naturwiss., 60:425.
Kurland, C. G., 1977, in: "Molecular Mechanisms of Protein Biosynthesis," H. Weissbach, and S. Pestka, eds., pp. 81-116, Academic Press, New York.
Lacey, J. C., Yuki, A., and Fox, S. W., 1979a, BioSystems, 11:1.
Lacey, J. C., Stephens, D. P., and Fox, S. W., 1979b, BioSystems, 11:9.
Lee, S. G., and Lipmann, F., 1974, Proc. Nat. Acad. Sci. U.S., 71: 607.
Lehninger, A. L., 1975, "Biochemistry," Worth, New York.
Lehninger, A. L., 1982, "Principles of Biochemistry," Worth, New York.
Lipmann, F., 1974, in: "The Origin of Life and Evolutionary Biochemistry," K. Dose, S. W. Fox, and G. A. Deborin, eds., pp. 321-330, Plenum Press, New York.
Matheson, A. T., Möller, W., Amons, R., and Yaguchi, M., 1980, in: "Ribosomes: Structure, Function and Genetics," G. Chambliss, G. R. Craven, J. Davies, K. Davis, L. Kahan, and M. Nomura, eds., pp. 297-332, University Park Press, Baltimore.
Matsuo, K., Fuke, M., Tsuboi, M., and Wada, A., 1969, Biochim. Biophys. Acta, 197:39.
Nadezhda, V. B., Gulnara, Z. T., and Spirin, A. S., 1982, BioSystems, 15:233.
Nakashima, T., Fox, S. W., and Wang, C. T. 1968, in: "2. Arbeitstagung über extraterrestrisch Biophysik und Biologie und Raumfahrtmedizin," H. Bücker, ed., pp. 223-224, ZLDI der Deutschen Gesellschaft für Flugwissenschaften e. V., München.
Nakashima, T., Lacey, J. C., Jungck, J. R., and Fox, S. W., 1970, Naturwiss., 57:67.
Nakashima, T., and Fox, S. W., 1972, Proc. Nat., Acad, Sci. U.S., 69:106.
Nakashima, T., Jungck, J. R., Fox, S. W., Lederer, E., and Das, B. C., 1977, Int. J. Quantum Chem., QBS4:65.
Nakashima, T., and Fox, S. W., 1980, J. Mol. Evol., 15:161.
Nakashima, T., and Fox, S. W., 1981, BioSystems, 14:151.
Nierhaus, K. H., and Dohme, F., 1974, Proc. Nat. Acad. Sci. U.S., 71:4713.
Nierhaus, K. H., 1980, BioSystems, 12:273.
Noller, H. F., and Woese, C. R., 1981, Science, 212:403.
Nomura, M., and Held, W. A., 1974a, in: "Ribosomes," M. Nomura, A. Tissières, and P. Lengyel, eds., pp. 193-223, Cold Spring Harbor Laboratory Press, Long Island, New York.

Nomura, M., Sidikaro, J., Jakes, K., and Zinder, N., 1974b, in: "Ribosomes," M. Nomura, A. Tissières, and P. Lengyel, eds., pp. 805-814, Cold Spring Harbor Laboratory Press, Long Island, York.
Oparin, A. I., and Gladilin, K. L., 1980, BioSystems, 12:133.
Otaka, E., and Kobata, K., 1978, Mol. Gen. Genet., 162:259.
Otaka, E., Higo, K., and Osawa, S., 1982, Biochemistry, 21:4545.
Ozaki, M., Mizushima, S., and Nomura, M., 1969, Nature, 222:333.
Perutz, M. F., Kendrew, J. C., and Watson, H. C., 1965, J. Mol. Biol., 13:669.
Pestaka, S., 1968, J. Biol. Chem., 243:2810.
Pestaka, S., 1972a, Proc. Nat. Acad. Sci. U.S., 69:624.
Pestaka, S., 1972b, J. Biol. Chem., 247:4669.
Ryan, J. W., and Fox, S. W., 1973, BioSystems, 5:115.
Spitnik-Elson, P., and Elson, D., 1976, in: "Progress in Nucleic Acid Research and Molecular Biology," J. N. Davidson, and W. E. Cohn, eds., Vol. 17, pp. 77-98, Academic Press, New York.
Stöffler, G., and Wittmann, H. G., 1977, in: "Molecular Mechanisms of Protein Biosynthesis," H. Weissbach and S. Pestka, eds., pp. 117-202, Academic Press, New York.
Stöffler, G., Bald, R., Kastner, B., Lührmann, R., Stöffler-Meilicke, M., and Tischendorf, G., 1980, in: "Ribosomes: Structure, Function and Genetics," G. Chambliss, G. R. Craven, J. Davies, K. Davis, L. Kahan, and M. Nomura, eds., pp. 171-205, University Park Press, Baltimore.
Tai, P. C., and Davis, B. D., 1974, Proc. Nat. Acad. Sci. U.S., 71:1021.
Traub, P., and Nomura, M., 1968, Proc. Nat. Acad. Sci. U.S., 59:777.
Waehneldt, T. V., and Fox, S. W., 1968, Biochim. Biophys. Acta, 160:239.
Weber, A. L., Caroon, J. M., Warden, J. T., Lemmon, R. M., and Calvin, M., 1977, BioSystems, 8:277.
Wittmann-Liebold, B., Geissler, A. W., Lin, A., and Wool, I. G., 1979, J. Supramol. Struct., 12:425.
Wittmann, H. G., 1982, Proc. Roy. Soc. Lond., B 216:117.
Wool, I. G., and Stöffler, G., 1974, in: "Ribosomes," M. Nomura, A. Tissières, and P. Lengyel, eds., pp. 417-460, Cold Spring Harbor Laboratory Press, Long Island, New York.
Wool, I. G., 1980, in: "Ribosomes: Structure, Function and Genetics," G. Chambliss, G. R. Craven, J. Davies, K. Davis, L. Kahan, and M. Nomura, eds., pp. 797-824, University Park Press, Baltimore.
Yuki, A., and Fox, S. W., 1969, Biochem. Biophys. Res. Commun., 36:657.
Zwieb, C., Glotz, C., and Brimacombe, R., 1981, Nucl. Acids Res., 9:3621.

PROTOCELL ACTION POTENTIALS: A NEW PERSPECTIVE OF BIO-EXCITATION

Wilford P. Stratten
Rose-Hulman Institute of Technology
Terre Haute, Indiana 47803

INTRODUCTION

The term bio-excitation generates many mental pictures ranging from the machinations of the brain, responsiveness to environmental stimuli, facilitated efferent motor responses, electrical discharges, ionic gradients, membrane structure, etc. At the cellular level, those familiar with excitation immediately retrieve one model or another upon which they have depended for mental imagery. Some begin with a precise phospholipid bilayer membrane peppered with protein gates or carriers with precision architecture. Others also consider the biochemical and energy transfer processes associated with active transport, and maintenance of membrane anisotropy. The eons of natural selection which have allowed the development of such sophisticated electrochemical processes concerns yet others. And there are many primarily concerned with the process by which external chemical, electromagnetic, or mechanical energies are converted into excitation. But what becomes of all of these images if it is suddenly found that simple heating and subsequent hydration of amino acid mixtures results in spherical structures manifesting all fundamental characteristics of excitation? There is no need for biological evolution. There is no need for phospholipids. There is no need for nucleotide control of protein precision. There is no need for ATP or other sophisticated metabolic processes. There is no apparent afferent-efferent link approximating *in vivo* excitation processes. There is no longer any intimate relationship to intercellular communications, or electrical induction of sophisticated intracellular enzymatic processes. All of these facts have profound impact upon our conceptualization of the process of excitation.

It is the purpose of this chapter to compare some of the popular imagery of excitation with the realities demonstrated by the discovery of excitation in proteinoid protocells. This will hopefully provide incentives to some investigators of excitation to approach its unanswered problems with a fresh perspective.

MODELS FOR MEMBRANE EXCITATION

Since the publiciation of the Hodgkin and Huxley theory (1952), based upon the concept of "gated" ionic flow, it has dominated the field of membrane excitation. This has been the case in spite of the fact that it gives little consideration to the physicochemical nature of the membrane. This theory has been followed by others portraying the process as dipole flip-flops (Wei, 1972), simple induction (Ling, 1982), ionic energy barrier transitions (Hille, 1975), membrane ion exchange (Tasaki, 1968), acetylcholine induced gating (Nachmansohn, 1970), or even hydrostatic induction of capillary action (Teorell, 1962). This chapter's limited review of the Hodgkin-Hurley, Tasaki, and Nachmansohn models and the lipid bilayer work of Mueller and Rudin will hopefully provide adequate example to indicate the potential impact of the protocell findings.

Hodgkin and Huxley

Using the results of voltage-clamp experiments on squid giant axon, Hodgkin and Huxley (1952) began with a mathematically simplifying assumption that the current flow through the membrane could be separated, i.e., the current of each ion could be considered to be flowing through its own selective path. The total current was the sum of that flowing through all of the parallel paths. Equations were then formulated describing the relationship of the ion conductance to the variables of time and membrane potential. The sum of potassium and sodium ionic flow explain most membrane responses. Of the many conductance-time empirical expressions which could be chosen to fit the sigmoid rise in potassium conductance or the transient rise-fall of sodium conductance, Hodgkin and Huxley (1952) used expressions portraying multi-particle movements. The concept of "movement" and membrane semipermeability allows the application of the Nernst equation and the assessment of the membrane potential contribution of each ion. The model fits the observed phenomenon in a broad set of conditions and the "gate" forms a simple mental picture. The combination of its fit, its simplicity, and its precedence have kept this model on the forefront of membrane excitation studies.

Tasaki

The Tasaki model (Tasaki, 1968; Singer and Tasaki, 1968) includes considerations of the physicochemical properties of the membrane--at least to the limited extent to which they are known. Using varied internal and external ionic milieu, membrane function can be explained by steric alterations of macromolecular structure induced by univalent/divalent cation interactions. The membrane current is not via simple flow-through "gates" but rather a membrane transfer of ions due to sequential exchange between the divalent and the univalent cations. The macromolecular cation "carriers" have two stable conformations. In the resting state, the macromolecules are occupied primarily by divalent cations which are not readily displaced as long as the membrane potential maintains its inside-negative polarity. Activation results from displacement of the divalent cation by univalent cations. In the "active" state the univalent current efflux soon decreases (sodium vs. potassium) and upon diminution the divalent cations once again manifest their stronger affinity for the membrane's external anionic sites (Singer and Tasaki, 1968). Although some claim that the Tasaki model is not significantly different from Hodgkin-Huxley's (cf. Mueller and Rudin, 1969a), there exist several phenomenon which cannot be explained by the Hodgkin-Huxley model, but seem to be adequately explained by the Tasaki model: (1) Effects of dilution of external univalent cations can be mimicked by concentration of external divalent cations. (2) The resting potential resulting from external potassium ion dilution does not obey the Nernst equation (its elimination will not alter the resting potential). (3) An action potential spike can still be manifest when internal and external sodium concentrations are equal. (4) Tetrodotoxin, a presumed sodium "gate" blocker, blocks action potential and inward current in sodium-free media (Singer and Tasaki, 1968).

Fundamental to most of the disagreement between the Tasaki and Hodgkin-Huxley models is the indication that there exists no ionic specificity, i.e., to sodium and potassium. Rather, a lyotropic sequence of favorability is supported. The depolarizing cations such as potassium can actually be substituted for sodium in its hyperpolarizing role. In fact, the external univalent cation can be eliminated altogether, given the proper balance of lyotropic substitutes. Even the anionic lyotropic sequence influences excitability, fluoride being the most favored, indicating a competition between the membrane and media anions. The cation fixation to membrane "carriers" can account for these lyotropic sequences of the internal anions, but it cannot be explained by the Hodgkin-Huxley model (Singer and Tasaki, 1968).

Nachmansohn

The model of Nachmansohn illustrates membrane excitation as a sequence of events beginning with a stimulus induced release of acetylcholine (ACh) from its membrane-bound loci. The "free" ACh quickly associates with a neighboring protein ACh receptor, inducing a conformational change in the protein. The conformational change results in the release of calcium ions which interact with other membrane macromolecules, altering the membrane's ion permeability. Membrane associated acetylcholinesterase quickly hydolyzes the ACh and the processes reverse, reestablishing the ionic barrier (Nachmansohn, 1970). The membrane potential is represented by the sum of active and passive transport. The metabolic energy exchanging enzymes essential for the active processes are also thought to be an intrinsic component of the excitation complex. This model is based largely upon the research on fish electric organs, but Nachmansohn refers to it as the "unified concept of the role of AcCh." He claims that "the elementary processes that change ion permeability are essentially the same in the axonal, in the nerve terminal, and in the postsynaptic membrane" (Nachmansohn and Neumann, 1974). This claim is supported by a variety of circumstantial evidences such as the finding of Ach, and acetylcholinesterase in a wide variety of excitable tissue; however, there exists considerable evidence against this view (Koelle, 1966). This model represents a growing notion that the excitation process is a result of interaction of a sophisticated set of highly evolved interacting enzymatic components, implying that the process of excitation may be a newcomer in the evolutionary scheme (Neumann et al., 1973). If one is to assess the process of membrane excitation, from an evolutionary perspective, this model should be included even though it may only apply to a specialized tissue.

Black Lipid Membranes

Although not proposing a unique model *per se*, Mueller and Rudin (1968, 1969a,b) have made significant contribution to our understanding of membranes by their work using bimolecular membranes formed from cellular lipids. These membranes are often referred to as black lipid membranes (BLM) due to their failure to reflect visible light. The properties of the artificial membranes in terms of thickness, water permeability, surface tension and passive electrical characteristics are very similar to those of cell membrane (Mueller and Rudin, 1969b). The BLM electrical properties are all passive and linear in nature until supplements are added. Those supplements which convey alteration in electrical and/or chemical affinity characteristics are referred to by Mueller and Rudin as translocators. Some of the translocators impart simple increases in conductances which display Ohmic linearity, while others display non-linear potential-conductance relationships resulting in activity similar to action potentials (Mueller and Rudin, 1969a).

Excitability inducing material, EIM, a cytolytic bacterial endotoxin, was the first compound found to induce a voltage dependent conductance in the lipid membranes. The EIM translocated species are cations and the electrical dynamics are similar to those of potassium in nerve (Mueller and Rudin, 1969b). The addition of protamines or polyamines to the EIM membrane changes the translocated species to anions. The proper titering of EIM and protamines will display bimodal translocation, cationic and anionic, which manifests action potentials exactly like those of the alga Nitella, and similar to nerve, albeit it is an anion-cation selectivity instead of sodium-potassium. Monazomycin another voltage dependent translocator possesses bimodal translocation properties, carrying hydrogen ions in one conformation and potassium or sodium ions in the other. This alleviates the necessity for two separate channels (Mueller and Rudin, 1969a).

STRUCTURE OF EXCITABLE MEMBRANE

Protein-Lipid Cooperation

The general view of cell membrane structure has changed little over the past nine decades. In 1895 Overton (cited in Troshin, 1966) described the membrane as a matrix of nonpolar lipids supporting arrays of localized polar carriers or ion channels. A later refinement, the Danielli-Davson model, is a phospholipid bilayer "unit membrane" to which proteins are attached (Nachmansohn, 1970). The model represents the lipid component as a passive structural barrier whose functional loci are various adsorbates, most likely proteins. It is these proteins which impart the properties of semipermeability, excitation, metabolism, transport, and enzyme activity to the membrane (Mueller and Rudin, 1969a). More precisely, since the proteins themselves are modified by the forces of the lipid-water interface, the membrane appears to be comprised of functional lipoprotein complexes with the lipid-modified protein structure serving as the functional core of each complex (Nachmansohn, 1970). The view that these functional complexes are "fixed" within the lipid matrix of the membranes is not supported by recent work on artificial membranes (Mueller and Rudin, 1969a).

Macrostructure

Numerous studies have indicated asymmetric membrane structure, attributable to the protein components. Evidence to support this includes (1) the external application of proteases to squid giant axon has no effect but internal application is damaging, (2) the membrane exterior is cation sensitive while the interior is not, (3) the interior is anion sensitive while the exterior is not, and (4) differences exist as to the interior vs. exterior sensitivity to alteration of heavy metal, organic ion, or hydrogen ion concentrations (Singer and Tasaki, 1968).

Lipids

The intra-membrane arrangement of lipid molecules has not been totally resolved. The formation of the lipid bilayer is energetically strongly favored; however, the work with artificial membranes indicates that there is no energetic preference between the formation of bilayer spheres, spheres of many bilayers, small micelles, or even an extended smectic form. Mueller and Rudin (1969a) have found, however, that the multiple layering and smectic phases are broken up by the addition of proteins, leaving a likelihood of only the bilayer of micellar forms as being the biological conformation. Their arrangement in a bilayer is supported by electron microscopy, x-ray electron density, x-ray diffraction, birefringence, and thermal phase transition. BLM are of known lamellar conformation; and their properties of membrane resistance, lipid/water solubility ratio, protein association, excitability, and scanning calorimetry are found to be the same as those of natural membrane. However, Sjoestrand and Barajas (1968) using techniques designed to preserve protein structure, have found micellar structures in electron micrographs of the membrane.

BLM have high electrical impedance, five to seven orders of magnitude higher than cell membranes (Mueller and Rudin, 1969b). It is only with the addition of proteins or other conductive adsorbates that the membrane displays electrical conductance of the magnitude displayed by biological membranes. Little is known about the degree to which the lipids are involved in the facilitation of the adsorbate conduction increase. There seems to be no general dependence of the adborbates on the lipids, but with artificial membranes one adsorbate, the polyenes, require cholesterol to be active (Mueller and Rudin, 1968). In at least one case the function of the adsorbate changes with a change in the lipid structure. Dipicrylamine transfers dynamically from one surface to the other of one membrane type and acts as the charge carrier. In another lipid (lecithin) the dipicrylamine acts as an ion carrier for potassium or hydrogen ions (Mueller and Rudin, 1969a).

Protein

Ions are accepted to be the carriers of membrane currents. The mechanism accounting for the rapid dynamics of action potentials is the primary riddle for the electrophysiologists today. The answer to the riddle seems to lie hidden with other protein properties (Nachmansohn, 1970).

The great diversity of membrane function, specificity, and efficiency is attributable more to the proteins than to the phospholipid composition. The precision stereo selectivity, and the effect of sulfhydryl-blocking and disulfide-reducing agents on excitation receptors support the notion of their protein identity (Karlin and Bartels, 1966). There are also indications that cell membranes are highly ordered dynamic structures, the architecture of which facilitates the membrane's intrinsic activities (Neumann et al., 1973). One underlying question of many studies is that of how much "order" is essential for any given membrane function. Complexity appears in most fundamental models, e.g., Hodgkin and Huxley were able to quantitatively fit their analytical data only after assuming that a fixed multiple of charges were associated with each equilibrium state (Mueller and Rudin, 1969a). Kennedy et al. (1977) have formed ionic channels in membranes by the application of simple synthetic polypeptides; however, these polypeptides were modeled after biological structures known to influence membrane permeability. Even these simple polypeptides were found to induce permeability changes only upon hexameric formation.

Protein Complexes

The more advanced models of functional membranes are even more replete with the notion that complex organized protein structures are essential to excitation. Macromolecules and their multiple internal or external linkages are essential to the model portrayed by Tasaki (1968). The model of excitation via field-induced conformational changes in biopolymers causing highly selective alteration in ion permeability, alludes to highly specialized macromolecular mediation (Nachmansohn and Neumann, 1974). The occurrence of extrinsically gated translocators (possibly the fundamental mechanism of chemoreceptors) implies a heterotrophic allosteric system (Mueller and Rudin, 1969a). Nachmansohn (1970) claims that all sites of excitation are complex membrane structures including a multiprotein receptor, a deactivating acetylcholinesterase, and other macromolecular components. The Katchalsky model (Neumann et al., 1973) also views excitation units as being multiple protein gateway complexes possessing a structure and charge array which is dynamically altered with change in field potentials.

Ions and Cations

Ions also have structural implications via their membrane interactions. The work of Tasaki with squid giant axon links divalent cations with the manifestation of excitability. Ionic affinities indicate that the external surface of biomembrane possesses an excess of intrinsic negative charge (carboxylate) and the inner surface shows an excess of positive charge (phosphate) (Singer and Tasaki, 1968). External divalent cations, are suspected of contributing to the maintenance of membrane structure by bonding adjacent negative macromolecule side groups (Schellman and Schellman, 1964). This is supported by the finding that when the external medium is limited to univalent cations, the axons become inexacitable. The cation exchange process between divalent and univalent cations is the key to the function of the "two-stable-state" membrane model of Tasaki (1968).

The most functionally relevant divalent cation seems to be calcium. The monitoring of osmotic coefficient by Katchalsky (1964) indicates the 99% of the calcium is bound to the resting biomembrane. This is in agreement with the prediction of Tasaki, who contends that excitation would accompany the release of the calcium (Singer and Tasaki, 1968). The notion of protein conformational changes upon calcium release is supported by the known potency of calcium in inducing such changes. This process is especially dramatic on surfaces possessing abundant negative charge (Nachmansohn and Neumann, 1974).

Protein Basics

In spite of the many complex models, few ask the basic question: "What are the minimum requirements for membrane excitation?" It must be remembered that most proteins are ampholytes with the N-terminal being positive and the C-terminal negative (Singer and Tasaki, 1968). This common property of all amino acids added to the varied polar and apolar properties of the various R-groups leads one to believe that there are many likely protein conformations able to (1) change conformation within an electrical field, (2) carry ions, and (3) possess sufficient lipophilicity to function in association with lipids. The sum of these attributes makes the possessor a likely candidate as a translocator of non-lipid-soluble inorganic ions which are normally restricted to the aqueous phase (Mueller and Rudin, 1969a).

Protein Membrane

Since the role of lipids appears to be that of a passive barrier to charge flow, and since some proteins also possess strong apolar characteristics, it might be guessed that membranes could also be formed from proteins alone. Under select conditions such have been formed and have dimensions similar to lipid bilayers; however, they are found to be "leaky" in relation to the lipids (Mueller and Rudin, 1969a; Anderson et al., 1953; Fox and Dose, 1977).

PROTEINOIDS, PROTEINOID MEMBRANES, AND PROTOCELLS

Proteinoids

Proteinoids are synthetic copolyamino acids formed by heating various mixtures of amino acids. The polymers are self-ordering (Fox and Dose, 1977). This is likely due to the "ordering" propensity of the precursor amino acids (Fox, 1978). The ordering tendency is carried beyond the intraproteinoid level to ordering for populations of proteinoid molecules to form membranes and homogeneous sphere populations. This macro-ordering is an indication of repetition of order from one proteinoid or proteinoid array to the next (Fox and Dose, 1977). In addition to the self-ordering, these polymers have many characteristics fundamental to life processes, including the many structural and functional aspects outlined in the following paragraphs.

Kept aseptic, the proteinoids are indefinitely stable within a broad range of pH, temperature, and hydration conditions (Fox and Dose, 1977).

Proteinoids abundant in hydrocarbon-rich amino acids display lipid-like properties, and are among those which combine most readily with lecithin and other lipids. It is these proteinoids which have the highest probability of displaying properties of excitation (Ishima and Fox, 1973). Electrical excitation requires the ambivalence of electrical insulation and conductivity. Since electrical excitation is manifest in some proteinoid complexes lacking any phospholipid, the proteinoids must possess both properties (Przybylski et al., 1982). The presence of both hydrophobic and amphophilic regions along the surface of the polymers allows such to be possible. The most strongly conductive products found amongst proteinoids and their fractions are also oil-soluble. The balance between these two properties is likely to be essential to the manifestation of excitability (Grote et al., 1978).

A broad array of catalytic activities have been discovered for the proteinoids. These include various forms of hydrolysis, decarboxylation, amination, oxidoreductions, photoactivated decarboxylation, hormonal interactions, and synthesis of internucleotide and peptide bonds with ATP (Fox et al., 1978; Fox and Dose, 1977).

The addition of either proteinoid polymers or polymer fractions increases the conductance of chloroform solutions indicating these thermal products to be current carriers (Grote et al., 1978). These findings are not unexpected since thermal copolyamino acids have ESR densities of 10^{18}/g (Bone et al., 1978) allowing visualization of the proteinoids as shuttling electrons or holes. Those proteinoids or fractions which are the most effective in solutions are also the most effective in enhancing phospholipid membrane conductance. This circumstantial association may imply that the current of the thin layer membrane may be "carrier-mediated" as it is in the bulk solvent, but does not eliminate the possibility of other ionophoric or porous modalities (Grote et al., 1978).

In summary, the proteinoids appear to be able to stand as a "one man band" having diverse properties and functional capacities of several chemical classes including those of lipids, proteins, and even the nucleic acids. The possession of this diversity may obviate the need for communion of many complex molecular classes in the manifestation of primitive life function (Przybylski et al., 1982).

Proteinoid Membranes and Protocells

Structure. Proteinoids form membranes, either of the spontaneous spherical form upon hydration (Fox and Dose, 1977), or black bilayers using special techniques similar to those of Mueller and Rudin (1969b). Electron micrographs show the proteinoid sphere membrane to be a double layer, but the membranes of many are thicker than the BLM. These proteinoid membranes are "leaky" as were the protein membranes mentioned previously, but possess many properties of "tight" membranes (Przybryski et al., 1982; Fox and Dose, 1977).

Many proteinoids have lipid-like properties, these properties being attributed to non-polar amino acid sidechains (Lehninger, 1975). It is the proteinoids which have these lipid-like properties which form membrane and spheres (p-protocells) (Przybylski et al., 1982). Adding lecithin during hydration results in the formation of electrically excitable proteinoid-lecithin spheres (p-l-protocells). The apolar side chains seem to be important, possibly for an essential complexing with the lecithin in the sphere formation. Three amino acids with apolar side chains, leucine, proline, and threonine, are of special interest (Ishima et al., 1981).

The coordination of ions by the translocator polypeptide complex of Kennedy et al. (1977) is thought to mediated by hydroxyl groups of serine and to be essential to electrical activity. Comparatively, it is possible that threonine hydroxyls are the "ion coordinators" in the protocell membrane since they seem essential to excitability (Ishima et al., 1981).

Properties. The p-protocells maintain the several properties just outlined for the proteinoids alone, but include new properties related to its various attributes of membranicity, and ordered macrostructure (Przybylski et al., 1982; Fox and Dose, 1977).

In addition to the attributes of stability attributed to the proteinoids alone, the protocells maintain remarkable durability. Visual and electrical examination reveal tolerance to extremes of pH (Snyder and Fox, 1975), temperature and dehydration (Fox and Dose, 1977; Ishima et al., 1981). Mechanical resilience is demonstrated in the self-sealing of black membranes following puncture, a property matching that of lipid bilayers or biological membranes (Fox et al., 1978).

The incorporation of the proteinoid catalytic activity into the protocells is indicated by the glucose catalysis (Fox and Krampitz, 1964), peroxidase, and phosphatase activities of the spheres (Hsu and Fox, 1976). The spheres also have the ability to synthesize peptides in water (Nakashima and Fox, 1980).

The spheres manifest junction formation and visibly participate in intercellular activities (Hsu et al., 1971).

The protocells manifest selective permeability (Fox, 1969) and the osmotic properties which accompany this membrane property

(Fox et al., 1978). Being selective, they are effective in maintaining anisotropy for long periods of time (Ishima et al., 1981).

PROTOCELL EXCITATION

Introduction

A number of proteinoid and proteinoid-lecithin combinations manifest nerve-like electrical properties (Ishima and Fox, 1973; Ishima et al., 1981; Przybylski et al., 1982; Stratten, 1982). The electrical properties vary with the identity of the proteinoid. Four of one hundred proteinoids tested with lecithin were found to manifest significant characteristics of excitability under given conditions. Expectedly, variables other that proteinoid identity influence excitation. Of the many solutions evaluated, Ishima et al. (1981) found artificial pond water to be optimum. Subsequent work by Przybylski et al. (1982) with p-protocells indicates transmembrane potassium ionic disequilibrium as a minimum requirement for membrane excitability. Spiking can be induced in active protocells by shifts in external ionic concentrations. For p-1-protocells the external medium must be adjusted to levels hypoosmolar to the internal medium to induce the emergence of a thin, transparent, spherical membrane from the p-1 crust; there are increases in membrane stability with incubation; and incubation periods for p-1 protocells of 2 hours to 2 days increase the yields of units manifesting excitability. Increased concentration of calcium ion concentrations in the external fluid decreases the spontaneity of the membrane electrical activity as it does for neurons (Ishima et al., 1981).

Proteinoid-Lecithin Protocells

The electrical properties of the p-1 protocells seem to be provided by a conductive component of proteinoid and an electrically insulating component of phospholipid. This is equivalent to the neuronal component properties (Nachmansohn and Neumann, 1974). The p-1-protocells manifest five potential types: (1) a steady-state resting membrane potential, lying between -20 and -70 mV with the mean at -44 mV, (2) a flip-flop activity similar to that expected in the "two-stable-state" hypothesis of Tasaki (1968), (3) singular spikes resembling neuron action potentials, (4) bursts of spikes, or (5) miniature activity at the flopped (hyperpolarized) phase, comparable to miniature end-plate potentials of neuronal soma. Further data has been acquired by alteration of external potassium concentrations and application of transmembrane square wave current pulses. Elevation of potassium results in membrane depolarization. The current pulses induced Ohmic changes in the membrane potential at low currents, but spiking at the higher currents (Ishima et al., 1981).

Proteinoid-only Protocells

The membrane potentials and excitation manifest in p-l-protocells are also manifest in p-protocells. Although the latter manifest higher conductance and greater instability, only quantitative differences in electrical properties are indicated between the two (Przybylski et al., 1982).

Application of Hodgkin-Huxley Model

An evaluation of protocell electrical activities by application of the Hodgkin-Huxley model reveal many similarities in biological systems.

The resting membrane potential is determined by the membrane conductance to potassium and sodium. The sodium conductance is higher than that of neurons, but potassium plays the prime role (Stratten, 1982).

The elevation of external potassium ion reduces the membrane potential and increases potassium conductance (Ishima et al,, 1981). Application of the Nernst equation indicates that the potassium conductance increases are associated with an opposing decrease in sodium conductance (Stratten, 1982). The increase in potassium conductivity is similar to that found with nerve axon membrane when the membrane potential is voltage clamped at lower membrane potentials; however, the biological tissue manifests no steady state decrease in the conductance of sodium as does this preparation.

Spike generation appears to result from a shift in one or more of the ionic conductances. Calculations indicate an increase in sodium permeability during the spike rising phase, the increase being disproportionate to that of potassium. However, the increase in sodium conductance may be less than twofold, quite miniscule in comparison to the 500-fold increase seen in nerve cells. The spike shift in sodium conductance is comparable to that of nerve axon membrane in its brevity (approximately 2 msec) (Stratten, 1982). Application of successful dynamic translocators such as EIM have all been quite slow in comparison to bioresponses (Mueller and Rudin, 1969a).

The RC constant of the spike recovery phase indicates total recovery of the sodium and potassium conductances to the resting levels. The repolarization is attributable primarily to potassium conductance, as is the case with nerve cell spike recovery (Stratten, 1982).

Summary

The following are some of the similarities and differences between the protocell and the excitable nerve cell. It might be noted that most of the similarities are qualitative, whereas most of the differences are quantitative. Protocell-neuron similarities include: (1) RC time constants, (2) spiking depolarization attributable to increased sodium conductance, (3) spike duration, (4) spike recovery attributable to potassium conductance, (5) negative resting potential attributable to sodium and potassium conductance, and (6) stabilization by external calcium ions. Protocell-neuron differences include: (1) variable resting membrane potential, (2) spiking thresholds, (3) greater RC instability, (4) much less change in sodium conductance with spiking, (5) shifting conductance with steady-state depolarization, (6) higher resting sodium conductance, and (7) the membrane environment in which spiking occurs (Ishima et al., 1981; Stratten, 1982).

ENERGETICS AND EXCITATION

Metabolism

Anisotropism is essential for electrical excitation. The second law of thermodynamics demands a continual degradation of this anisotropy; therefore, it is assumed that this degradation must be compensated by endergonic processes. This need has evolved into a contest between those who feel the membrane potential is coupled with metabolism (Nachmansohn and Neumann, 1974; Hill, 1958) and others who contend the metabolic process is not necessary to the presence of a membrane potential (Troshin 1966; Tasaki, 1968). The evidence supporting either view is largely circumstantial.

Metabolism-Coupled Excitation

The metabolism-coupled view is supported by the high levels of heat production and absorption accompanying electrical activity, implying a tight association with chemical energetics processes, i.e., metabolism (Nachmansohn and Neumann, 1974; Hill, 1958). Nachmansohn (1970) has also found the energy production and absorption, "metabolism", of the *Electrophorus* to be low except in the membrane.

Non-Coupled Excitation

The non-coupled view is exemplified in the work of Tasaki (1968) who, with others, has shown the squid axon to be capable of conducting impulses for days with most of the cellular axoplasm removed. Although some of the characteristics may be due to residual internal and external components, since soluble metabolites are gone in a few minutes, matabolism, if any, must be restricted to within the membrane. These findings in combination with those of BLM research implies that the metabolism, if coupled to excitation, must be catalyzed by the very substances which are the membrane translocators. The artificial membrane studies indicate only dissipative energy transduction mechanisms, alleviating such a necessity (Mueller and Rudin, 1969a). Tasaki (1968), further, finds that the exothermia associated with the initiation of an action potential is adequately explained by the accompanying displacement of divalent by monovalent cations. Similarly, the repolarization is endothermic due to a reverse in this cation exchange.

Energy Conversions

More generalized research on membrane energetics indicate biological membranes to be capable of interconverting between bonding, redox, and ion gradient energy forms. Of the six conversions most biomembranes possess only a few. The protocell excitation dramatically illustrates, at the very least, an example of STATE I > STATE II > STATE I. Such a sequence demands use of at least two conversions, one exergonic and one endergonic. The energy source, i.e., bond, redox, or ion gradient, is not yet obvious.

The work with BLM and certain translocators, e.g., monazomycin, reveals dynamic ionic affinities coupled with a sequential double gating effect. Jardetsky (1966) indicates such a combination as essential to pumping action. This sequential-valve/ion-dissociation pump could presumably provide the endergonic service of maintaining membrane anisotropy by coupling with a net exergonic "excitation" while incurring only minimal secondary energy losses. Redox agents could presumably act to provide electron energy to these pumps as easily as to the postulated oxidative phosphorylation processes. Mueller and Rudin (1969a) have gone so far as to say: "Many properties of the gated translocators resemble those attributed to active ion transport mechanisms. . . ."

Membranes may be involved in energy trapping and conversions including various phophorylations with the polyphosphates, the pyrophosphates, or even ATP (Lipmann, 1965). Such has been demonstrated in the synthesis of amino acid polymers by ATP-protocell combinations (Nakashima and Fox, 1980).

Photo-Excitation

Photosensitivity has been found to induce "spontaneous" electrical activity of some protocells (Przybylski et al., 1982). Chromophores (pteridines and flavins) formed as byproducts of the formation of the copolyamino acid polymers or chromatic proteinoids themselves may act as the pigments channeling the photon energy toward enhancing membrane excitability (Heinz et al., 1979).

Although there is not yet a clear link between this photon-chemical energy coupling and the energy coupling of excitation, the enhancement of excitation with illumination arouses stimulating ideas associating the two.

Similar photon influences have been noted in biological membranes following addition of pigments. Singer and Tasaki (1968) found addition of pigments to allow light-mediated induction of excitation. In addition, electroplax, following addition of the pigment p-phenylazophenyl-trimethylammonium chloride, demonstrate wavelength sensitive potential changes (Nachmansohn, 1970). Although the latter may be associated with shifts in the cis-trans equilibrium, the mechanism of photon-excitation coupling is far from clear.

EVOLUTION OF MEMBRANE AND EXCITATION

The various enzymatic, permeability, and excitation properties of membrane are more likely physicochemical properties of protein than phospholipid (cf. Nachmansohn, 1970). The numerous roles of accretion, enzymatics, selective permeability, dynamic "gating", active transport, and maintenance of anisotropy are all functions which are presumably filled by proteins and have been shown to be filled by proteinoids. Only the last, a passive process, can be accomplished by the lipids. In spite of this, there are some investigators who insist that since "lipids are required for the formation of the only known examples of cellular membranes" it follows that lipid structures preceded those of protein, i.e., proteins are "extrinsic" additions (Mueller and Rudin, 1969a). It seems far more likely that a system first acquires the multiple selective advantages of molecular order and ordering via a primal proteinoid composition and only subsequently eases its burden of anisotropy maintenance by acquisition of the lipid. The argument that "the protein membranes are more permeable . . . therefore are not able to provide an efficient barrier structure" (Mueller and Rudin, 1969a) is considerably weakened by the myriad of stable proteinoid protocell types displaying many life-essential characteristics (Fox and Dose, 1977).

The argument "lipid first" is as untenable as the "DNA first" notion, which has been peddled as "truth" in most biology texts for the past two decades. In fact proteinoids and their bilayer protocells can be synthesized without DNA (Fox and Dose, 1977), but the opposite, i.e., DNA without protein, has never been shown. Similarly, the proteinoid membranes manifest essential characteristics of membranicity sans lipid, but the opposite has never been documented. The Miller-Urey-like experimental synthesis of hydrocarbons and long-chain fatty acids by Wilson (1962) and Allen and Ponnamperuma (1967) respectfully, does not alter the intrinsic lipid inadequacies. It simply implies that lipids may have been available in early stages of evolution to associate with and stabilize the protein structures.

There indeed would have been a selective advantage of those proteinoid-lipid protocells which had struck the optimal balance between the advantages and disadvantages of leakiness. The maintenance of some leakiness has advantages (Kuhn, 1976) including optimizing acquisition of molecular intermediates for structure, metabolism, etc., until the protocells acquire abilities to synthesize these from smaller components. Conversely, too great a leakiness would result in osmotic, anisotropy, and molecular intermediate retention problems (Mueller and Rudin, 1969a). One might conjecture that the "best of all worlds" would be the ability to control the moments of leakiness, in particular if they could be timed to correlate with the availability of beneficial sustances. This may, indeed, be the earliest of selective advantages provided by the process we now call excitation. In summary, a likely sequence on the evolutionary path begins with proteinoid spheres which are eventually displaced by an advantageous proteinoid-lipid combination.

The finding that artificial fossilization of proteinoid spheres results in structures indistinguishable from fossilized algae (Barghoorn and Tyler, 1965) and that some active spheres are stable at high pH similar to that of the conjectured primordial alkaline ocean (Snyder and Fox, 1975) stand as circumstantial reinforcement to the "proteinoid first" notion.

Some of the current models of bioexcitation declare highly evolved molecular complexes (Neumann et al., 1973; Nachmansohn and Neumann, 1974), to be essential to the manifestation of excitation. The fact that such is the case for some types of excitation is likely, but arguments such as "it would be difficult to reason that the enzyme . . . is essential for electrical activity in some, but not others" (Nachmansohn, 1970) are inadequate. In fact, the manifestation of excitation in BLM by addition of some simple translocators seems to preclude such complexity, especially any essential linkage to cellular metabolism. However, the need for highly evolved structures for excitation was not totally obviated by this work, since the effective translocators have primarily been biological macromolecules with clear membrane-linked function which is highly selective and probably a result of natural selection. The most effective translocators isolated from tissue have been small molecular structures which appear to possess the prime function of cell membrane degradation, e.g., antibiotics, bacterial toxins, and the immune system's complement. A function which they seem to hold in common is the facilitation of membrane transfer of substances (e.g., ions) which do not normally cross the lipid barrier (Mueller and Rudin, 1969a).

Proteinoid excitation is the first indication of excitation occurring in macromolecular structures totally independent of biological natural selection. Even the peptide used by Kennedy et al. (1977) was modeled after life structures.

CONCLUSIONS

These findings indicate that the protein-lipid stereodynamics allowing for electrical activity of a very selective nature may not be processes attained only by biological natural selection, but rather functional molecular configurations which form due to the non-random ion-proteinoid interaction in non-biological milieu.

The high incidence of excitability with proteinoids indicates a possible selective advantage of those molecules over the molecular arrangements which lack such dynamic two-stable-state characteristics.

The comparable properties of excitable tissue and protocells, even to selection of similar ionic conductances and inhibition by calcium, points to possible functional commonality. The effect of calcium and the improbability of precise protocell membrane architecture forces an increased consideration of excitation as being a colloidal function with a "carrier" as opposed to "gating" mechanism.

There should be a careful assessment of the energetic relationships since the molecular evolution of these particular species implies the manifestation of significant functional "ordering" within a high entropy albeit high energy system. The implication is that "excitation" may itself be an energy transfer system allowing structural maintenance by the utilization of surrounding entropic processes. This link between excitation and energetics may be what Mueller and Rudin (1969a) are observing when they note common properties between "gated translocators" and active ion transport mechanisms.

REFERENCES

Allen, W. V., and Ponnamperuma, Co., 1967, Current Mod. Biol., 1:24.

Anderson, T. F., Rappaport, C., and Muscaline, N. A., 1953, Ann. Inst. Pasteur, 84:5.

Barghoorn, E. S., and Tyler, S. A., 1965, Science, 147:563.

Bone, S., Lewis, T. J., Pethig, R., and Szent-Gyorgyi, A., 1978, Proc. Nat. Acad. Sci. U.S., 75:315.

Fox, S. W., 1969, in: "Physical Principles of Biological Membranes," F. Snell, J. Wolken, G. J. Iverson, and J. Lam, eds., pp. 417-432, Gordon and Breach, New York.

Fox, S. W., 1978, in: "The Nature of Life," W. Heidcamp, ed., pp. 23-92, University Park Press, Baltimore, Maryland.

Fox, S. W., and Dose, K., 1977, "Molecular Evolution and the Origin of Life," 2nd edn., Marcel Dekker, New York.

Fox, S. W., and Krampitz, G., 1964, Nature, 203:1362.

Fox, S. W., Adachi, T., Stillwell, W., Ishima, Y., and Baumman, G., 1978, in: "Light Transducing Membranes: Structure, Function, Evolution," D. W. Deamer, ed., pp. 61-75, Academic Press, New York.

Grote, J. R., Syren, R. M., and Fox, S. W., 1978, BioSystems, 10:287.
Heinz, B., Ried, W., and Dose, K., 1979, Angew. Chem. Int. Ed., 18:478.
Hill, A. V., and Howarth, J. V., 1958, Proc. R. Soc., B149:167.
Hsu, L. L., and Fox, S. W., 1976, BioSystems, 8:89.
Hsu, L. L., Brooke, S., and Fox, S. W., 1971, Current Mod. Biol., 4:12.
Ishima, Y., and Fox, S. W., 1973, Soc. Neurosci. Abstr., 3rd Ann. Mtg., p. 172.
Ishima, Y., Przybylski, A. T., and Fox, S. W., 1981, BioSystems, 13:243.
Jardetsky, O., 1966, Nature, 211:969.
Kardin, A., and Bartels, E., 1966, Biochem. Biophys. Acta, 126:525.
Katchalsky, A., 1964, Biophys. J., S4:9.
Kennedy, S. J., Roeske, R. W., Freeman, A. R., Watanabe, A. M., and Besch, Jr., H. R., 1977, Science, 196:1341.
Koelle, G. B., 1966, in: "Nerve as a Tissue," J. Rodahl, and B. Issekutz, Jr., eds., pp. 287-292, Harper and Row (Hoeber), New York.
Kuhn, H., 1976, Naturwiss., 63:68.
Lehninger, A. L., 1975, "Biochemistry," Worth, New York.
Lipmann, F., 1965, in: "The Origins of Prebiological Systems and of their Molecular Matrices," S. W. Fox, ed., Academic Press, New York.
Mueller, P., and Rudin, D. O., 1968, Nature, 217:713.
Mueller, P., and Rudin, D. O., 1969a, in: "Current Topics in Bioenergetics," D. R. Sanadi, ed., Vol. 3, pp. 157-249, Academic Press, New York.
Mueller, P., and Rudin, D. O., 1969b, in: "Laboratory Techniques in Membrane Biophysics," H. Passow, and R. Stamfli, eds., pp. 141-156, Springer-Verlag, New York.
Mueller, P., Rudin, D. O., Tien, H. T., and Wescott, W. C., 1962, Nature, 194:979.
Nachmansohn, D., 1970, Science, 168:1059.
Nachmansohn, D., and Neumann, E., 1974, Ann. N.Y. Acad. Sci., 227:275.
Nakashima, T., and Fox, S. W., 1980, J. Mol. Evol., 15:161.
Neumann, E., Nachmansohn, D., and Katchalsky, A., 1973, Proc. Nat. Acad. Sci. U.S., 70:727.
Przybylski, A. T., Stratten, W. P., Syren, R. M., and Fox, S. W., 1982, Naturwiss., 69:561.
Schellman, J. A., and Schellman, C., 1964, in: "The Proteins," H. Neurath, ed., Vol. 2, pp. 1-137, Academic Press, New York.
Singer, I., and Tasaki, I., 1968, in: "Biological Membranes: Physical Fact and Function," D. Chapman, ed., pp. 347-410, Academic Press, New York.
Sjoestrand, F. S., and Barajas, L., 1968, J. Ultrastruct. Res., 25:121.
Snyder, W. D., and Fox, S. W., 1975, BioSystems, 7:222.
Stratten, W. P., 1982, Soc. Neurosci. Abstr., 8:253.

Teorell, T., 1962, Biophys. J., 2:27.
Troshin, A. S., 1966, "Problems of Cell Permeability," Hell, M. G., trans., W. F. Widdas, ed., Pergammon Press, London.
Wei, L. Y., 1972, Biophys. J., 12:1159.
Wilson, A. T., 1962, Nature, 196:11.

PHYSICAL BACKGROUND OF EXCITABILITY: SYNTHETIC MEMBRANES AND EXCITABLE CELLS

Aleksander T. Przybylski

NFCR Laboratory at the Institute for Molecular and Cellular Evolution, University of Miami
Coral Gables, Florida 33134

INTRODUCTION

The value of biomimetric models is unquestionable. Beyond their heuristic role, they have also a practical aspect. The studies of reassembled biological material models preceded those of fully artificial materials. We shall in the first three sections review very briefly the irritability (excitability), which is common not only in the animal realm but also in living matter in general, and then discuss the details of our new findings on excitability of synthetic artifical cells.

The electrophysiological properties manifested by plant protoplasmic droplets were very helpful in finding a way to answer whether artificially assembled cells possess excitability. The answer was positive (Takenaka et al., 1971). About one hour after their formation, reassembled protoplasmic droplets displayed membrane and action potentials.

In 1973 the first findings on excitability of synthesized proteinoid cells were reported (Ishima and Fox, 1973). Three findings were further expanded (Ishima et al., 1981; Przybylski et al., 1982). selective permeability and osmotic properties (Fox et al., 1969; Fox and Nakashima, 1980) and bilayer membranes made of proteinoids (Fox et al., 1978) give us an approach which allows an understanding of the origin of excitability. The present paper aims to point out some data obtained on artificially produced proteinoid cells with an attempt to find the very basic physical laws underlying the origin of excitability.

EXCITABLE MEMBRANES

The membranes fulfill two main functions: (a) creation of the barrier between the cell's internal contents and the external environment, and (b) generation and propagation of excitation. Both of these functions are intrinsic in understanding the origin and function of the first living entity. Excitable membranes, beyond natural ones, comprise two types of artificially made micelles: laminar (bilayer molecular membranes) and spheroidal (vesicles or microspheres). The underlaying mechanism of the origin of excitable membranes made of such compounds as phosphatidylcholine (lecithin), cholesterol, retinal, etc., under conditions of mechanical agitation (Goldacre, 1958; Rutten, 1971; Tien, 1974) is their energetically favorable interface orientation.

A decrease in bifacial energy takes place during bilayer membrane formation, and, therefore, the micelle is more stable than a separate molecule. The corresponding entropy change ΔS is given by:

$$\Delta S = d(\Delta F_i)/dT_i$$

where

F_i : standard free energy,

T : absolute temperature (Tien, 1974).

A membrane potential in a bilayer membrane-liquid system is due to at least one of the following potentials:

distribution potential
Donnan potential
redox (electrostenolytic) potential
electrokinetic potential
photoelectric potential

and depends on the chemical composition of the membrane and the liquid phase of the system. The contribution of each of the above potentials is different. Physical factors resulting in membrane conductance comprise a series of events: interphase surface charge, ion activity, transmembrane ion gradient, oriented dipoles at membrane surface, dielectric constants, and intramembrane ion mobility. In turn, these conductance properties of the membrane with the associated ionic gradients determine the steady-state potential as well as dynamic potential changes, either induced (e.g., by current stimulation) or spontaneous. The spontaneous potential oscillations are observed under strict conditions of membrane composition, preparation, salt gradient of the liquid phase of the system, pH, and temperature. If photosensitive dyes are present in the membrane, the electrical behavior of the membrane is determined also by factors responsible for trapping of photons, formation of carriers and charge separation.

MECHANISM OF POLARIZATION CHANGE OF THE MEMBRANE

The membrane resistance is the first contributing factor involved in the observed membrane polarization changes.

Resistance of the Membrane

The resistance of the membrane is its crucial feature, both in regard to barrier and permeability function as well as to charge separation. The above mentioned factors contributing to membrane polarization are evidently involved in determination of the membrane resistance. The membrane resistance is

$$R_m = E_m R_i / E_i - E_m,$$

where

R_i is the series resistance, E_m--the voltage across the membrane, and E_i--the calibrated input voltage. The membrane capacitance (of oxidized cholesterol BLM, e.g.) is related to voltage (White, 1970)

$$C_m = C_o + bV_a^2 ,$$

where

C_o : the capacitance at zero voltage,
b : temperature constant,
V_a : applied voltage.

The change of BLM capacitance is dependent on electrolyte concentration. Its minimum has been detected for uni-valent ions in electrolytes such as KCl, NaCl, at a concentration of 0.1 M, and in the case of di-valent electrolytes such as $MgSO_4$ it has been found at 0.025 M concentration (Rosen and Sutton, 1968).

The bilayer capacitance C_m can be expressed also as a dependence:

$$C_m = t/R_p \ln (E_o/E_t),$$

where

t : time (sec),
E_o, E_t : voltage at $t = 0$, and t, respectively,
R_p : leakage resistance, equals ($R_m R_i / R_m + R_i$).

When $E_m/E_o = 1/q = 0.37$ (q : charge) the time constant is $\tau = R_p C_m$.

This constant is a useful measure of the temporal membrane characteristics.

The resistance of the membrane is linear when the concentration of electrolytes across the membrane is the same. If different, both types of membranes (laminar and spheroidal) display a nonlinear characteristic with a negative resistance.

Because of the high dielectric constant of the bilayer membrane and very small interlayer distance, it can hold a high charge (of an order of 10^5V/cm). The lipid-like substances are characterized by high resistance, and protein presence of lipid-like membranes diminishes resistance considerably. It has been found that an addition of natural protein (excitation-inducing material) reduces membrane resistance to as low as 10^1 ohm/cm^2 from an initial value of 10^6 - 10^{10} ohm/cm^2 (Mueller and Rudin, 1968).

The Electronic Process in Membranes

The possibility that mechanism analogical to electron conduction in proteins might play a role in biological processes was considered by Szent-Györgyi over 40 years ago (1941).

Studies of such processes as vision, photosynthesis, chemical respiration, carcinogenesis and nerve excitations on submolecular and quantum levels are being better understood in terms of the underlying mechanism of these processes.

We can consider BLM as an organic semiconductor; however, the current carriers involved in the dark conductivity of unmodified BLM are most likely ions rather than electrons and holes. But the conductivity of BLM can be raised several orders of magnitude.

If sufficient energy is absorbed by the membrane, electrons and holes can be generated. To emphasize similarities between the living system and the semiconductor device, an analogy between biological semiconductors and inorganic semiconductors is given in Table 1.

There are more considerations and experimental data on electronic processes in the living system. Several possible mechanisms such as electron tunneling and hopping are discussed.

The high probability and significance of the tunneling mechanism in biological processes is within the range of 10^9 to 10^{-9} s^{-1} with slight changes only in interstate distance or energy between states (Lewis, 1982). This means that the occurrence of an event or its absence is conditioned by slight changes of energy or geometry of the molecule.

Table 1. An Analogy Between Biological and Inorganic Semiconductors

Component	Inorganic semiconductor	Biological semiconductor
Base material	Covalent-bonded Crystalline phase (e.g., germanium crystal)	Hydrophobic hydrocarbon phase (e.g., lipid bilayer)
Electron donor	Group V elements (e.g., As, Sb)	Bio-reductants (e.g., cytochromes, ferrous ions, and H_2O)
Electron acceptor	Group III elements (e.g., Ga, In)	Bio-oxidants, (e.g., ferredoxins, quinones, and H_2O)
Electron pathway	Crystal proper	Conjugated hydrocarbon chain and ring systems
Connector	Metallic wire	Electrolyte solution

(Tien, 1974)

The role of the membrane is significant because of its ability to orient active molecular groups and direct charge transfer and electron transport.

These processes are determinated by a descending sequence of energy levels and the existence of an electric field. This includes the energy of water solvation. Eley et al. (1979) established that in a dry state conduction is determined by holes and by electrons as water absorption occurs. The water-protein interaction involving charge carriers determines the electrical properties of the membrane. In the case of the artificially assembled protocell of thermal polymers of amino acids there seems to be channel properties of the membrane.

Channeling Through Membranes

Both natural and artificial membranes contain a large number of discrete gating site (Hodgkin and Huxley, 1952) which open and close at random as revealed by fluctuation analyses (Stevens, 1977) and single channel recordings (Neher and Sakmann, 1976).

Channels undergo structural transitions between conducting and nonconducting states. A charge and ion transfer through the membrane and kinetics of this process seem to be common in bilayers, spheroids and natural membranes. It has been shown (Kennedy, 1967) that synthesized peptides of the sequence (Leu-Ser-Leu-Gly), having the helical structure, also form ion channels across the lipid bilayer membrane.

Nernst and Hodgkin-Huxley Formulas

Despite some criticisms (Habib and Bockris, 1982) the Nernst and Hodgkin-Huxley equations maintain broad acceptance as the explanation for the phenomenon of excitability. Their simplicity may be one reason for this.

The gradient of ions, both in concentration and in charge across an asymmetric membrane are the basic and sufficient factors explaining membrane excitability. The charge q of n ion moles is

$$q = z\,n\,F,$$

where

z : valency,
F : Faraday constant.

The differential of this value, in the case of the presence of more than one kind of ion, is

$$dq = \sum_{i=1}^{m} z_i\,F\,d\,n_i.$$

Diffusion of ions through a membrane also means simultaneous transfer of charge. This process, depending on the ratio of ion charge to membrane polarization and to its dielectric property, is governed by the diffusion force moving ions, and by an opposite electrophoretic force inhibiting ion movement.

If the diffusion of ions is inhibited, a state of equilibrium is reached. This state is described by the Nernst formula. It has been derived also from a very fundamental theoretical background. The chemical potential is

$$\mu = \mu^o + R\,T\,\ln p,$$

where

μ^o : reference chemical potential,
p : pressure,
R : gas constant.

In the case of dissolved substances there is a ratio of ions expressed by a_1 and a_2, the log of which when taken and multiplied by RT/zF gives the equilibrium potential P_{eq} as stated by the formula

$$P_{eq} = \frac{RT}{zF} \ln \frac{a_1}{a_2} .$$

It is worth emphasizing here the similarity of this formula with the formula for oxidation-reduction potential P_{or}:

$$P_{or} = E_o + \frac{RT}{zF} \ln \frac{A^{n-1}}{B^{m+1}} ,$$

where

E_o : standard electrochemical potential,

A^{n-1} : oxidized compound,

B^{m+1} : reduced compound.

The actual value of equilibrium potential across the membrane, estimated by the Nernst equation, in the living cell may be compensated and overcompensated by the metabolically driven potential. The intracellular versus extracellular proton concentration may be considered here as an equivalent ratio in the Nernst equation. The water environment of the cell, the role of water in electronic properties of peptides and proteins in membrane function seem to be evidence of the importance of protons in sustaining the membranes polarization and its dynamic properties during de- and repolarization (and generation of action potentials). In this case we can consider even a metabolically driven proton battery under ΔpH difference with an electromotive force E_p (Glaser, 1971)

$$E_p = 0.059 \log \Delta pH .$$

One unit of pH difference across the membrane is equivalent to 0.059 V of membrane potential. The resulting electrical potential across the membrane $\Delta\psi$ and the pH difference constitute the Mitchell proton motive force P_{mf}

$$P_{mf} = 0.059 (\Delta pH) + \Delta\psi .$$

Because of metabolism and its consequence upon the pH value, the actual voltage in the living cells exceeds P_{eq} value, and it follows that the membrane current I_m (Palti, 1971) is given by the following formula:

$$I_m = g_m (E_m - P_{eq}) ,$$

where

g_m : membrane conductance,
E_m : membrane potential,
P_{eq} : equilibrium potential.

When $E_m = P_{eq}$ the transmembrane current is zero. Such a situation takes place in artificial membranes after a certain time lapse, unless they are not charged.

COMPARATIVE DATA ON EXCITABLE MEMBRANES

Under strict conditions of preparation of bilayer membranes, and addition of KCl and Excitability-Inducing-Material they display electrical properties such as membrane potential, and spontaneous (Fig. 1), as well as current-induced electrical discharges (Fig. 2).

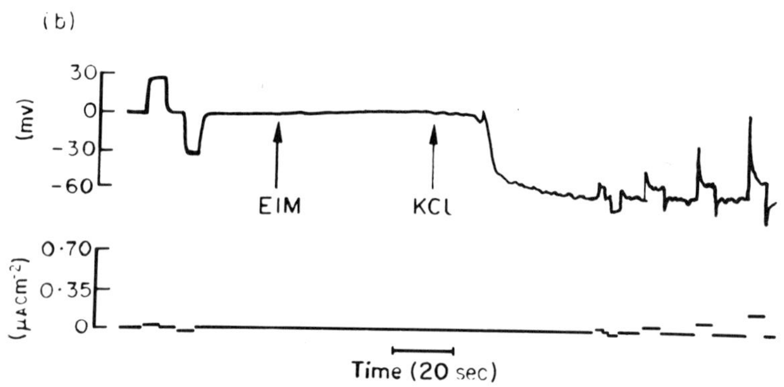

Fig. 1. Membrane and current-induced potentials of the bilayer lipid membrane (upper record) due to electrical stimulation (lower record) (Mueller and Rudin, 1968).

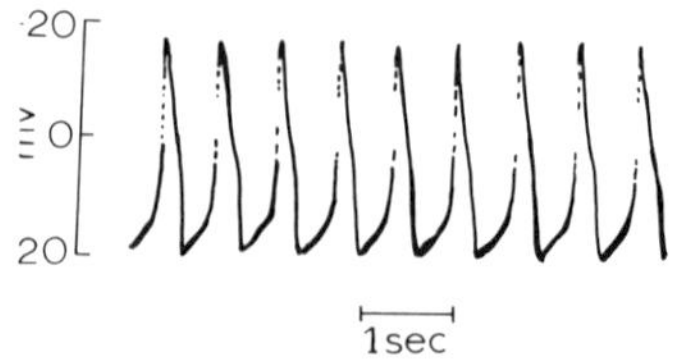

Fig. 2. Spontaneous electrical oscillations of the bilayer lipid membrane (Mueller and Rudin, 1968).

Very similar responses were observed in proteinoid spherules (Figs. 3-5). The difference was the presence in the first case of EIM and its absence in the proteinoid spherules.

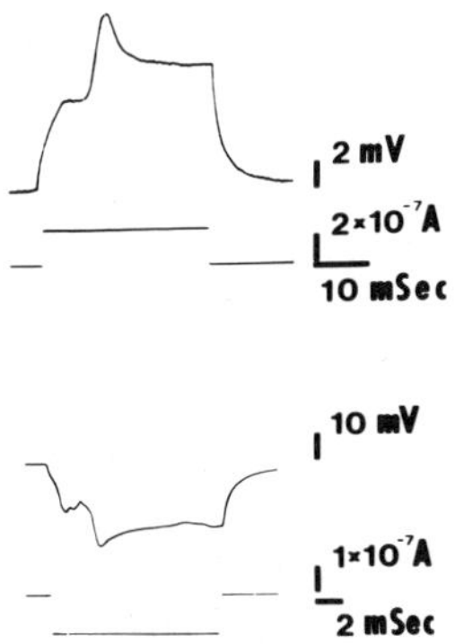

Fig. 3. Induced potentials (upper record) of the spheroidal membrane during electrical stimulation (lower record).

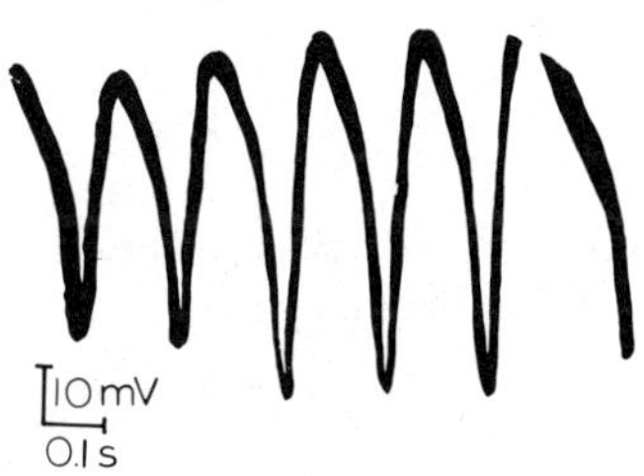

Fig. 4. Spontaneous electrical oscillation of the poly (Asp:Glu) proteinoid-lecithin artificial cell.

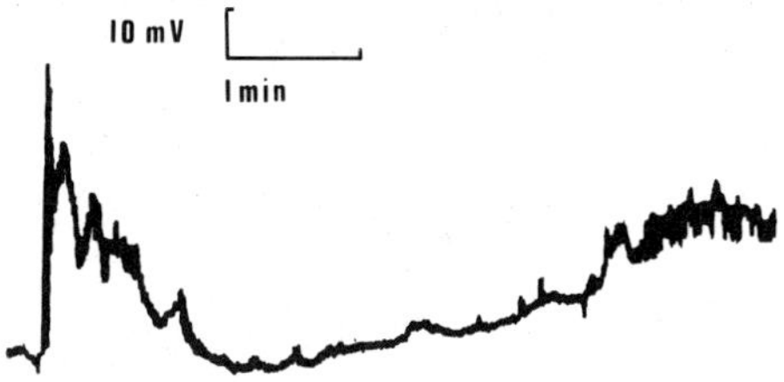

Fig. 5. Recovery of the membrane potential following mechanical injury of the proteinoid cell due to its microelectrode impalement.

There is also a high degree of similarity of the pattern of electrical discharges of the bilayer membrane and the spheric membrane of the proteinoid cell. Various patterns observed in proteinoid cells are different in regard to the shape and gradient of discharges (Figs. 6-8). Some highly resemble natural neuronal spiking (Fig. 6). The repeatable examples of electrical discharge display very similar patterning among spheres made of the same material (Figs. 6, 7). However, similar patterning of electrical discharges may be observed in spherules made of different polymers, although there are indications that patterning is connected with the chemical composition of microspheres. The resultant discharge pattern is conditioned by, at least, several factors such as membrane thickness, its ion permeability, etc. Because of this, the influence of composition of the cell

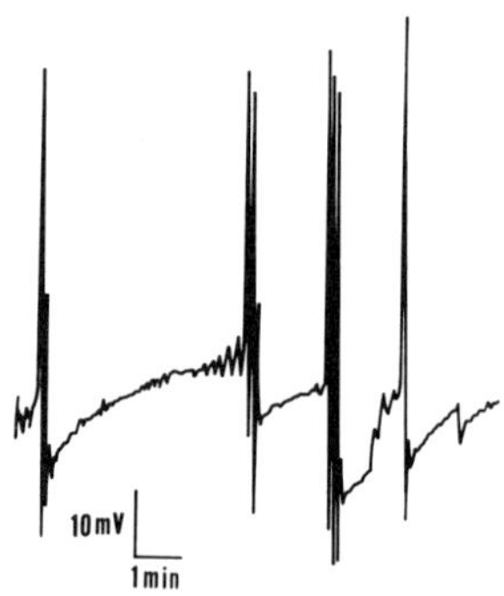

Fig. 6. Pattern of electrical discharges of the proteinoid cell made of the 2:2:1 proteinoid.

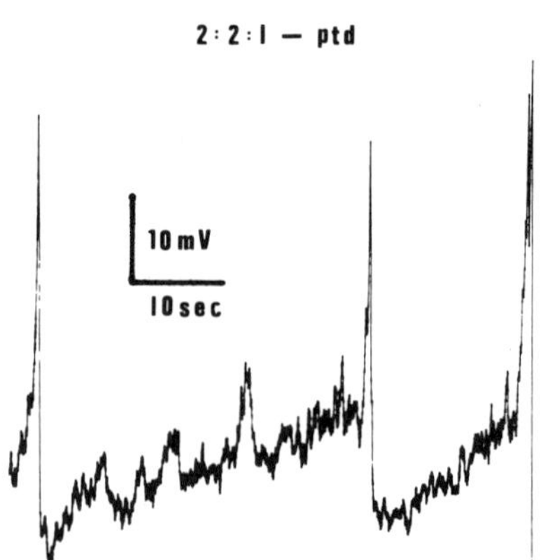

Fig. 7. Similar pattern of discharges of the proteinoid cell made of the 2:2:1 proteinoid several months later.

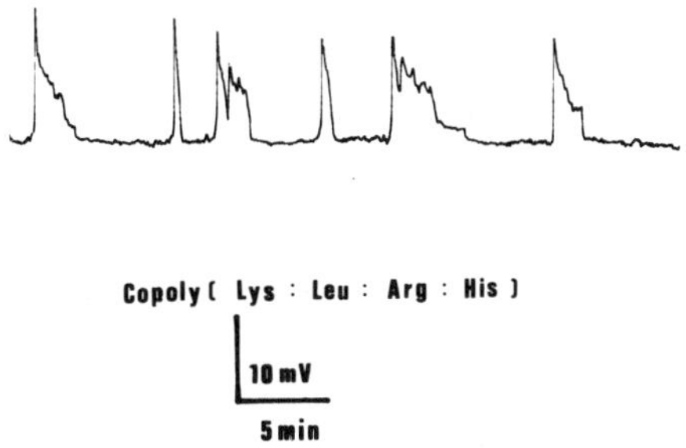

Fig. 8. Pattern of electrical discharges of the proteinoid cell made of copoly (Lys:Leu:Arg:His).

and, in particular, of the membrane cannot be simply related to its electrical properties.

The very characteristic feature of spheroidal membranes is their recovery, both of membrane polarization due to microelectrode impalement (Fig. 5) and external KCl treatment of the cell (Fig. 9). This implies an intrinsic homeostatic mechanism.

All the above seems to indicate not only a superficial but rather a deeper analogy between these artificial entities and natural excitable cells, including perhaps also the very mechanism underlying membrane polarization and its discharges.

An application (Stratten, 1982) of the Hodgkin-Huxley formalism in analysis of the proteinoid cell indicates common ionic mechanisms

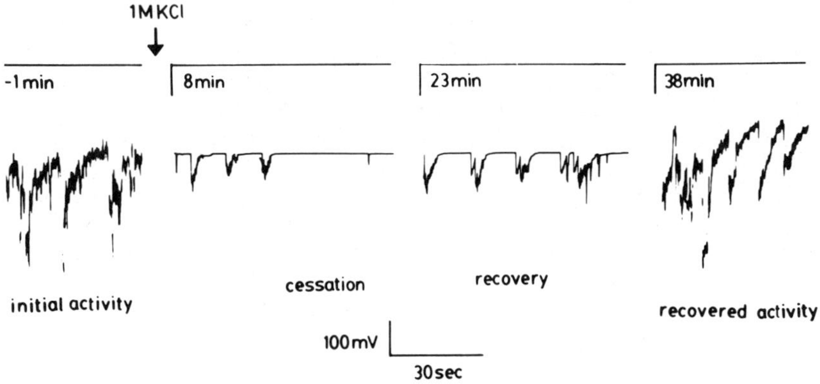

Fig. 9. Homeostatic recovery of the electrical discharges of the proteinoid cell (2:2:1-proteinoid) due to external KCl treatment of the cell.

of excitation in proteinoid and natural excitable cells. Evaluation of membrane potential E and its conductance g with intrasphere electrodes and current clamping with varying external potassium ion (K) concentrations clearly indicate a significant contribution of g(K). An induced reduction in E is accompanied by an increase in g(K) as is the case with nerve membrane, but it may also be accompanied by a decrease in g(Na) which is not manifested with nerve membrane.

There is more than a simple analogy between proteinoid excitable cells and neuronal or muscle cells. The similarities recognized so far are as follows:

RC membrane characteristics,
electrical stabilization by calcium ion,
transient depolarizing spiking,
spike recovery phase,
apparent homogeneity of channels,
negative inside resting potential.

Differences are rather secondary.

A comparison of some physical characteristics of artificial and natural membranes (Table 2) allows us to search for closer simistanding of construction and function of natural membranes and excitable cells.

In both cases of artificially made membranes (bilayer and spheriodal) there is, except light and ion concentration gradient across membrane, no energy source driving the electrical discharges. The absence of electrical discharges in darkness seems to be strong enough proof that illumination of the preparation even by day-light or white microscope-illumination light is the primary energy source of electrical discharges observed both in bilayer and spheroidal membranes. In regard to proteinoids, photoeffect is displayed (Przybylski and Fox, 1982). This property is due to the presence in the polymer of flavin and pterin chromophores (Heinz and Ried, 1981). These chromophores would be, hence, responsible for trapping of photons and electron release, whereas the membrane structure would be responsible for charge separation, i.e., holding the potential. As the consequence, periodic electrical discharges of the membrane would take place with a time constant determined by its resistance and capacitance. Additionally, in the case of the ionic gradient across the membrane, and due to the presence of the channel-like structure of the membrane, the ionic flow through the membrane would be influenced by changes in the membrane potential. These two factors seem to be responsible for the observed electrical phenomena in artificial membranes.

Table 2. Physical Characteristics of Bimolecular Lipid Membranes, Spheric and Natural Membranes

Membrane	Natural Membranes	BLM	Spheric Membranes
Thickness (Å)	40–130	60–90	10000
Potential difference (mV) (resting)	10–88	0–140	10–1000
Resistance (Ω-cm^2)	10^2–10^5	10^3–10^9	10^4–10^8
Capacitance (μF/cm^2)	0.5–1.3	0.3–1.3	0.7
"Excitability"	Observed	Observed	Observed
Ion selectivity and specificity	Observed	Observed	Observed
Excitation by light	Observed	Observed	Observed

CONCLUSION

In conclusion we can say that the experimental proteinoid protocell model, characterized by one or more proteinoid or proteinoid-lecithin membrane layers and by a proteinoid core with electrolyte, is also a model of an excitable cell.

The excitable artificial cell possesses many features characteristic of natural neuronal cells such as: membrane potential, all-or-none spontaneous and induced electrical discharges, hypo- and hyper-polarization type of membrane potential changes, asymmetric permeability, channeling phenomena, current-voltage characteristics with negative resistance, intra- and extra-cellular ionic influence upon the membrane potential including homeostatic recovery of discharge spiking and membrane potential.

The change of the membrane potential of the excitable cell is compatible with Hodgkin-Huxley equations. The light-sensitivity of the thermal amino acid polymer making up the artificial cell membrane. The flavin and pterin chromophores seem to be the energy trapping source which gives rise to the functions described.

ACKNOWLEDGMENTS

This study was also aided by grants from the National Aeronautics and Space Administration Grant No. NGR 10-007-008, and Mr. David Rose.

REFERENCES

Eley, D. D., Lockhart, N. C., and Richardson, C. N., 1979, J. Chem. Soc. Faraday Trans., 75:323.

Fox, S. W., McCauley, R. M., Montgomery, P. O'B., Fukushima, T., Harada, K., and Windsor, C. R., 1969, in: "Physical Principles of Biological Membranes," F. Snell, J. Wolken, G. J. Iversen, and J. Lam, eds., Gordon and Breach, New York.

Fox, S. W., Adachi, T., Stillwell, W., and Ishima, Y., 1978, in: "Light Transducing Membranes: Structure, Function, Evolution," D. W. Deamer, ed., Academic Press, New York.

Fox, S. W., and Nakashima, T., 1980, BioSystems, 12:155.

Glaser, R., 1971, "Einführung in die Biophysik," VEB G. Fischer Verlag, Jena.

Goldacre, R. J., 1958, in: "Surface Phenomena in Chemistry and Biology," J. F. Danielli, K. G. A. Pankhurst, and A. C. Riddiford, eds., Pergamon Press, London.

Habib, M. A., and Bockris, J. O'M., 1982, J. Bioelec., 1:289.

Heinz, B., and Ried, W., 1981, BioSystems, 14:33.

Hodgkin, A. L., and Huxley, A. F., 1952, J. Physiol. (London), 117:500.

Ishima, Y., and Fox, S. W., 1973, Abstract, Third Ann. Mtg. Soc. Neuroscience, 17.10:172.

Ishima, Y., Przybylski, A. T., and Fox, S. W., 1981, BioSystems, 13:243.

Kennedy, S. J., Roeske, R. W., Freeman, A. R., and Besch, Jr., H. R., 1977, Science, 196:1341.

Lewis, T. J., 1982, Phys. Med. Biol., 27:335.

Mueller, P., and Rudin, D. O., 1968, J. Theor. Biol., 18:222.

Neher, E., and Sakmann, B., 1976, Nature, 260:779.

Palti, Y., 1971, in: "Biophysics and Physiology of Excitable Membranes," W. J. Adelman, Jr., ed., p. 168, Van Nostrand Reinhold Co., New York.

Przybylski, A. T., Stratten, W. P., Syren, R. M., and Fox, S. W., 1982, Naturwiss., 69:561.

Przybylski, A. T., and Fox, S. W., 1982, in: "Alternative Energy Sources," Vol. 3, T. N. Vezirolgu, ed., Ann Arbor Science, Ann Arbor.

Rosen, D., and Sutton, A. M., 1968, Biochim. Biophys. Acta, 163: 226.

Rutten, M. G., 1971, "Origin of Life" Elsevier Press, Amsterdam.

Stevens, C. F., 1977, Nature, 270:391.

Stratten, W. P., 1982, Abstracts Soc. Neuroscience Ann. Mtg., Minneapolis, 66.12.

Szent-Györgyi, A., 1941, Nature, 148:157.

Takenaka, T., Inoue, I., Ishima, Y., and Horie, H., 1971, Proc. Japan Acad., 47:554.

Tien, H. T., 1974, "Bilayer Lipid Membranes," Marcel Dekker, New York.

White, S. H., 1970, Biophys. J., 10:1127.

THE GENETIC ANTICODE: THE ROLE OF THERMAL PROTEINOIDS IN THE DEVELOPMENT OF AN HYPOTHESIS

James C. Lacey, Jr. and Dail W. Mullins, Jr.

Department of Biochemistry
University of Alabama
Birmingham, Alabama 35294

PREFATORY REMARKS BY J. LACEY, JR.

I had the good opportunity to work as a postdoctoral fellow with Sid Fox for two years (1969 and 1970). It was during this period that my already existing interest in, and fascination with, the origin of the genetic code became firmly embedded in my soul. The permanancy was established by two aspect of my experience at Miami that were due, in the main, to Sid Fox's exceptional scientific intuition. On the one hand, Sid was convinced that an understanding of the origin of the genetic code necessitated the use of a dynamic system which is synthesizing peptides. He and his coworkers were studying peptide formation from aminoacyl adenylates when I arrived there (Krampitz and Fox, 1969), and I became a part of that effort (Nakashima et al., 1970). Furthermore, as still another application of thermal proteinoids, Sid had already begun studying the question of their interaction with polynucleotides and had established that about 15 mole % lysine in the proteinoid was required to initiate their binding to polynucleotides (Waehneldt and Fox, 1968), and that lysine-rich and arginine-rich proteinoids had different preference among the homopolynucleotides (Yuki and Fox, 1969).

Since those days, I have continued on the genetic code problem, and as we recount the story it will become obvious that we are still following up leads developed as a result of the work with Sid. Our efforts are still delineated into the two broad aspects (amino acid-nucleotide interactions and the chemistry of activated aminoacyl species) that his intuition and expertise had told him were important.

Whatever merit is found in the story that follows, a lot of credit must go to Sid Fox for the initial push and direction he gave me, and I am pleased to express my appreciation for that and for his friendship, and to his lovely wife Raia, who extended to us, along with Sid, the greatest hospitality during our Miami stay.

THE PROBLEM

In the exciting period from 1961-1966, molecular biologists, principally Khorana, Ochoa, and Nirenberg, succeeded in deciphering the genetic coding system. Excellent accounts of this marvelous experimental achievement are given in books by Woese (1967) and by Ycas (1969). Further study through the years has shown that the coding system is essentially universal, with only minor exceptions being detected recently in mitochondria and chloroplasts (Macino et al., 1979; Barrell et al., 1979). But knowing what a thing is, and understanding why it is as it is, are two entirely different questions. We know with as great a certainty as is presently possible in science, that there are two codons, UUU and UUC, for phenylalanine. What we do not yet understand, however, is why these two particular codons were assigned to phenylalanine. Indeed why is the entire genetic code as it is? Is there some discernible molecular basis (e.g., selective interactions) for the codon assignments, or did they evolve along with the process of protein synthesis through stochastic routes which might not leave traces that we can presently elucidate by experiment? As will become obvious, we are proponents of the idea that there was a definite molecular basis for the code origin.

While the question can be readily stated in a general way, the specific problem can only be stated in the context of contemporary protein synthesis. A simplified flow diagram for protein synthesis is shown in Fig. 1. From the knowledge of the steps in the process one can state three general requirements for any process of protein synthesis:

1. Each amino acid must be activated;
2. While activated, the amino acids must be organized into predetermined sequences; and
3. While organized, the amino acids must be polymerized into peptides.

Step one in the contemporary process is carried out by the amino acyl-synthetases, which simultaneously recognize each amino acid, activate it, recognize the cognate tRNA and esterify the amino acid to the 3' end of the tRNA. Thus, in the contemporary process, activation and recognition are carried out simultaneously and prior to the appearance of the codonic message. Attachment of the amino acid to tRNA now allows steps 2 and 3, organization into defined sequences, by base pairing of the anticodon loop to each codon. After alignment on the ribosomae, peptide bond formation ensues.

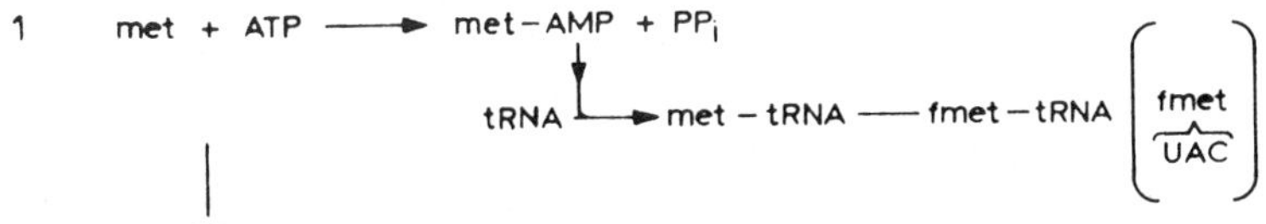

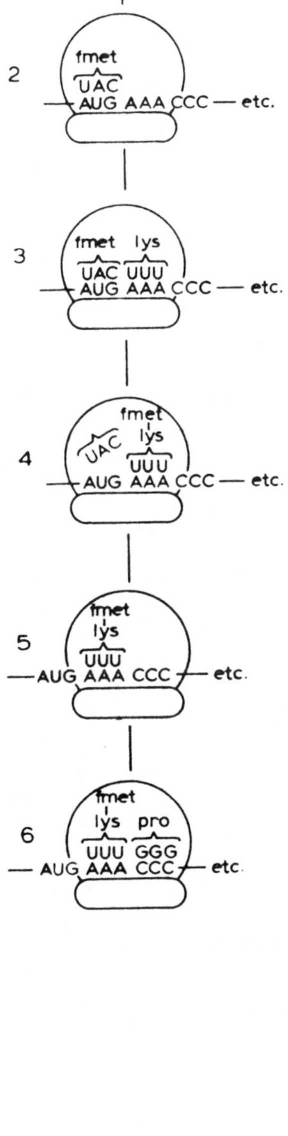

Fig. 1.
Highly simplified diagram of the process of translation (protein synthesis) which converts RNA nucleic acid language into protein language. 1. Each amino acid is activated by reaction with ATP forming a highly reactive aminoacyl adenylate anhydride (in this example methionyl adenylate anhydride). The amino acid is then passed to a special transfer RNA (tRNA) molecule which carries a decoding sequence for that particualar amino acid in its anticodon loop (-UAC in this example). This process of charging the amino acid on to its special tRNA is carried out by a special enzyme, methionyl-tRNA-synthetase. This enzyme recognizes the amino acid, ATP and the appropriate tRNA. In prokaryotes after the initiating amino acid, which is methionine, is attached to its tRNA, a formyl group is added to its α-amino group. The fromyl group is not added in eukaryotes. 2. The charged initiator tRNA then combines with the mRNA (utilizing the same base pairing specificities as in transcription and in forming the DNA double strand) in the peptidyl site on a ribosome to form an initiation complex. This initiation requires GTP, three protein initiation factors (IF-1, 2 and 3) and Mg^{2+}. 3. A second tRNA (here lys-tRNA) with its charged amino acid is brought into the amino acid site on the ribosome. This step requires GTP and two protein factors (EF-Tu and EF-Ts). 4. An enzyme (peptidyl transferase) in the larger part of the ribosome catalyzes the formation of the peptide bond, forming fmet-lys which is still attached to the lys transfer RNA. 5. An enzyme, EF-G, and GTP are required to translocate the mRNA over one triplet frame so that now the next codon CCC appears in the aminoacyl site and the fmet-lys-tRNA is now in the peptidyl site. Pro-tRNA now is bound to the CCC codon and we are ready to form the next peptide bond. This process is repeated over and over until a termination signal (UAA, UGA or UAG) is reached. The protein is then complete and the whole complex disassociates and the peptide hydrolyzes off the tRNA.

Since the amino-nucleotide relationships of the genetic code are not known to have significance outside the context of protein synthesis, it seems logical that whatever studies are carried out should simultaneously attempt to understand not only the origin of the coding mechanism, but also how it allowed the origin of a protein synthesizing system. This view of the problem has been espoused by Sid Fox and Tad Nakashima for many years, and has been favored by others, including Carl Woese. As our own thinking on the problem has matured through the years, we have also become convinced that the code and the process of protein synthesis must necessarily have coevolved (Lacey and Weber, 1977). Consequently, work on what might appear to be two separate problems actually concerns dual aspects of a single problem. Perhaps the most succinct statement of the problem can be asked as follows: "Is there a set of organizing principles which operated between amino acids and nucleotides, and which encourages the synthesis of selected peptides?" Based on our present information, we have generated the following working hypothesis as an answer to this question: genetic coding arose due to selective affinities between amino acids and their anticodonic nucleotides; this, in turn, allowed selected enhancement of reactions leading to peptide synthesis. By way of refinement of this statement, it seems likely that, while some genetic code assignments were made on the basis of direct affinities, others may have been assigned on a completely different basis.

The story that follows is essentially a recounting of the events and awareness that led us to the above hypothesis, with special notes on the role that Sid Fox played in the evolution of our working and ideas.

ACCUMULATION OF DATA AND IDEAS

It is probably an apt corollary to the scientific method that the less data one has on a particular problem, the wilder and more frequent are the speculations, especially when the problem is as fascinating as the origin of the genetic code. Our first work (Lacey and Pruitt, 1969) on the coding problem did consist of a fair amount of speculation, although arising out of some reasonably good data on the association of mononucleotides with poly-L-lysine. The results actually showed, as Woese et al. (1966) had found with poly-L-arginine, that the order of association of mononucleotides with basic polyamino acids was GMP>AMP>CMP>UMP. We concluded, based mainly on salt sensitivity, that the initial interaction between the nucleotides and polylysine was electrostatic in nature (negative nucleotides with positive polylysine), but that the above ordering was determined by the base stacking ability of the nucleotides themselves. Consequently, these data reflected mainly the self-ordering of the nucleotides, and did not directly relate to the coding problem. That first report also contained a model for reverse translation (i.e., using a peptide as a template for polynucleotide synthesis) and based on some simplified thermodynamic considerations, we proposed a selective association of the codonic nucleotides with their amino acids.

It was at this point in early 1969, that our collaboration with Sid Fox was begun. He and his coworker, Krampitz, were already studying the effect of homopolynucleotides on the formation of peptides from amino acyl adenylates made from mixtures of amino acids. The adenylate, being the cell's own activated intermediate in protein synthesis, was an obvious compound to use, and Sid's intuition was already leading us into the proper type of experiments involving peptide synthesis (i.e., kinetic studies instead of just static). Many experiments were done, but no definite and reproducible effect of polynucleotides on the composition of peptides from mixed amino acids was found. This introduction to the synthesis and properties of amino acyl adenylates, however, was to be important in our later work,

Sid was also aware that molecular interactions between amino acids and nucleotides must have played an important role in establishing the code, and he had us working on the interaction of thermal proteinoids with polynucleotides. He and Waehneldt (1968) had already shown that 15 mole % lys was required to initiate interactions of thermal proteinoids with polyribonucleotides, and that there were preferential interactions of Lys-rich or Arg-rich proteinoids with polypyrimidine and polypurines, respectively (Yuki and Fox, 1969). When such thermal proteinoids were mixed with polynucleotides under proper conditions, a second phase appeared in the form of fine droplets. One could monitor these interactions by turbidity or by centrifuging down the precipitate and analyzing it. The latter technique appeared to be the better one. In one of our early experiments we had prepared a Lys-rich proteinoid and were studying its interaction with the four homopolyribonucleotides. We were especially interested in the effect of added Mg^{++} because of the importance of Mg^{++} in maintaining ribosome structure. Those experiments (Fig. 2) (Fox et al., 1971) showed that with no Mg^{++} present the order of preference was poly U>poly C>poly A>poly G, but that as the Mg^{++} concentration increased, the amount of precipitate decreased with all polynucleotides. Presumably the positively-charged Mg^{++} was displacing the positively-charged Lys-rich proteinoid from the polynucleotide. These experiments unquestionably showed that at all Mg^{++} concentrations, poly U was preferentially precipitated with the Lys-rich proteinoid. To our knowledge, this experiment was the first to show a clear-cut preference of an amino acid (lys) for its anticodonic nucleotide, poly U. While we saw this preference as anticodonic, we had some reservations at overreacting to it because we felt that factors such as the rigidity of the poly G (for example) might have influenced its interaction with the proteinoid. Regardless, it remained a crucial experiment because it was so clear cut and our thoughts through the years would constantly hark back to it. It was soon after this experiment (1971) that Lacey left Miami and joined the faculty at the University of Alabama in Birmingham. Nevertheless, the communication and collaboration with Fox continued.

For a number of years, our work involved chemical studies,

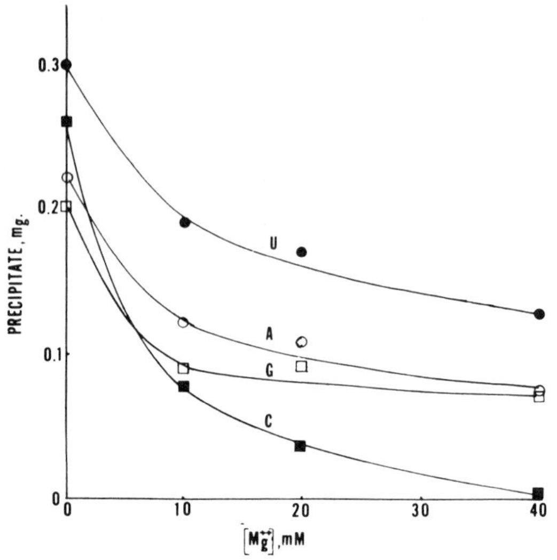

Fig. 2. Effect of [Mg^{++}] on the total wt. of precipitate from a mixture of thermal proteinoid 20Y (24.5 mole % lys) at 1.0 mg/ml concentration with polyribonucleotide (A, ○; U, ●; G, □; C, ■) at 0.1 μmole/ml on a residue basis in tris buffer, 0.05 M and pH 7.0 at room temperature (Fox et al., 1971).

i.e., aminoacyl transfer reactions and the problem of specificity was left hanging somewhat. It was in 1974, still remembering the Lys-poly U result and some additional preliminary experiments that also suggested and anticodonic preference of phe for poly A that Ania Mejido had done at Miami, that we went to the literature to look for additional data that might support the idea that there might be preferential association of amino acids with their anticodonic nucleotides. It was then that we found the data of Garel et al. (1973) who had determined the relative hydrophobicities of all the amino acids and nucleotides. Testing the question about the relationship between amino acids and anticodonic nucleotides, we plotted their data for the homocodonic amino acids (i.e., Phe, Pro, Gly, and Lys each has one homogeneous codon) versus the data for their anticodonic nucleotides (Fig. 3). An obvious direct relationship resulted. Furthermore, Art Weber (a former graduate student of Sid Fox's) taking the data for mononucleotides, constructed a three-dimensional plot in which one axis was the hydrophobicity of the anticodonic nucleotide and the other horizontal axis was the hydrophobicity of the second anticodonic nucleotide and the vertical axis was the hydrophobicity of the amino acid. He suggested that such a plot, in effect, estimated the hydrophobicity of the anticodonic dinucleotide; at least the plot did include the hydrophobicity of the two most important letters of the anticodon (i.e., the wobble nucleotide was not included).

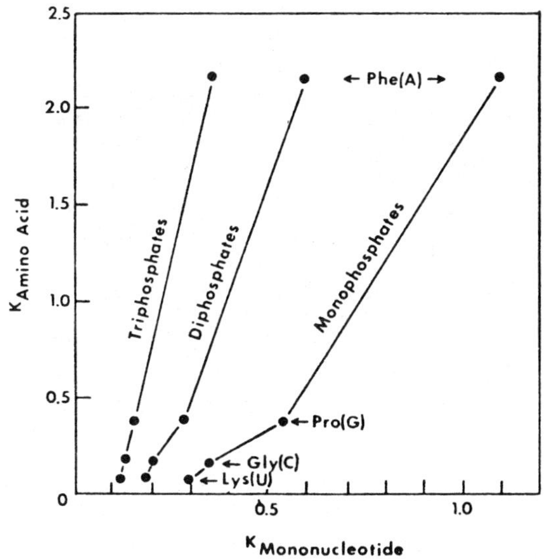

[1]lys, gly, pro, and phe each has one homogeneous codon

Fig. 3. Plot of relative hydrophobicity of the homocodonic amino acids versus that of their anticodon nucleotides (mono-, di-, and triphosphates), Data from Garel et al. (1973).

After constructing the 3-D plot, we could see that when viewed from a particular angle, most of the points fell in a curved plane. Fig. 4 is a photograph taken of that view. This rather remarkable outcome shows again a direct correlation of properties between all twenty amino acids and their anticodonic nucleotides. Points not falling on the curve are for Arg, Ser, Leu, and Trp. Interestingly, of these four, Arg, Ser, and Leu have six anticodons each. While one set CG (Arg), GA (Ser), and AA (leu) (5'-3') does not fit the correlation the other set CU (Arg), CU (Ser) and AG (Leu) does correlate. Trp is later shown not to correlate in other studies. The lack of correlation of these particular anticodons suggests that whereas the correlating ones might have been assigned on the basis of similarity of properties, the ones not correlating might have been assigned on some other evolutionary basis, e.g., mutation of an existing tRNA. There is even an obvious example of this in mitochondria in that Trp has taken over one of the terminator codons (Macino et al., 1979; Barrell et al., 1979). Ocher and amber mutations are obviously examples of just that same occurrence.

We then determined the relative hydrophilicities of the twenty amino acids and all sixteen possible dinucleotide monophosphates using high salt concentrations as eluents in paper chromatography

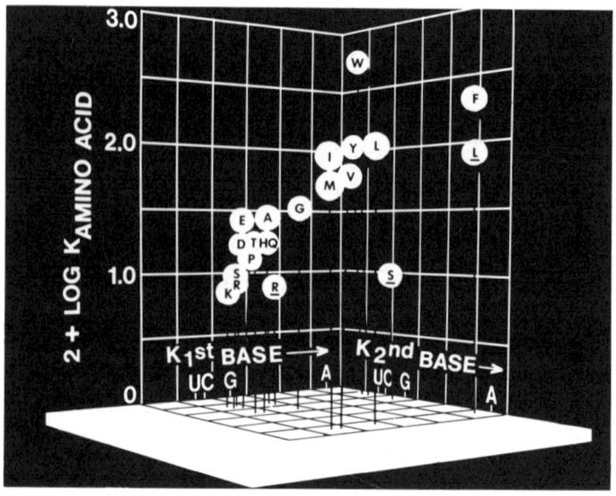

Fig. 4. Photograph of a three-dimensional plot of the partition coefficient data of Garel et al. (1973). The data are from mononucleotides and amino acids. The horizontal axis on the left represents the nucleotide at the 3' end of the anticodon, the horizontal axis on the right is the data for the nucleotide at the 5' end of the dinucleotide anticodon. The vertical axis is the partition coefficient data for the corresponding amino acids. The letters on the circles which are underlined represent amino acids that have a second set of anticodons which fit the correlation better.

(Weber and Lacey, 1978). In this case, the more hydrophilic a substance is, the better it competes with salt for water molecules and the further it moves up the paper. So the more hydrophilic a substance is, the higher the R_f. Again testing the anticodonic idea, we plotted (Fig. 5) the data for the homocodonic amino acids versus that for their anticodonic nucleotides, again obtaining a direct correlation, but naturally inverted from the hydrophobicity data. The plot of the data for all twenty amino acids and their anticodonic nucleotides gave Fig. 6. While this plot shows two sets of correlations, the dotted line being most of the hydrophilic amino acids, it is somehow less convincing than the data for the homocodonic amino acids. Nevertheless, John Jungck (1978) (another former graduate student of Sid Fox's) showed this correlation is statistically significant but in addition found that the properties of bulkiness and polarity also correlate anticodonically. Trp and Tyr did not fit these correlations at all. These two amino acids are the most hydrophobic of all but their anticodons are quite hydrophilic. We believe these assignments were made late in evolution not based on amino acid-anticodon relationships.

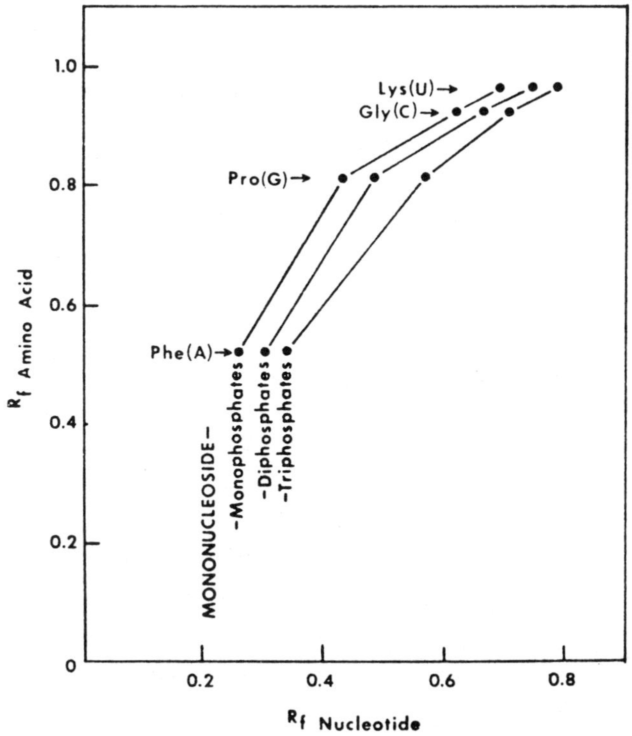

Fig. 5. R_f values of the homocodonic amino acids versus the R_f values of their anticodon mononucleotide mono-, di-, and triphosphates. These results were obtained with Whatmann 3 MM chromatography paper and 1.0 M ammonium acetate: saturated ammonium sulfate (10/90 v/v) at pH 7.0 and room temperature. All samples were run several times, each time with the four nucleotides on the same chromatogram. Although there was day-to-day variation in the absolute R_f values, the R_f ordering was invariably U>C>G>A.

While these correlations are indeed fascinating and seem to suggest that relationships between amino acids and anticodonic nucleotides are the basis for the codon assignments it would seem more satisfying if we had binding constants that told the same story. Inferring from the data that the correlation suggested preferential "binding" we first proposed that the data supported the idea of an anticodonic basis for the code in 1975 at a meeting in Czechoslavkia. This report was published as proceedings (Lacey and Weber, 1976). Dunnill (1976), Ralph (1968) as well as Nagyvary and Fendler (1974) had earlier suggested the anticodonic basis. While we are acknowledging the originators of various ideas and

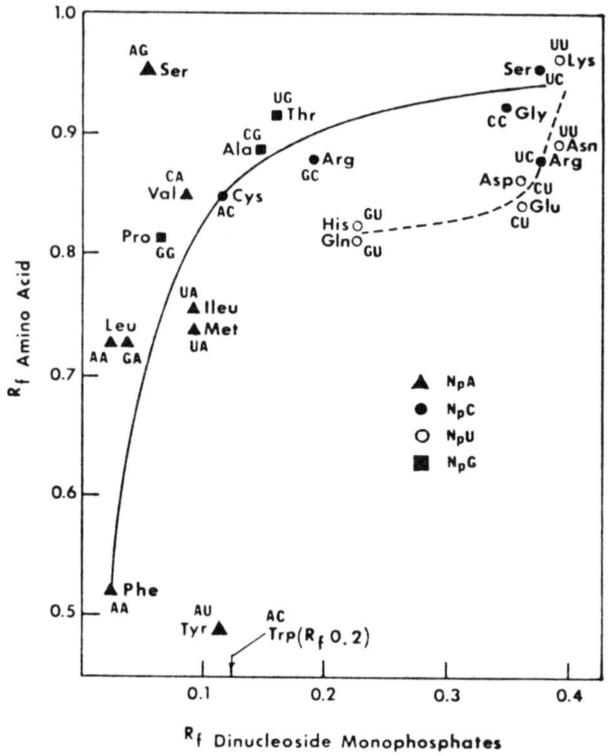

Fig. 6. R_f values (the solvent was 10 vol % saturated ammonium sulfate, pH 7.0) of amino acids plotted versus the R_f values of the dinucleoside monophosphates representing the first two letters (3' - 5' direction) of their anticodons.

awarenesses, Carl Woese et al. (1966) had been the first to point out the importance of hydrophobicity-hydrophilicity considerations. They did assign polarity values to the amino acids but didn't have data on the nucleotides.

Since the early data on Lys-rich proteinoids and poly U was the only set showing an anticodonic relationship, we decided to try making proteinoids rich in the other three homocodonic amino acids as well, but each containing enough lysine to initiate binding to the polynucleotides. Lacey returned to Miami in the summer of 1975 to do those experiments. Lys-rich (at several lysine concentrations) Gly-rich, Pro-rich and Phe-rich proteinoids were prepared, this time using ^{14}C amino acids. In each case, that amino acid was radioactive which was present in the greatest amounts (e.g., Gly-rich proteinoid contained ^{14}C gly). We could then quantitate the amount of proteinoid in the precipitate by scintillation counting and the polynucleotide by absorbance at 260 nm. The results of these experiments are in Fig. 7. These data again confirmed that Lys-rich proteinoids, especially in the presence of Mg^{++} do preferentially precipitate with poly U. Gly-rich preferentially precipitated with poly C and Pro-rich with poly G (Lacey et al., 1979). While those three are anticodonic preferences, Phe-rich preferentially

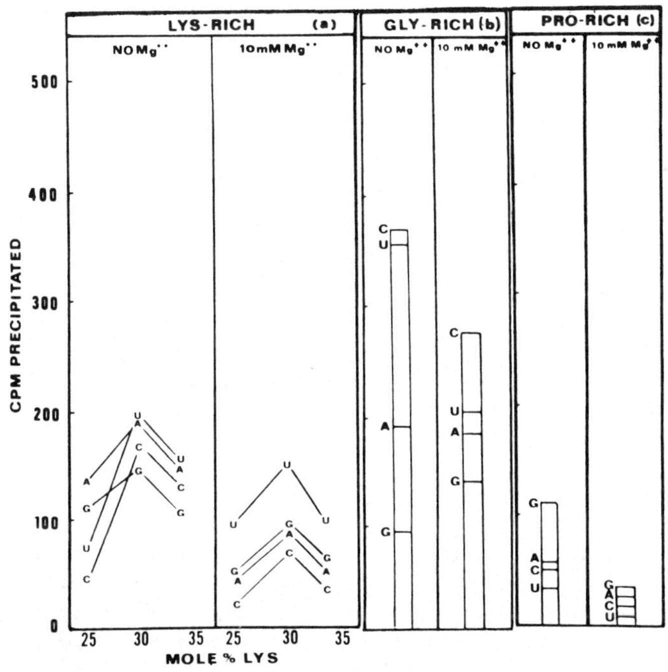

Fig. 7. Precipitation of radioactive thermal proteinoids with homopolyribonucleotides without and with 10 mM Mg^{++}. (a) Lys-rich at three Lys compositions; (b) Gly-rich; and (c) Pro-rich. One milliliter of solution of 0.18 μ mole/ml (mononucleotide residue basis) of polynucleotide in 0.045 tris buffer pH 7.0 was added to 1.0 ml of a solution of proteinoid 2mg.ml) also in 0.045 M tris buffer pH 7.0. After 10 min. the solution was filtered through Whatmann GF/C filter and washed with 5 ml of absolute ethanol. After drying, the filters were placed in borosilicate scintillation vials with 10 ml toluene containing 4 g/L PPO and 0.1 g/L POPOP. Triplicate samples were run on a Beckman LS-100 scintillation counter (Avg. standard deviation 10 cpm on triplicates). Values from blanks containing no polynucleotides were subtracted from sample values (Lacey et al., 1979).

precipitated with poly U, a codonic preference. This seeming anomaly was possibly explained by supposing that the Phe residues were folded inward as in a globular protein, leaving the more hydrophilic residues outside and causing the proteinoid to appear rich in hydrophilic amino acids. If this were so, we should be able to run the experiment in a solvent that would unfold the proteinoid and change the preference to poly A. This was found to be the case, if 25% ethanol was used as solvent, poly A was preferred. This work with the thermal proteinoids was actually done in 1975 before the correlations data. Publication of the proteinoid data was delayed hoping that we could somehow clarify the data from the Phe-rich proteinoid. In the meantime the correlations data were published in 1977 (Lacey and Weber) in a review and more formally in 1978 (Weber and Lacey) and the proteinoid data in 1979 (Lacey et al.). Actually, there was considerable overlap in the time frame of the correlations data and proteinoid data, but without question, it was the early results with the proteinoids which were accumulated at Miami with Sid Fox, which bent our path in the anticodonic direction and as Robert Frost might say "that has made all the difference."

But as Crick (1968) noted years ago, what we need are binding constants between amino acids and nucleotides. That has not been an easy matter, principally because the binding constants are quite low, usually being 10 M^{-1} or less. NMR seems the method of choice for such studies and Reuben and Polk (1980) had studied the interaction of all amino acids with 5'-AMP with an eye to explaining the code origins. They showed quite definitely that there are differences in the binding constants of all amino acids for 5'-AMP, the more hydrophobic ones having the higher binding constants. But we needed to resolve the preference of Phe for the various nucleotides. Fortunately, M. A. Khaled resolved the problem for us. Using the methyl ester of Phe along with the 5' mononucleotides, he observed that as the concentration of the methyl ester increased, the protons of the nucleotide bases moved upfield suggesting a ring-ring interaction as shown in Fig. 8 (Khaled et al., 1983). Using the changes noted, we then plotted that data in a Scatchard type plot yielding the straight lines in Fig. 9. The slopes of these lines are the binding constants. There seems little doubt from the binding constants in Table 1, that Phe prefers its anticodonic nucleotide, AMP. When one couples this NMR data with the thermal proteinoid precipitation data, we believe we can conclude qualitatively that the homocodonic amino acids do preferentially interact with their anticodonic amino acids. We need the binding constants of Pro, Gly, and Lys for the various nucleotides before we can make that a quantitative statement.

The early proteinoid-polynucleotide data coupled with the correlations data had led us to organize the anticodonic assignments as in Table 2 based principally on ordering as to hydrophobicity. This arrangement gives more reasonableness to the code, e.g., those

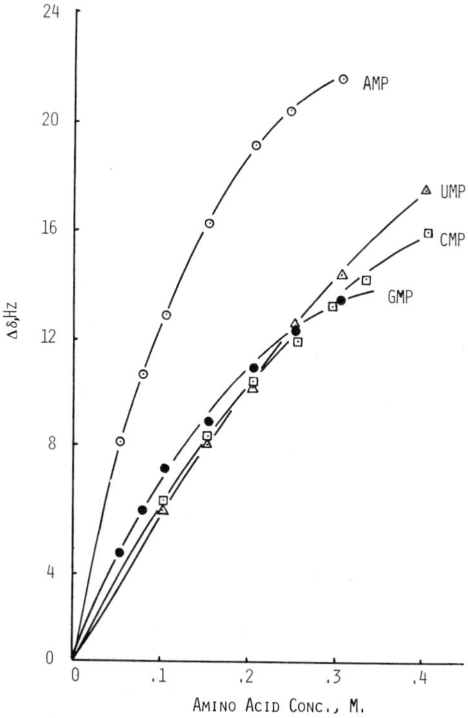

Fig. 8. Variation in chemical shift (ΔδHz) of nucleotide proton signals as the concentration of phe-O-Me is increased. Nicolet 300 MHz spectrometer, pD 7.0, 25°C operated in fourier pulse mode.

amino acids having the most hydrophobic nucleotide, A, as their middle (and most important) anticodonic letter are all hydrophobic amino acids. Those having U as their middle letter are all hydrophilic amino acids or derivatives of hydrophilic amino acids. Tyr is an exception, but as mentioned earlier Trp and Tyr do not fit any correlations and we believe were given assignments late in evolution.

We have continued the study of selective affinities using various methods including NMR, UV Spectroscopy and continue to find selective affinities of amino acids for their anticodonic nucleotides. Furthermore, we have continued our studies of the chemistry of reactions related to protein synthesis and the mechanisms are becoming clarified, but the most fascinating findings are that rates of reactions appear to be directly related to affinities between reacting molecules (Lacey and Mullins, 1981).

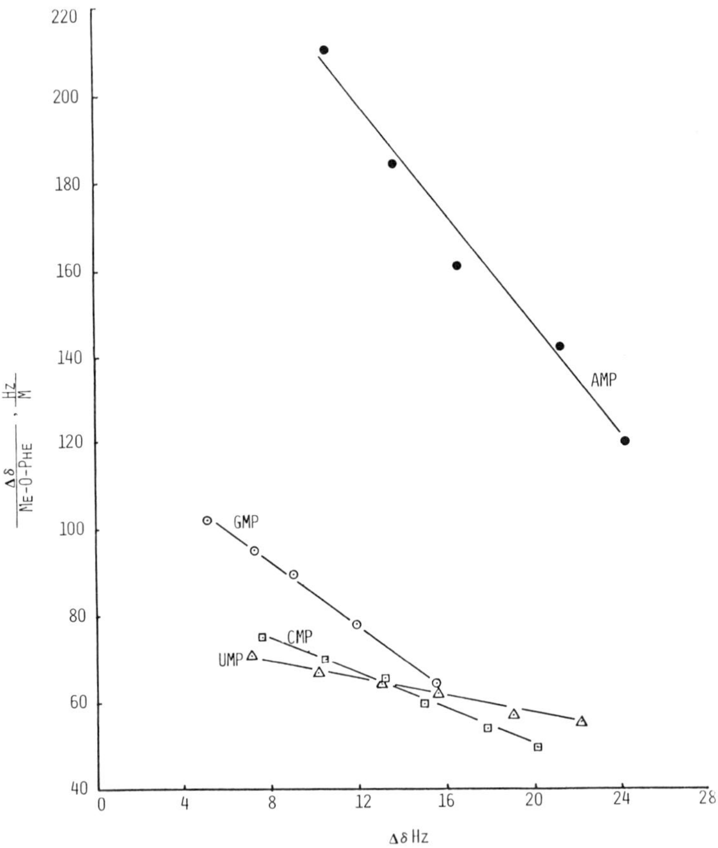

Fig. 9. Plot of change in chemical shift, ΔδHz, divided by the concentration of methyl ester vs. the change in chemical shift, ΔδHz. The slope of the lines calculated from the least squares best fit gives the binding constant in reciprocal moles, M^{-1}.

Table 1. Binding Constants of Me-O-Phe for Mononucleotides

Nucleotide	Binding Constant, M^{-1}
5'-AMP	6.6
5'-GMP	3.5
5'-CMP	2.1
5'-UMP	1.1

Table 2. The Genetic Anticode

Decreasing Hydrophobicity

3' end		middle letter A	G	C	U	5' end
A	Decreasing Hydrophobicity	phe	ser	cys	tyr	A
		phe	ser	cys	tyr	G
		leu	ser	try	term	C
		leu	ser	term	term	U
G		leu	pro	arg	his	A
		leu	pro	arg	his	G
		leu	pro	arg	gln	C
		leu	pro	arg	gln	U
C		val	ala	gly	asp	A
		val	ala	gly	asp	G
		val	ala	gly	glu	C
		val	ala	gly	glu	U
U		ile	thr	ser	asn	A
		ile	thr	ser	asn	G
		met	thr	arg	lys	C
		ile	thr	arg	lys	U

The genetic anticode is presented 3' → 5' so that the anticodons can be more easily imagined as base pairs of their codon equivalents, i.e., the codon-anticodon strands pair in an antiparallel fashion.

In summary we can say that Sid Fox's original emphasis on molecular interactions (using thermal proteinoids) and systems actively synthesizing peptides (either from ATP or starting with adenylates) has led us to continuing experiments in both areas which in turn have led us to our present working hypthesis: The genetic code is based on selective affinities between amino acids and their anticodonic nucleotides. These selective affinities in turn resulted in selective reactions leading to selective peptide synthesis. It was this combination of affinities and reactivities that allowed the origin of genetic coding and the process of protein synthesis.

ACKNOWLEDGMENT

This work has been supported throughout by the National Aeronautics and Space Administration (present grant NGR 01-010-001).

REFERENCES

Barrell, B. G., Bankier, A. T., and Drouin, J., 1979, Nature, 282: 189.

Crick, F. H. C., 1968, J. Mol. Biol., 38:367.

Dunnill, P., 1966, Nature, 210:268.

Fox, S. W., Nakashima, T., and Lacey, Jr., J. C., 1971, in: "Nucleic Acid Protein Interactions," D. W. Ribbon, and J. F. Woessner, eds., p. 113, North-Holland, Amsterdam.

Garel, J. P., Filliol, D., AND Mandel, P., 1973, J. Chromat., 78: 381.

Jungck, J. R., 1978, J. Mol. Evol., 11:211.

Khaled, H. A., Mullins, Jr., D. W., and Lacey, Jr., J. C., 1983, J. Mol. Evol. (in press).

Krampitz, G., and Fox, S. W., 1969, Proc. Nat. Acad. Sci. U.S., 62:399.

Lacey, Jr., J. C., and Pruitt, K. M., 1969, Nature, 223:799.

Lacey, Jr., J. C., Stephens, D. P., and Fox, S. W., 1979, BioSystems, 11:9.

Lacey, Jr., J. C., and Weber, A. L., 1976, in: "Protein Structure and Evolution," L. Fox, Z. Deyl, and A. Blazej, eds., p. 213, Marcel-Dekker, Basal.

Lacey, Jr., J. C., and Weber, A. L., 1977, Precamb. Res., 5:219.

Lacey, Jr., J. C., and Mullins, Jr., D. W., 1981, in: "Origins of Life," Y. Wolman, ed., p. 447, D. Reidel, Dordrecht, Holland.

Macino, G., Coruzzi, G., Nobrega, F. G., Li, M., and Tzagoloff, A., 1979, Proc. Nat. Acad. Sci. U.S., 76:3784.

Nagyvary, J., and Fendler, J., 1974, Origins Life, 5:357.

Nakashima, T., Lacey, Jr., J. C., Jungck, J. R., and Fox, S. W., 1970, Naturwiss., 57:67.

Ralph, R. K., 1968, Biochem. Biophys. Res. Comm., 33:213.

Reuben, H., and Polk, F., 1980, J. Mol. Evol., 15:103.

Waehneldt, T. B., and Fox, S. W., 1968, Biochim. Biophys. Acta, 160:239.

Weber, A. L., and Lacey, Jr., J. C., 1978, J. Mol. Evol., 11:199.

Woese, C., 1967, "The Genetic Code," Harper and Row, New York.

Woese, C. R., Dugre, P. H., Saxinger, W. C., and Dugre, S. A., 1966, Proc. Nat. Acad. Sci. U.S., 55:966.

Ycas, M., 1969, "The Biological Code," North-Holland, Amsterdam.

Yuki, A., and Fox, S. W., 1969, Biochem. Biophys. Res. Comm., 36:657.

RELATIONSHIP BETWEEN BIOLOGICAL INFORMATION AND THE ORIGIN OF LIFE

Barry Commoner

Center for the Biology of Natural Systems
Queens College, CUNY
Flushing, New York 11367

The question of how life arose on the Earth and the earliest stage of its evolution has been of considerable, but largely theoretical, interest. Theoretical considerations initiated by Haldane (1938) and Oparin (1938) have argued persuasively that the appearance of organic compounds preceded the first living thing, thus establishing the environmental milieu in which the latter emerged. There remained, however, the basic question of what changes within this primitive "organic soup" could lead to a system that possessed the fundamental properties which distinguish living systems from non-living ones: replication (with mutation at a low frequency), growth and metabolism.

The view that DNA is a "self-duplicating molecule"--i.e., that DNA contains within its molecular structure all the properties required for the production of a replica molecule--implies one answer to this question. It suggests that once even a single molecule of DNA appeared in the organic soup, it could replicate itself and establish this new "self-duplicating" entity as a growing component of the primitive milieu. Thus, according to Ohno (1970):

> Since the complementary replication of nucleic acid make use of certain structural characteristics inherent in the bases themselves, it is almost certain that reasonably accurate self-replication of polynucleotides is possible in the absence of enzymes. This means that self-replication of polynucleotides also began to occur in the primeval soup long before the emergence of the first life on this earth.

Moreover, since modern DNA theory also holds that its molecular structure governs that of RNA and of proteins, it would appear that such a primitive self-duplicating DNA system could accrete to itself systems which accomplished the synthesis of RNA and protein, thus achieving the attributes of a second fundamental biological property, growth. And finally, since such a system could produce proteins with specified amino acid sequences, which are a requirement for the catalytic properties of enzymes, it would possess as well the property of generating the specified, catalyzed system of chemical reactions that comprise metabolism.

There are two basic difficulties which militate against accepting this approach to the origin of the first living system. One difficulty is that the spontaneous appearance of the entire system, or even of a molecule capable of autonomously replicating its own molecular structure in the chaotic milieu of the "organic soup," is exceedingly improbable. Indeed, confronted with this improbability, Crick (1981) has suggested that the first life form arrived on Earth from some extra-terrestrial source.

A second difficulty is that the synthesis of the basic polymers--DNA, RNA, and protein--from their constituent monomers is an endergonic process which requires, at some point, an input of free energy. In living cells, this is accomplished by elevating the free energy content of the monomers by converting them to phosphorylated derivatives, such as nucleotides or amino-acyl-AMP. The latter are then able to polymerize by means of a condensation that requires no free energy input. In living cells, the free energy required for polymerization is generated by cellular oxidation-reduction processes; they yield "energy-rich" ATP, which is in turn capable of reacting with the monomeric constituents to produce equally "energy-rich" phosphorylated derivatives. Thus, ATP mediates the link between the exergonic cellular metabolic processes and the necessarily endergonic process that synthesize DNA, RNA, protein and, indeed, all cellular polymers from their constituent monomers.

These energetic relationships must have obtained as well in the original organic soup. Thus, if in fact the first step involved the replication of an original molecule of DNA, then the synthesis of the replica must have been based on "energy-rich" nucleotides. Similarly, the synthesis of primitive proteins must have involved nucleotide-amino acid precursors. Thus, the availability of "energy-rich" nucleotides appears to be a necessary condition for the polymerization of DNA, RNA, and protein, and therefore of the postulated "self-duplication" of DNA.

While these relations are common to the synthesis of nucleic acids and proteins, in one crucial respect they differ sharply. In protein synthesis the nucleotide serves as an "energy-rich" carrier of the amino acid monomer, and it is regenerated--albeit in a form which is at a lower energy level--during the polymerization process. Hence, as long as the organic soup includes processes which can once more raise the nucleotide to its high energy (for example, a photochemical process which derives the necessary energy from UV or visible solar energy), protein synthesis is compatible with ongoing polymer synthesis. These relationships may be symbolized by the following diagram in which ATP is used, generically, to signify an "energy-rich" nucleotide, and in which the intermediation of RNA in protein synthesis is omitted:

Hence, even if the organic soup contained a limited supply of nucleotides (which is likely given the probably slow rate of their formation from the soup's initial, relatively simple ingredients), protein synthesis can continue as long as the photochemical process regenerates ATP from ADP.

In contrast, the synthesis of nucleic acids does not permit the regeneration of ATP and the other "energy-rich" nucleotides essential for polymerization, *for these monomers are themselves incorporated into the polymer*. Thus, in this case:

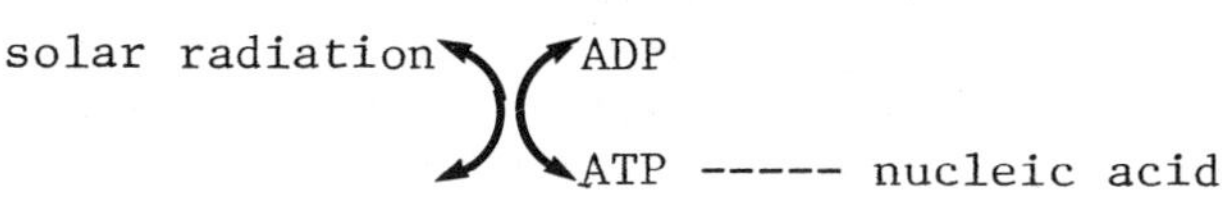

In effect, the synthesis of nucleic acid *sequesters* the nucleotides which are essential to carry the free energy needed to drive the polymerization of nucleic acids, proteins and indeed all other polymers. For that reason, in contrast to the synthesis of proteins, nucleic acid synthesis is a self-limiting process.

If we return now to a consideration of the conditions which mediated the formation of the ingredients of the organic soup, it becomes possible to evaluate the relative probabilities that DNA or protein represented the first-formed polymer. The classical Urey-Miller experiment has shown that amino acids are quite readily produced when a mixture of methane, ammonia, and water is exposed to radiant energy. Given the complexity of the nucleotide molecule, in comparison with amino acids, the formation of nucleotides would appear to be less probable than the formation of amino acids. It seems likely then, that in the period which preceded the first appearance of polymers the organic soup contained a significantly higher concentration of amino acids than nucleotides.

If, now, the next event was the appearance (by whatever means, including an extra-terrestrial intrusion) of a "self-duplicating" molecule of DNA, it would instigate a self-limiting process. As DNA replicated, it would remove free nucleotides from the soup, and, given the considerable stability of the polymer, bind them so tightly in its structure as to sharply reduce the already relatively low concentration of nucleotides. The net result would be the conversion of the soup's free nucleotides into DNA, removing the necessary condition for the further synthesis of all polymers, including DNA. Thus, such a process, if it did occur, would block the further events that are essential to the actual formation of a living system: the synthesis of protein and the continued synthesis of nucleic acids. On these grounds it seems reasonable to reject the view that the process which led to the emergence of a living system in the primitive organic soup was initiated by the appearance of DNA, or, for that matter, of RNA.

When such considerations are applied to the proposal that the life-initiating event was the appearance of protein, for example, by the mechanism proposed by Fox, in the organic soup, the results are very different. In contrast with nucleic acid synthesis, protein synthesis does not sequester nucleotides and therefore does not inhibit the essential energy-transfer mechanisms which they mediate. Moreover, again in contrast with nucleic acids, proteins as a generic group have catalytic properties--that is, some of them are enzymes. Consequently, there is inherent in the synthesis of proteins the possibility that some of them will contain an amino acid sequence capable of providing an active catalytic site capable of enhancing the metabolic processes that lead to polymer synthesis. And once enzymes capable of catalyzing anaerobic oxidation/reduction processes appeared, the free energy needed to produce "energy-rich" nucleotides could be obtained from the soup's store of reduced compounds (e.g., sugars) instead of through the random occurrence of photochemical conversions.

In sum, while nucleic acid synthesis tends to inhibit the further synthesis of the essential biopolymers, protein synthesis is capable of accelerating these processes. On these grounds, the initial synthesis of protein appears to be compatible with the further steps that could lead to the formation of a living system, whereas the initial synthesis of DNA (or of RNA) is not.

It should also be noted that these relationships--i.e., the participation of nucleotides in the enzymatic processes that derive free energy from reduced metabolic substrates and transfer it to the free energy-dependent polymer synthesis--are ubiquitous in present-day life. Accordingly, they are to be regarded as primitive and therefore closely linked with the origin of life.

The foregoing considerations suggest the following sequence of early events in the origin of life:

1. Synthesis of proteins with varied (random) amino acid sequences (given that the sequence-specificity now imposed by the structure of nucleic acid--i.e., the "genetic code"--was absent), with the required free energy available from nucleotides, for example, is the form of amino acyl derivatives;
2. The occurrence among the proteins of structures that catalyze oxidation/reduction of soup constituents (e.g., sugars) and the resultant non-photochemical formation of "energy-rich" nucleotides;
3. Synthesis of nucleic acids initially with random nucleotide sequences, and eventually with sequences specified by a nucleic acid template;
4. Establishment of the relationships among the DNA template, RNA, and the synthesis of specific proteins (i.e., containing specified amino acid sequences) characteristic of present living cells.

These steps, together with processes that gave rise to membrane structures that established a boundary between the foregoing processes and the surrounding environment, can be reasonably regarded as leading to the origin of the first living cells.

Step (2) encompasses a primitive form of the exergonic metabolic processes which in all living cells support the synthesis of biopolymers. In this primitive system, there is a positive feedback between the exergonic metabolic process and protein synthesis, i.e., as the synthesis of proteins, including metabolic enzymes, progresses, the exergonic process and therefore protein synthesis will tend to accelerate. On the other hand, there is a negative feedback between the exergonic process and nucleic acid synthesis. Since nucleic acid sequesters the nucleotides essential to catalysis of polymer synthesis, it tends to decelerate the exergonic process, and nucleic acid synthesis as well.

As indicated earlier, one of the requirements for a primitive living system is that it should persist and provide the starting point for an ongoing evolutionary process. This requires that the various chemical processes that constitute a living system should be in equilibrium. From the foregoing considerations, it would appear that the opposite relations of nucleic acid and protein synthesis to exergonic metabolism represents an important aspect of such an equilibrium.

All of these considerations are, of course, speculative and not directly subject to experimental test. There are no fossil remnants of the primitive organic soup and, while test-tube versions of the processes involved in it, such as the Miller/Urey experiment, or those of Sidney Fox, provide important data, they are only fragmentary evidence of the process as a whole. However, at least one of the novel concepts that plays a role in these considerations--nucleotide sequestration--is subject to test in present living systems. The concept suggests that one of the roles that nucleic acids could play in living cells is that their synthesis withdraws free nucleotides from participation in various metabolic processes, thereby exerting an influence on the rate of these processes.

The basic relevant data are the following:

1. Nucleic acids are synthesized from triphosphonucleotides, which, in the process, lose their terminal pyrophosphate groups and become the residues that make up the covalent backbone of nucleic acids.

2. ATP is a crucial carrier of phosphate and pyrophosphate groups in a series of energetic processes. The free energy derived from carbohydrate oxidation/reduction, for example, is incorporated in these groups and is the source of the free energy required to drive the endergonic processes which synthesize biopolymers.

3. A series of enzyme processes transfer the ATP phosphate group, and its inherent free energy content, to the diphospho derivatives of the three other nucleotides characteristic of RNA, forming UTP, CTP, and GTP. Similar reactions form the triphospho desoxynucleotides characteristic of DNA: dATP, dCTP, dGTP, and dTTP. UTP and ATP provide the "high-energy" phosphate groups required for the endergonic synthesis of polysaccharides. GTP and ATP perform this function in the synthesis of proteins, and CTP and ATP in the synthesis of lipids. Thus the energy required for the synthesis of RNA and DNA is carried into the process in the "high-energy" phosphate groups of their respective nucleotides.

4. The foregoing relationships between exergonic metabolism, which is catalyzed by free nucleotides, and the synthesis of biopolymers establishes two feedback processes, one positive and the other negative. Since protein synthesis regenerates ADP and produces, among others, the enzymes of oxidative metabolism, it is related to that process by a positive feedback. In contrast, nucleic acid synthesis sequesters free nucleotides, and since these are in equilibrium with the free nucleotides involved in oxidative metabolism, nucleic acid synthesis has a negative feedback effect on that process. The synthesis of polysaccharides (and lipids), in which ADP is regenerated, has no feedback effect on oxidative metabolism. These relations are diagrammed in Fig. 1.

5. The amount of DNA synthesized per cell is a fixed species-specific characteristic; the amount of DNA per cell in a given species is therefore an inherited characteristic. This is a consequence of the mechanism of DNA replication in which a new strand of DNA is synthesized on the template provided by pre-existing

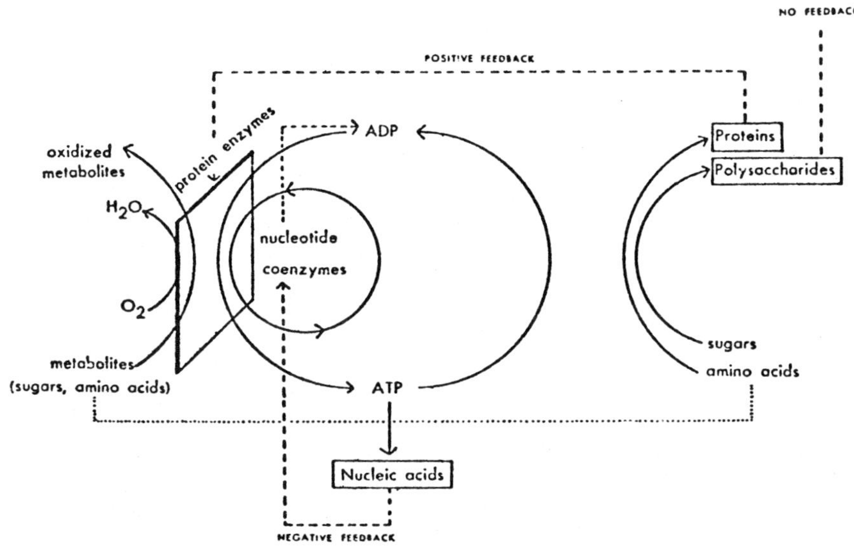

Fig. 1. Diagram to illustrate interrelationships between oxidative metabolism, and the synthesis of polysaccharides, proteins, and nucleic acids respectively. The left-hand part of the diagram symbolizes the activity of the enzyme system of oxidative metabolism, enzymes, generically, being represented by the vertical surface. Synthesis of polysaccharides and proteins is coupled to oxidative metabolism by the ADP/ATP system; in both case, synthesis of the polymer requires ATP, but ADP is generated in the process. Since ADP concentration regulated the rate of oxidative metabolism, which in turn determines the ATP level, the ADP/ATP balance is maintained when these polymers are synthesized. Protein synthesis has a positive feedback relationship to oxidative metabolism, since the enzymes necessary for the latter are proteins. In contrast, nucleic acid synthesis not only depends on ATP (and other nucleotides), but also results in the sequestration of the entire nucleotide residue. Hence, oxidative metabolism is coupled to nucleic acid synthesis by a negative feedback relationship. Therefore, synthesis of a very stable nucleic acid, especially DNA, may be expected to lower the levels of free nucleotides, and with it the rate of oxidative metabolism characteristic of a specific cell.

strands of DNA. During mitosis, the DNA content of the cell is exactly doubled and then evenly divided between the two daughter cells.

6. The observations summarized in (4) and (5) above comprise a mechanism of inheritance which is quite distinct from the conventional template mechanism in which DNA nucleotide sequence, through RNA intermediates, governs protein amino acid sequence, thereby determines enzyme specificity and the presence in the cell of a particular enzymatic product. In contrast, the second mechanism of inheritance is based on nucleotide sequestration. In this mechanism, the amount of DNA synthesized per cell, which is inherited, governs the availability of free nucleotides, which in turn regulates the cellular rate of oxidative metabolism. In other words, the mere synthesis of DNA, without regard to nucleotide sequence, can be expected to govern the inheritance of those cellular characteristics which are mediated by the intracellular concentration of free nucleotides. The latter includes such very general inherited characteristics as the rate of oxidative metabolism and cell size (which is generally inversely related to metabolic rate).

These considerations, which were originally advanced by the author in 1964 (Commoner, 1964), are subject to the test of experimental evidence. To begin with, they help to explain a number of observations regarding cellular DNA content which do not conform to expectations based on the conventional template theory. It has been known for some time that in eukaryotic organisms a considerable part of the chromosomal DNA, generally classified as heterochromatin, does not carry typical Mendelian genes, but does have certain generalized inherited effects--on metabolic rate and cell size, for example--which are compatible with the nucleotide sequestration theory. Modern techniques of DNA analysis have confirmed that eukaryotic cells contain nucleotide sequences which are repeated 100 to 10,000 times (Lehninger, 1975a). Since conventional genetic theory requires the presence of only one copy of each gene, the redundant copies have no "informational" function. Such redundant sequences and the simple repetitive sequences of satellite DNA, which is regarded as an "informationless spacer," constitute more than half the DNA content of eukaryotic cells (Lehninger, 1975a). If such DNA was in fact functionless in inheritance, there is no reason to expect it to be retained, as it has been, during the course of evolution. The nucleotide sequestration hypothesis suggests that such "informationless" DNA does in fact carry out an important function in inheritance, by determining species characteristics such as metabolic rate and cell size, and thus helps to explain the retention, in evolution, of DNA that does not participate in the formation of template genes.

Another major discrepancy between the data on cellular DNA content and the template theory occurs in relation to evolutionary development. If genes based on a DNA segment with a characteristic nucleotide sequence carry all inherited information, then the amount of DNA per cell is expected "to be in proportion to the complexity of the cell and thus to the amount of genetic information it contains." (Lehninger, 1975b) This leads to the further expectation that the cellular DNA content will increase with a progressively higher position on the evolutionary scale.

This expectation is sharply contradicted by the actual data. For example, most amphibia contain considerably more DNA per cell than mammals. Indeed, within Amphibia there are differences in cellular DNA content very much larger than the difference between man (6.8 pgm/cell) and molluscs (about 0.7 pgm/cell). Thus, anuran amphibia such as Rana and Bufo have cellular DNA contents of 15 and 7 pgm/cell respectively, whereas urodeles such as Amphiuma and Necturus have cellular DNA contents that range from 48-168 pgm/cell.

While such gross differences are inexplicable in terms of conventional template theory, they conform quite well to expectations based on the nucleotide sequestration theory. Anuran amphibia undergo metamorphosis, in which the respiratory rate of the larval form increases sharply to that characteristic of the adult. However, urodeles are characterized by the phenomenon of neoteny in which metamorphosis and its attendant increase in respiratory rate does not occur. In neotenous species, such as Amphiuma (168 pgm/cell), the adult is simply an enlarged larval form, and retains the latter's very low metabolic rate; the adult is characteristically sluggish in behavior. Another exceptional species--the lungfish, which despite its primitiveness has a cellular DNA content of 220 pgm/cell--also conforms to this expectation, for the lungfish has an exceptionally low metabolic rate and a sluggish behavior. Thus, differences in metabolic rate appear to be better explanations than position on the evolutionary scale for such differences in cellular DNA content.

When the nucleotide sequestration theory was first advanced, an effort was made to test it by examining the relation between cellular DNA content and the characteristic metabolic rates of different organisms. Considerations advanced earlier by Krebs (1950) suggest that such comparisons of metabolic rate ought to be based on the rate of respiration of a particular tissue common to different species. Erythrocytes provide a convenient basis for this comparison. As reported earlier (Commoner, 1964), from the relatively few reports in the literature of erythrocyte rates of respiration for species of known cellular DNA content, it appeared that the two values were, in fact, inversely related. However, since the available values comprised only three species of birds, three reptiles, and four mammals, this relationship was suggestive rather than conclusive.

In 1968, in collaboration with Dr. James M. Bennett, a series of analyses of both cellular DNA content and erythrocyte respiration rate were carried out on 35 species of wild and domesticated reptiles, birds, and mammals (most of them wild). The results, originally published in Dr. Bennett's Ph.D. thesis (Bennett, 1968), are summarized in what follows.

In each case, blood was obtained by heart puncture and the erythrocytes were separated by a series of centrifugations. A measured volume of the washed erythrocytes (which was also counted) was suspended in Krebs (1950) medium and the respiratory rate determined at a series of temperatures in a Warburg respirometer. An aliquot of the washed, counted erythrocytes was also used to isolate nuclei from which DNA was extracted and determined by the diphenylalanine procedure. (In the case of mammalian species the nuclei were isolated from liver, kidney, or spleen tissue.)

The results are summarized in Fig. 2. It is evident that the expected inverse relationship between cellular DNA content and erythrocyte respiratory rate does in fact occur in the case of mammals and reptiles. Mammalian DNA values range from 5.0 to 7.7 pgm/cell, and erythrocyte respiratory rates vary from 0.14 to 0.18 μℓ/mg/hr. Reptilian DNA values range from 2.75 to 5,1 pgm/cell, and erythrocyte respiratory rates vary from 0.14 to 0.4 μℓ/mg/hr. Both relations are linear, but the slope for reptile values is three times that of the mammalian slope.

The bird values represent an apparent exception in that erythrocyte respiratory rate appears to be largely independent of DNA content, which varies only from 1.5 to 3.1 pgm/cell. However, this relationship in fact serves to further confirm the nucleotide sequestration hypothesis. This theory is based on the mechanism of oxidative phosphorylation process originally described by Lehninger, in which the energy released from reduced metabolites by DPN/DPNH is taken up by the conversion of ADP to ATP. If the oxidation process is uncoupled from the phosphorylation process (for example, by dinitrophenol and other drugs), the energy is released as heat; in effect, coupling suppresses oxidative heat production. Birds maintain a particularly high body temperature, so that a relatively large proportion of the energy derived from respiration must be released as heat rather than incorporated in ATP. This suggests as well a considerable degree of uncoupling, and consequently a weaker relation between DNA synthesis and respiratory rate.

One of the most revealing aspects of the relationships shown in Fig. 2 is that they provide, for the first time, an explanation of the variation of DNA/cell within each of the three phyla. As noted earlier, the template theory implies that those organisms that are more advanced on the evolutionary scale and are therefore more complex ought to contain more DNA per cell than more primitive, less complex organisms. However, this relationship holds only with respect to that part of the DNA which in fact serves as conventional Mendelian genes. It does not apply to that part of the cell's

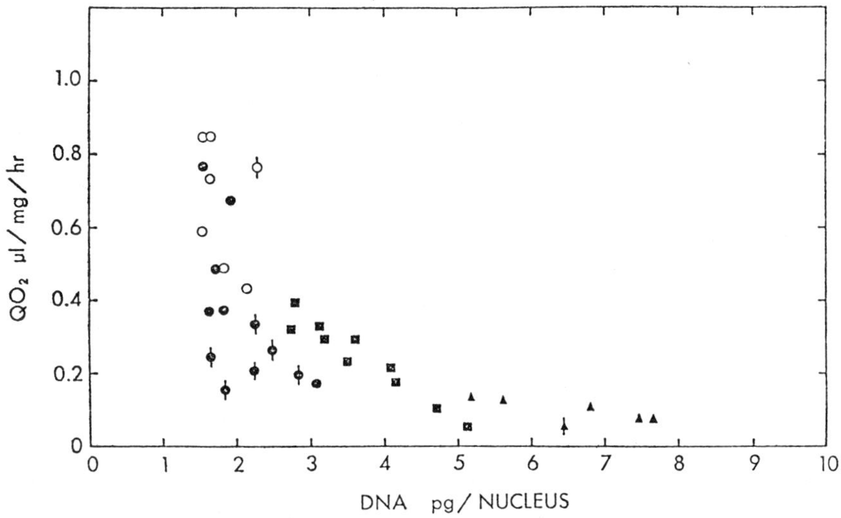

Fig. 2. DNA content per cell and erythrocyte rate of oxygen consumption for various species of mammals (triangles); reptiles (squares); nonpasserine birds (filled circles); passerine birds (open circles). A vertical line through a point signifies a domesticated species; the indicated domesticated mammal is the laboratory rat.

total DNA that does not carry conventional genetic information and is concerned only with nucleotide sequestration. If DNA content is regarded as solely a reflection of Mendelian genetic content, then it is difficult to explain why the alligator has a cellular DNA content (5.11 pgm/cell) about equal to that of Peromyscus (a wild mouse) (5.16 pgm/cell). However, when respiratory rate is taken into account, this contradiction is resolved, for the alligator has the lowest respiratory rate of the reptiles, while Peromyscus has the highest respiratory rate of the mammals. Thus, the alligator has a relatively large amount of non-template DNA while Peromyscus has a relatively small amount; hence, the alligator must have the smaller amount of template DNA, as expected.

If the regression lines for reptiles and mammals are extrapolated to zero rate of respiration, it appears that the maxmimum possible DNA content per cell is about 5.2 pgm for reptiles and about 9 pgm for mammals. It is of interest to note (Rendel and Kellerman, 1955), in this connection, that the largest cellular DNA content reported for a mammal is, in fact, 9.3 pgm/cell (for the Bandicoot, Parameles Nasuta, an Australian marsupial). This relationship suggests, too, that if the effect of respiratory rate on DNA content is taken into account, then as expected, the more advanced group, mammals, has a larger cellular DNA content than the less advanced group, reptiles.

These data seem to confirm the conclusion reached on speculative considerations of the relationships among energetic matabolism, polymer synthesis and replication. Given that energetic metabolism is a necessary prerequisite for polymer synthesis and given its opposite relations to the synthesis of protein and nucleic acid, it appears that protein synthesis, rather than nucleic acid synthesis or replication, is an initial step in the origin of life. In this sense, these considerations support the conclusion, which derives from the work of Sidney Fox, that structures based on protein played an early role in the origin of life.

REFERENCES

Bennett, J. M., 1968, "Comparative Studies of Cellular Metabolic Rate and DNA Content of Birds, Reptiles and Mammals," Doctoral Dissertation, Washington University, St. Louis.

Commoner, B., 1964, Nature, 206:486; Amer. Sci., 52:365.

Crick, F., 1981, "Life Itself, Its Origin and Nature," Simon and Schuster, New York.

Haldane, J. B. S., 1938, "Science and Life," Macmillan, New York.

Krebs, H., 1950, Biochim. Biophys. Acta, 4:249.

Lehninger, A. L., 1975, "Biochemistry," 2nd edn., p. 884, Worth, New York.

Lehninger, A. L., 1975a, "Biochemistry," 2nd edn., p. 884, Worth, New York.

Ohno, S., 1970, "Evolution by Gene Duplication," p. 6, Springer, Berlin.

Oparin, A. I., 1938, "The Origin of Life," Macmillan, New York.

Rendel, J. M., and Kellerman, G. M., 1955, Nature, 186:829.

THE STEREOCHEMICAL APPROACH TO STUDIES OF THE ORIGIN OF THE GENETIC CODE

Cyril Ponnamperuma and Mitchell K. Hobish

Laboratory of Chemical Evolution
University of Maryland
College Park, Maryland 28742

Since Wohler's synthesis of urea in the early nineteenth century, organic molecules have been displaced from their previously exalted position as the result of some special vitalistic process, and have taken their rightful place in the toolbox of the chemist and biologist. In studies of the origin of life, it becomes of paramount importance to determine the molecular parameters which define living systems from nonliving ones.

One such defining parameter is the ability of living systems to convey complex information to succeeding generations. Years of research have been dedicated to elucidation of the steps necessary for faithful transmission of hereditary characters in the process of cell division.

A major step in our understanding of this intricate process of information transfer came in the mid-1940s, with the experiments of Hershey and Chase which demonstrated that DNA, not protein, is the chemical entity which carries the genetic information. Subsequent work by many investigators led to the elucidation of the structure of DNA by Watson and Crick (1953), with its concomitant implications concerning the mode of its own replication. Next to be determined were the processes whereby the information coded in DNA as a linear array of bases is transmitted to the proteins, which are linear arrays of amino acids. However, it should be noted that a significant milestone in this research was provided by the work of Nirenberg and Matthei (1961). Their data demonstrated the basis for the transfer of information, and led to a description of the nature of the early conceptions of the "genetic code": sets of triplets of nucleotides along the nucleic acids (codons) correspond to one amino acid in the protein molecule.

While we understand the process of protein synthesis in great detail, there are still several questions that remain unanswered. Perhaps the most basic is: What is the nature of the "genetic code" that results in such specificity, with nonrandom combinations of three nucleotides coding for specific amino acids? Several authors have presented theories and data in support of stochastic models, while others support models that account for the observed specificity on physico-chemical bases.

There are, however, data available which makes possible assessment of these models. Several groups have demonstrated that there are indeed specific, measurable associations between the code carriers and the code products. These data provide strong support for the stereochemical basis of the genetic code. In the following pages, we will describe what is currently known about the genetic code, and summarize some of the data which bear on this issue.

NATURE OF THE PRESENT-DAY GENETIC CODE

The main features of the present-day genetic code may be summarized as follows (Baricelli, 1977; Crick, 1966; 1968; Epstein, 1966).

1. The genetic code mediates information transfer between nucleic acids and proteins with an alphabet of 4 bases and 20 amino acids.

2. Each codon is a triplet of bases, as defined by the sequence found on mRNA. In this light, the information-containing strand of DNA transcribed into that RNA stores the information as the anti-codon sequence.

3. The list of all possible triplet combinations, of which there are 64 codes, in a nonrandom way, for the twenty amino acids. These associations may be described as follows:

3.1. XYC and XYU always code the same amino acid.
3.2. XYA and XYG usually code the same amino acid. There are exceptions. For example, methionine and tryptophan do not conform to this paradigm.
3.3. XY* generally codes a specific amino acid, where "*" means any of the four bases may occupy that position.
3.4. UXY and CXY code for one amino acid, leucine.
3.5. CXY and AXY code for one amino acid, arginine.
3.6. In general, triplet combinations which code for single amino acids start with the same two bases. Exceptions are arginine, coded for by CG* and AG(A or G), leucine, CU* and UU(A or G), and serine, UC* and AG(U or A).

3.7. In cases where G's or C's are the first two bases, all codons code for the same amino acid. This is the basis of the so-called "wobble hypothesis," which proposes that the third base in the triplet is not as important as the first two.

4. All of the above demonstrates that the code is degenerate. This degeneracy, however, is not uniform (Table 1).

5. Allocation of triplet codes to amino acids is nonrandom.

5.1. all XUY codons code for hydrophobic amino acids;
5.2. polar amino acids are coded for by the more hydrophilic codons;
5.3. codons with a central pyrimidine tend to code for hydrophobic amino acids. Codons with a central purine tend to code for the hydrophilic amino acids.

6. The code is without punctuation; there are no commas between codons to separate their codes. One implication of this finding is that a proper reading frame must be established at the onset of translation. The "start" codon, AUG, establishes this reading frame.

7. The code is "universal," in that, with a few modifications, this genetic code is found in all terrestrial organisms.

With all that we have discovered about the nature of the genetic code, we have yet to determine its origin. As mentioned earlier, theories abound which purport to describe these origins. We shall now describe several of these models, many of which posit a probabilistic origin, and lay the groundwork for a discussion of physicochemical models.

Table 1. Degeneracy of the Genetic Code

Amino Acid	Degeneracy
LEU, SER, ARG, VAL, THR, ALA	6
PRO, GLY	4
ILEU	3
TYR, GLU, ASP, LYS, GLN, ASN, HIS, PHE	2
TRP, MET	1

MODELS FOR THE ORIGIN OF THE GENETIC CODE

The Frozen Accident Theory

Several stochastic models have been proposed. Their common feature is that development of the genetic code was the result of a series of interactions which were random, and that reinforcement through function evolving over geological time led to the current code. Francis Crick (1967) has been the leading exponent of this class of hypothesis, presenting the basic idea as early as 1967. His approach is perhaps the most probabilistic of the stochastic models, in that it suggests that the existing coding system was fixed at some early point in evolution, and remained so defined because any change in codonic assignment would affect the sequence of so many highly-evolved proteins that lethality would be unavoidable (Hoffman, 1974). One conclusion to be drawn from such a situation would be that the original assignment of codons to amino acids was just a matter of chance. A major criticism of this approach is that it alone cannot explain the exclusion of nonprotein amino acids (Rohlfing and Saunders, 1978).

Mikelsaar (1978) compiled data on the abundances of amino acids in vertebrates, microorganisms, thermal proteinoids, and products of simulated prebiotic syntheses. His results led him to suggest that the structure of the early genetic code was determined solely by the ratios of the amino acids then present. As with so many of the nonstructural origin theories, it suffers from being experimentally untestable.

Vocabulary-Expansion Theory

There are two paradigms employed in support of this theory. The first is the stipulation that the sixty-four triplet codons initially coded for only a small number of amino acids. Any new amino acids entering the scheme would therefore, of necessity, share codons with existing amino acids.

The second variation on this theme assumes initial specificity for a limited number of amino acids, mediated by a triplet code where fewer than the three positions have informational content (Jukes, 1967; Hartman, 1978). A result of these assumptions is that proteins coded for in this fashion were limited in compositional and functional complexity. Expansion of the vocabulary to yield more capable proteins occurred when new but related amino acids began to use a codon. Some of the original codons would retain their specificity; the expansion occurs by addition of an additional informational position in the triplet. The problem with this approach is that it appears to have, at its basis, some inherent specificity, such that the primordial genetic code allowed for the pairings that it has. The stipulation that vocabulary expansion would add an informational position to the codon triplet implies that there is something about the association between the bases and the amino acid being coded for. This brings this variation of the vocabulary expansion theory into the structural origins camp. It therefore has the advantage of being testable experimentally.

The Ambiguity Reduction Theory

The basis of this approach is the assumption that there arose classes of related codons which coded for classes of related amino acids. This would result in highly ambiguous proteins, which contained any member of a class of amino acids in the position coded for by the appropriate codon. Gradual reduction of the ambiguity may have been mediated by additional informational position within the triplet when the ability to recognize specific amino acids became a selective advantage. The stochastic origins of this theory have been discussed (Fitch, 1966). The degree of specificity initially required is smaller than that required by the previously-described theories. It may be coupled with direct interaction theories (see below) to account for the exclusion of nonprotein amino acids. Cell lines using nonprotein amino acids were likely selected against, since lines using protein amino acids had demonstrated superiority, as witnessed by their existence in present-day proteins. Support for this approach came from the work of Jukes (1974), who stated that arginine was a recent recruit to the roster of protein amino acids, and that it may have replaced ornithine. The high pKa of arginine relative to ornithine gave it greater functional variability. Such an explanation could also be used to explain the use of methionine in contemporary proteins rather than the structurally similar norleucine (Smith et al., 1968; Dickerson and Geis, 1969). Methionine has a hydrophobic portion, but also has the ability to form bonds with metal ligands. In addition, it can function as a donor of methyl groups.

The Lethal Mutation Theory

The theory requires an early specificity of most codons for amino acids. The codon assignments were such that a single base change probably resulted in insertion of the same or a similar amino acid. Thus, lethal mutations were minimized (Sonneborn, 1965).

The Co-Evolution Model

Starting with the five most easily synthesized amino acids (GLU, ASP, ALA, SER, GLY), and assuming the equivalent of a precursor-product relationship, an evolutionary scheme may be mapped out (Wong, 1975, 1976; Wong and Bronskill, 1979). This scheme assumes a single base change per step, and results in the generation of the genetic code for all twenty protein amino acids. Thus the genetic code evolved in tandem with the alphabet of amino acids used to form the proteins. An attractive feature of this model is that the amino acids it stipulates as the Founding Five are just those which were likely the most abundant on the primitive Earth (Miller, 1955; Palm and Calvin, 1962). Other attractive features include a plausible explanation of the observed degeneracy of the genetic code.

Egami (1964) has found a correlation between the first two letters of the codons and the length of the carbon skeleton of the amino acids. His approach to the coevolution of the genetic code involves this observation coupled with the observed biosynthetic pathways of the bases and the protein amino acids.

As varied as the models and theories for the origin of the genetic code are, many of them have, at their base, certain assumptions, either implied or stated, concerning the nature of the association between a specific amino acid and its coding triplet. In most cases where this is an issue, there is required a structural basis for the coding specificity. Physical biochemists have approached this question using some of the most common techniques, as well as some of the more sophisticated.

Direct-Interaction, Physico-Chemical, and Structural Models

The phenomenon of complementarity is the foundation upon much of biochemical specificity is based. Not only are "lock-and-key" arrangements found in enzyme-substrate interactions, but all information-transferring steps in the replication of DNA, and transcription and subsequent translation of RNA use complementarity as the guiding force. An early proponent of complementarity between amino acids and nucleic acids was Gamow (Pauling and Delbruck, 1950; Gamow, 1954). The "direct interaction theory" suggests differential affinity between the amino acids and their coding sequences due to differences in their physicochemical properties. One result of this approach is that the assignment of codons to specific amino acids is deterministic, and is therefore subject to experimental verification.

Large amounts of data have been collected in support of the direct interaction theories. The problem is that the data do not support any one model of interaction unequivocally. The models described by the extant data fall into roughly five categories:

1. amino acid-nucleotide interactions;
2. amino acid-codon/anticodon interactions;
3. amino acid-anticodon/discriminator site interactions;
4. amino acid-codon interactions;
5. amino acid-anticodon interactions.

In order for any of these models to be realistic, first must be demonstrated that there are, indeed, physical, measurable interactions between amino acids and nucleic acids and/or their constituent parts. This has been done (Leng and Felsenfeld, 1966; Wagner and Arav, 1968; Woese et al., 1966; Yuki and Fox, 1969). It is now quite clear that such interactions do take place. What is not as clear, however, is the exact nature of these interactions, and, as a result, which of the previously-listed structural models is most accurate. Part of the problem may be the low affinities that have been measured. Without careful collection and numerical analysis of the data, subtle differences in association constants may be missed. We will address this subject more, later.

Amino Acid-Nucleotide Interactions

Most of the studies in this area have been performed using nucleotides and basic polypeptides (Wagner and Arav, 1968; Woese, 1968; Yuki and Fox, 1969). Woese measured changes in turbidity of solutions of poly-L-arginine and nucleoside monophosphates. His data, which demonstrated a preferential affinity of guanylate for poly-L-arginine, led him to speculate that such interactions may have provided a rudimentary translation system.

5'-ribo- and deoxyribonucleoside monophosphates were used by Wagner and Arav (1968). The association between these molecules and poly-L-arginine and poly-L-lysine were measured using equilibrium dialysis. They found that the nature of the base was a determining factor in the specificity they observed. Superimposed on the nonspecific electrostatic interactions that would exist between the charged functions on the amino acids and the nucleotides, was a more specific interaction which depended on the nature of the base. The pyrimidine bases showed little or not selectivity, whereas the purine bases demonstrated significant selectivity between themselves, and when measured relative to the pyrimidines. In addition, they found a variation in binding constant with poly-L-arginine which correlates well with the tendency of the bases to stack. Little or no such variation was found with poly-L-lysine.

Nucleic acids, specifically DNA's, were used by Lang and Felsenfeld (1966) who showed that poly-L-lysine bound preferentially to AT-rich samples, and that poly-L-arginine bound more tightly to GC-rich samples.

The interaction between thermal proteinoids and polyribonucleotides was investigated by Yuki and Fox (1969), who found that lysine-rich proteinoids preferentially precipitate polypyrimidines, whereas arginine-rich proteinoids precipitated polypurines.

Several other groups have investigated the interactions between polynucleotides and basic polypeptides (Chargaff et al.,

1953; Spitnick et al., 1955; Johns and Butler, 1964; Akinrimisi et al., 1965; Tsuboi et al., 1966; Sober et al., 1966). The general conclusion drawn from these studies is that there is indeed some selectivity in the interactions, in that the association is influenced by the charge state of the molecules, and the relative amounts of purine and pyrimidine in the polynucleotides.

A significant advance in this area was provided by the work of Saxinger and Ponnamperuma (Saxinger et al., 1972; Saxinger and Ponnamperuma, 1971; Saxinger, 1972). Their use of immobilized amino acids are also consistent with selectivity in binding. Their results were described qualitatively in terms of a site-binding model. According to this model, there are several aspects to the observed selectivity. For example, superimposed on the electrostatic interaction between the positively-charged ammonium group and the negatively-charged phosphate group are (1) direct interactions between the base and the amino acid side-chain; (2) stacking interactions between bases; and (3) hydrophobic, hydrophilic, van der Waals and induced dipole interactions. Saxinger and Ponnamperuma also used nuclear magnetic resonance (NMR) spectroscopy to demonstrate that the chemical shifts of the heterocyclic ring protons of the bases move upfield with increasing amounts of added amino acid. This work was the first to demonstrate this phenomenon.

Subsequent work by several groups has also relied on NMR to investigate these interactions. The first quantitative effort was provided by Raszka and Mandel (1971a,b). They studied the interaction between the twenty protein amino acids and poly-A, and a selected subset of this group with poly-I and poly-U. The observed changes in chemical shifts of the protons on carbons 2 and 8 (C-2 and C-8) of the adenine moiety of poly-A were consistent with destacking of the partially-stacked polynucleotide chains, due, they speculated, to the intercalation of the amino acid side chains. There was an observed correlation between the degree of association and the size or hydrophobic character of the amino acids. The interactions with poly-U were measured by observation of the C-1 and C-5 protons of uridine. Here, also it was found that the hydrophobic amino acids, and especially phenylalanine, were most effective in producing the upfield shifts of these proton resonances. They also observed that phenylalanine bound more tightly to poly-A and poly-U than did lysine.

The interactions between two amino acid derivatives and nucleoside monophosphates were studied by Helene and Dimicoli (Helene et al., 1971a; Helene et al., 1971b). Using traditional binding theory, they measured association constants in the range 8.9 M^{-1} for 5'-AMP/tyramine, to 6.4 M^{-1} for AMP/tryptamine. Another study provided the association constants for tryptamine binding to GMP, AMP, TMP, UMP, and CMP as: 13.1, 18.4, 4.4, 2.9, and 2.4 M^{-1}, respectively. These results have been interpreted (Lawaczeck and Wagner, 1979) in terms of stacking specificity and induced-dipole interactions.

Reuben (1978) used NMR to study the binding between tryptophan methyl ester and nucleotides. He also reported a selectivity in the association. This work was extended to include nonaromatic amino acids by Reuben and Polk (1980). This was an indirect measurement, however, as they measured the effect of increasing amounts of the nonaromatic amino acid on the association between AMP and tryptophan methyl ester. Selectivity was also demonstrated in this study.

The effect of increasing the chain length of the amino acids to a dipeptide was investigated by Dasgupta and Podder (1979). Their dipeptides were of the form gly-X (X = tyr, gly, leu). The binding data for associations with AMP, UMP, CMP, and GMP also showed some selectivity.

With all these available data, it has become clear that there are, indeed, some kinds of associations between amino acids and mono- and polynucleotides. More refined questions arise from this conclusion, however. For example, is the association between a given amino acid and nucleotides in any form mediated by interaction with the sequences comprising the codon or the anticodon? Are there effects due to neighboring nucleotides? Neighboring amino acids? There is a plethora of theories and models which purport to predict the basis of the code interactions. Indeed, there are data in support of both the codon and anticodon models. How, then, to determine which is the correct one? We shall now present some of the details of several of these models, along with some of the data (where available) which bear on this issue.

Amino Acid/Codon-Anticodon Interactions

Central to the process of protein synthesis is the recognition of a codon triplet on mRNA by the anticodon triplet on tRNA. It is, therefore, not unreasonable to postulate a primitive system wherein a codon and an anticodon are both involved, or, at least, some analogue of this complex. Thomas (1976) has identified a site on such a complex where an amino acid may be found to associate. The major features of this model are: (1) the codon-anticodon pair provides a groove which is the correct size and shape for the appropriate amino acid. This structure provides the proper geometry for the positioning of the hydroxyl groups on the ribose moieties to complex with the carboxyl group on the amino acid to hold the amino acid in a rigid position; (2) short side chain amino acids, such as glycine and alanine, cannot extend over more than two base pairs; (3) longer amino acid side chains (e.g., met, trp) may be affected by the differences between A and G as the third letter in the code; (4) amino acids selected when A:U is the middle base pair are long and polar; (5) G:C and C:G middle pairs change the conformation of the complex, and require that either very short amino acids or long and awkward amino acids may bind; (6) A:U and U:A sites are flat, and would therefore select amino acids with flat aromatic rings, or, perhaps, methylene chains. Discrimination is further provided by this model in that the first condon letter determines the carboxylate angle, and that steric hindrance is provided by a pyrimidine carbonyl on the third codon letter. There is, unfortunately, no direct data on this topic. However, data has been obtained by Grosjean et al. (1978) on recognition interactions between two tRNA anticodon loops. Temperature jump techniques were used to determine the lifetimes of anticodon-anticodon complexes. Incorrect pairings had a shorter lifetime than correct pairings, by two orders of magnitude (Grosjean et al., 1976). They suggested that the present-day system, using the ribosome, is designed to amplify these lifetime differences.

Amino Acid/Anticodon/Discriminator Site Interactions

Crothers et al. (1972) has suggested that there are two sites on a tRNA molecule that are specific for an amino acid: the anticodon loop and the discriminator base located in the fourth position from the 3'-end of the nucleotide chain. Shimizu (1982) has proposed a model, called the C_4N complex, wherein he claims it is possible to combine the anticodon loop and the discriminator site of tRNA. As a result of this complexation, a pocket is formed which is specific for the appropriate amino acid. Furthermore, he claims that the possible pockets provide only those conformations which will bind the twenty protein amino acids. Stacking interactions and electrostatic attractions are required for the specificity, as are hydrophobic and van der Waals interactions. There is some provision for the enfolding of a "wrong" amino acid, but Shimizu contends that the nature of the side chains would be similar. Model building, which is well-known to be fraught with problems, has demonstrated that the interactions described above are indeed possible.

Amino Acid/Codon Interactions

An early model for this interaction was proposed by Lacey and Pruitt (1969). Using the assumption that amino acid polymers would be more likely to form under prebiotic conditions than nucleotide polymers, their model, based, again, on the use of CPK model structures, demonstrated that, when a polypeptide is in the α-helical form, it is possible to form a complementary strand of mononucleotides. In this model, the amino acid side chain is inserted between the first and second nucleotides of the codon, and that it is the total of all possible interactions which determines maximum stability and codon specificity. These interactions include: (1) ring stacking, electrostatic, or charge-transfer interactions between the amino acids and the nucleotides; and (2) interactions between amino acids and nucleotides due to hydrogen bonding between side chains of the amino acids and the phosphate groups of the nucleotides. This model was used to successfully predict the glycine codon.

A more generalized amino acid/codon interaction model has the following properties (Reuben and Polk, 1980; Woese et al., 1975; Woese and Fox, 1977; Woese, 1967): (1) the geometry of the amino acid/codon complex is similar in all possible pairs; (2) the base at position two of the codon interacts most strongly with the side chain of the amino acid, and is probably the most important "letter" in determining the amino acid assignment; (3) the base in position one actually acts as a perturbing factor in the interaction with the base in position two; (4) the base in position three interacts least strongly of all three; (5) all amino acids within the same classification (e.g., polar, hydrophobic, etc.) have the same energy of interaction with their respective codons.

Some evidence in support of this model came from the work of Nakashima and Fox (1972), in which was studied the incorporation of AMP-anhydrides of each of the four homocodonic amino acids (e.g., phenylalanine, glycine, proline, lysine) into nucleoproteinoid microparticles composed of each of the four homopolynucleotides. A codonic preference was observed.

Perhaps the strongest evidence in favor of this model comes from the data of Woese et al. (1966), who used partition paper chromatography to determine that the ordering of the amino acids with respect to their polarity was highly correlated with similar ordering of their codon triplets.

In general, the following observations support the amino acid/codon model: (1) exchanging one purine for another, or one pyrimidine for another in position three of the codon almost never changes the amino acid assignment; (2) changing a pyrimidine to a purine in position three usually does not change the amino acid assignment. In those cases where a change does occur, the two amino acids are related by their polarity. (3) A change of a base at position one has a greater effect than that observed in (2), but the changes still occur within a closely-related group.

Amino Acid/Anticodon Interactions

In the process of protein synthesis found in present-day organisms, the amino acid is first selected and attached to a polynucleotide bearing the anticodon sequence of nucleotides, the tRNA. This seems to indicate that recognition processes center around the anticodon (Dunnill, 1966; Nagyvary and Fendler, 1974;

Ralph, 1968). Several lines of evidence are now converging which indicate that the origin of the genetic code had as its basis a specific relationship between amino acids and their anticodon nucleotides (Lacey and Fox, 1972; Lacey and Weber, 1976; Woese, 1973).

One approach to this area has been taken by Nagyvary and Fendler who undertook to explain the origin of the genetic code on the basis of physical and chemical conditions that were likely to have existed on prebiotic earth. This entails the selective uptake of amino acids and nucleotides into compartments formed by colloidal aggregates. This was based on a suggestion for the composition of the primitive ocean (Lasaga et al., 1971), and recognized properties of micelles. Depending on the polarity of the media, the aggregates would have had either hydrophilic or hydrophobic interiors. Hydrophilic species, such as lysine and uridylate would have been segregated together, as would have phenylalanine and adenylate, but in a hydrophobic environment. This would provide for the observed genetic code, as mediated by an amino acid/anticodon interaction. This model has the additional advantage of providing a sequestered set of molecules that could have later given rise to condensation products, i.e., polymers of amino acids and nucleotides. In a sense, then, this is similar to the co-evolution model described earlier in this chapter. It is impossible to test this model *per se*; however, the observed correlations between polarities augurs well for this model.

As was the case for the amino acid/codon models, more work has gone into describing the basis of the interactions in terms of stereochemical associations. One of the more comprehensive presentations of this sort has been that of Melcher (1970). He suggested that the association of an amino acid through intercalation with nucleotide sequences comprising the anticodon would allow several interactions to occur. These are: (1) hydrogen bonding between the aliphatic atoms of the amino acid and the π-electrons of the bases. Discrimination between amino acids might be based on the different strengths of the hydrogen bonds (Lin and Tsay, 1970). Binding strength would increase with increasing positive charge of the hydrogen atom and increasing electron density of the bases. (2) The amino group of the amino acid could hydrogen bond with the lone pair electrons of the amino groups of the bases. Similar bonding with the keto groups is not possible because they lie in the plane of the aromatic rings of the bases. (3) The side chain of the amino acid is available for bonding with the phosphate groups of the phosphodiester backbone. Again, using models (Kendrew Atomic models), structures may be built in which a cage is formed by three nucleotides. These cages seem to be specific for phe, tyr, trp, his, and met.

Can models of this sort be tested experimentally? The answer is most assuredly, "Yes," and have been carried out in several cases (Fendler et al., 1975; Garel et al., 1973; Jungck, 1978; Lacey et al., 1979a,b; Raszka and Mandel, 1971; Weber and Lacey, 1978). Much of the evidence in favor of the amino acid/anticodon interaction has come from chromatographic and spectroscopic studies on the monomers alone, and in various combinations.

The first statement of the correlation between hydrophobicities of the amino acids and nucleotides was that of Woese et al. (1966). Fendler et al. (1975) and Armstrong et al. (1977) studied the affinities of amino acids and nucleotides for micelles. A general tendency for anticodon mediated origins may be inferred from their data.

The partition of amino acids and nucleotides between an oil phase and an aqueous phase was studied by Garel et al. (1973). The ratio of concentrations in the oil phase and that in the aqueous phase was defined as the partition coefficient. The data showed excellent correlation between the hydrophobicities of the amino acids and their anticodon nucleotides.

Using the same experimental protocols, Lacey and Weber (1977) organized the genetic code as the genetic anticode, based on hydrophilic and charge considerations. This was later rearranged and updated by Lacey and Mullins (1981), to provide three general sets of assignments. Their data led them to state that AC and AU were probably not initially usable as anticodons, and that the amino acids trp, tyr, cys, and perhaps ser were late evolutionary developments. Other work by Lacey and Weber (1977) show excellent correlation between amino acids that had homoanticodonic sequences and the monomers comprising those sequences. It was not immediately clear, however, how to extrapolate those data to heterogeneous anticodons.

Garel et al. (1973) used a three-dimensional plot to extrapolate these correlations to dinucleotides. There were several exceptions to the correlations predicted: leu/AA, trp/AC, and ser/AG. It would appear, then, that amino acids with adenine at the 3'-end of the anticodon do not correlate well.

Another study undertaken by Weber and Lacey (1978) demonstrated high correlation between the R_f values of the amino acids and dinucleotide monophosphates. These data also support the amino acid/anticodon model.

In fact, a systematic survey of sixteen properties of amino acids and nucleotides undertaken by Jungck (1978) bears out this contention. In forty of forty-five pairs, the correlation was greater for amino acid-anticodons than for the amino acid-codons.

His compilation of the data of Weber and Lacey showed the highest correlation, while the data of Barzilay et al. (1973) and Garel et al. (1973) were not as definitive, but were consistent with the anticodon model.

Spectroscopy has been used to great advantage in examining the interactions between amino acids and nucleotides. There exist considerable data which demonstrate that there are indeed associations. Furthermore, more and more data is being accumulated by several groups which further supports the anticodon model of association (Lacey, 1983).

Studies underway in our laboratory confirm this hypothesis, with some interesting and heretofore unexamined phenomena. We have recently undertaken a rigorous and (we hope) complete study of these interactions, employing NMR, and using a two-pronged approach. On the one hand, we are using single amino acids, and measuring their associations with increasing chain lengths of nucleotides. Conversely, we are keeping the chain length of the nucleotides fixed, and increasing the number of amino acids.

The binding of phenylalanine methyl ester to 5'-AMP was greater than that to 5'-UMP by about a factor of two (5.4 M^{-1} vs. 2.9 M^{-1}). Using a dipeptide, lys-phe, increases the binding to AMP to 125 M^{-1}. When dinucleoside monophosphates were used, phenylalanine methyl ester again bound more tightly to the anticodon dimer sequence (ApA) than to its codon dimer (UpU). Indeed, the binding to UpU was considerably looser than to UMP. At the present time we have no explanation for this observation. Attempts to understand the nature of the binding between Phe and ApA led us to build models (CPK). The result of this exercise was a prediction that the use of the enantiomer, D-phe would change the way the resonances were observed to shift. When we performed those experiments, in fact the resonances did change, but not in the manner we had predicted. D-phe binds just as tightly to ApA as does the L-enantiomer, but the mode of its interaction is quite different.

Perhaps the most fascinating observation to arise from these studies is that use of dinucleoside monophosphates which employ a 2'-5' phosphodiester linkage absolutely precludes binding of L-phe of any kind, whether to the codon or anticodon dimers. Experiments are currently being designed to further investigate these phenomena.

The low affinities being measured have been criticized by some investigators. Binding phenomena found in present living systems (e.g., organophosphate binding to hemoglobins) are often five to ten orders of magnitude greater than those that have been reported for these interactions (Hobish and Powers, 1983).

Nevertheless, the data reported are reproducible from group to group. Furthermore, seen in light of the chromatographic data described above, the results are also logical and self-consistent. This is not to say, however, that the numbers themselves are absolute. Indeed, unless one has a large number of data points to describe an interaction, determination of a binding constant is extremely difficult, and may result in large confidence limits. Only the use of good data, coupled with sophisticated numerical analysis can enable accurate and precise determinations. The advent of low-cost computational facilities and sophisticated software now makes this possible (Turner et al., 1962).

CONCLUSIONS

It is clear that there are, indeed, measurable interactions between amino acids and nucleotides in solution. In the light of the physical/chemical properties of these species, it is strongly suggestive that the genetic code has its foundation upon interactions between amino acids and nucleotides comprising their anticodon sequences. Other factors undoubtedly must play a part. Some combination of anticodon mediation with codon/anticodon interactions may be at work here. Suffice it to say that we seem to be dealing with a real phenomenon. Clearly, more work is required. The sensitivity and breadth of information available through spectroscopic studies make this a method of choice for further investigations.

ACKNOWLEDGMENTS

We would like to acknowledge the support of NASA grant #NGR21-002-317,514 to C.P. We would like to thank Nalinie Senaratne, Binny Cherayil, Sanjay Tyagi, and Masayatsu Aotani for sharing their thoughts on this subject with us, and Dr. Yiu-Fai Lam for his aid in the NMR studies.

REFERENCES

Akinrimisi, E. D., Bonner, J., and Tso, P., 1965, J. Mol. Biol., 11:128.

Armstrong, D. W., Seguin, R., and Fendler, J. R., 1977, J. Mol. Evol., 10:241.

Baricelli, N. A., 1977, J. Theor. Biol., 67:85.

Barzilay, I., Sussman, J. L., and Lapidot, Y., 1973, J. Chromat., 79:139.

Chargaff, E., Cramplan, C. F., and Lipshitz, R. L., 1953, Nature, 172:289.

Crick, F. H. C., 1966, J. Mol. Biol., 19:548.

Crick, F. H. C., 1967, Nature, 213:119.

Crick, F. H. C., 1968, J. Mol. Biol., 38:367
Crothers, D. M., Seni, T., and Soll, D. G., 1972, Proc. Nat. Acad. Sci. U.S., 69:3063.
Dasgupta, D., and Podder, S. K., 1979, Ind. J. Biochem. Biophys., 16:316.
Dickerson, R. E., and Geis, I., 1969, "The Structure and Action of Proteins," 20:16, Harper and Row, New York.
Dunnill, P., 1966, Nature, 210:1267.
Egami, F., 1964, in: "The Origin of Prebiological Systems and of Their Molecular Matrices," S. W. Fox, ed., p. 187, Academic Press, New York.
Epstein, C. J., 1966, Nature, 210:25.
Fendler, J. H., Nome, F., and Nagyvary, J., 1975, J. Mol. Evol., 6:21
Fendler, J. H., Nome, F., and Nagyvary, J., 1975, J. Mol. Evol., 6:215.
Fitch, W. M., 1966, J. Mol. Biol., 16:1.
Gamow, G., 1954, Nature, 173:318.
Garel, J. P., Filliol, D., and Mandel, P., 1973, J. Chromat., 78:381.
Grosjean, H. J., Henau, S., and Crothers, D., 1978, Proc. Nat. Acad. Sci. U.S., 75:618.
Grosjean, H. J., Soll, D., and Crothers, D. M., 1976, J. Mol. Biol., 103:499.
Hartman, H., 1978, Origins of Life, 9:133.
Helene, D., Dimicoli, J. L., and Brun, F., 1971, Biochemistry, 10:3802.
Helene, C., Montenay-Garestier, T., and Dimicoli, J. L., 1971, Biochim. Biophys. Acta, 254:349.
Hobish, M. K., and Powers, D. A., submitted to J. Biol. Chem., 1983.
Hoffman, G. W., 1974, J. Mol. Biol., 86:349.
Johns, E. W., and Butler, J. A. V., 1964, Nature, 204:853.
Jukes, T. H., 1967, Biochem. Biophys. Res. Comm., 27:573.
Jukes, T. H., 1974, Origins of Life, 15:331.
Jungck, J. R., 1978, J. Mol. Evol., 11:211.
Lacey, J. C., 1983, personal communication.
Lacey, J. C., and Fox, S. W., 1972, "Theory and Experiments in Exobiology II," pp. 35-63, Waters-Nordhoff, Groningen.
Lacey, J. C., and Mullins, W. D., 1981, Origins of Life, 11:447.
Lacey, J. C., and Pruitt, K. M., 1969, Nature, 223:779.
Lacey, J. C., Stephens, D. P., and Fox, S. W., 1979a, Biosystems, 11:9.
Lacey, J. C., and Weber, A. L., 1976, in: "Protein Structure and Evolution," p. 213, Fox et al., eds., Marcel Dekker, New York.
Lacey, J. C., and Weber, A. I., 1977, Precam. Res., 5:1.
Lacey, J. C., Yuki, A., and Fox, S. W., 1979b, Biosystems, 11:1.
Lasaga, A. C., Holland, H. D., and Dwyer, M. J., 1971, Science, 174:53.
Lawaczeck, R., and Wagner, K. G., 1974, Biopolymers, 13:2003.
Leng, M., and Felsenfeld, G., 1966, J. Mol. Biol., 15:155.
Lin, W. C., and Tsay, S. J., 1970, J. Phys. Chem., 74:1037.
Melcher, G., 1970, Biophysik, 7:25.

Mikelsaar, H. N., J. Theor. Biol., 50:203.
Miller, S. L., 1955, J. Amer. Chem. Soc., 77:2351.
Nagyvary, J., and Fendler, J. H., 1974, Origins of Life, 5:357.
Nakashima, T., and Fox, S. W., 1972, Proc. Nat. Acad. Sci. U.S., 69:106.
Nirenberg, M., and Matthei, J., 1961, Proc. Nat. Acad. Sci. U.S., 47:1588.
Palm, C., and Calvin, M., 1962, J. Amer. Chem. Soc., 84:2115.
Pauling, L., and Delbruck, M., 1950, Science, 92:77.
Ralph, R. K., 1968, Biochem. Biophys. Res. Comm., 33:213.
Raszka, M., and Mandel, M., 1971a, First European Biophysical Congress, Baden, September.
Raszka, M., and Mandel, M., 1971b, Proc. Nat. Acad. Sci. U.S., 68:1190.
Reuben, J., 1978, FEBS Lett. 94:20.
Reuben, J., and Polk, F. E., 1980, J. Mol. Evol., 15:103.
Rolfing, D. L., and Saunders, M. A., 1978, J. Theor. Biol., 71:487.
Saxinger, W. C., and Ponnamperuma, C., 1971, J. Mol. Evol., 1:63.
Saxinger, W. C., Ponnamperuma, C., and Gillespie, D., 1972, Proc. Nat. Acad. Sci. U.S., 69:2975.
Saxinger, W. C., Ponnamperuma, C., and Woese, C. R., 1972, Nature New Biology, 234:172.
Shimizu, M., 1982, J. Mol. Evol., 18:297.
Smith, E. L., Okunuki, K., Kamen, M. D., and Sekuzu, I., eds., 1968, "Structure and Function of Cytochromes," p. 282, University of Tokyo Press, Tokyo.
Sober, H. A., Schlossman, S. F., Yaron, A., Latt, S. A., and Rushizky, G. W., 1966, Biochemistry, 5:3608.
Sonneborn, T. M., Bryson, V., and Vogel, H., eds., 1965, "Evolving Genes and Proteins," p. 377, Academic Press, New York.
Spitnick, P., Lipshitz, R. L., and Chargaff, E., 1955, J. Biol. Chem., 215:765.
Thomas, B. R., 1976, "Bulletin of Molecular and Structural Biochemistry," #1, Lund, Sweden.
Tsuboi, M., Matsuo, K., Tso, P. O. P., 1966, J. Mol. Biol., 15:256.
Turner, B. W., Pettigrew, D., and Ackers, G. K., 1982, Meth. Enzym., 76.
Wagner, K. G., and Arav, R., 1968, Biochemistry, 7:177.
Watson, J., and Crick, F. H. C., 1953, Nature, 171:737.
Weber, A. I., and Lacey, J. C., 1978, J. Mol. Evol., 11:199.
Woese, C., 1967, "The Genetic Code," Harper and Row, New York.
Woese, C., 1968, Proc. Nat. Acad. Sci. U.S., 59:110.
Woese, C., 1973, Naturwiss., 60:447.
Woese, C., Dugre, D., Dugre, S., Kondo, M., and Saxinger, C., 1966, Cold Spring Harbor Symp. Quant. Biol., 31:723.
Woese, C., Dugre, D. H., Saxinger, W. C., and Dugre, S. A., 1966, Proc. Nat. Acad. Sci. U.S., 55:966.
Woese, C., and Fox, G. E., 1977, J. Mol. Evol., 18:1.
Woese, C., Fox, G. E., Zaclen, L., Ucida, T., Boner, L., Dechman, K., Lewis, B. J., and Stahl, D., 1975, Nature, 254:83.

Wong, J. T., 1975, Proc. Nat. Acad. Sci. U.S., 72:1909.
Wong, J. T., 1976, Proc. Nat. Acad. Sci. U.S., 73:2336.
Wong, J. T., and Bronskill, P. M., 1979, J. Mol. Evol., 13:115.
Yuki, A., and Fox, S. W., 1969, Biochem. Biophys. Res. Comm., 36:657.

REALISM AND THE ORIGIN OF THE GENETIC CODE

Lawrence S. Dillon

Department of Biology
Texas A & M University
College Station, Texas 77843

With the development of the proteinoid microsphere in the laboratories of Dr. Sidney W. Fox, whose own origins we honor in this book, a convincing model of preliving particles leading to an actual early protobiont has been created. Not only was this model originated in those laboratories but subsequently was perfected, through persistent experimental and constructionistic approaches, to the point where first generations--and even partial second generations--have been produced through self-replicatory processes (Fox, 1968, 1969). In fact, it would appear that little of major proportions remains to be elaborated further on this representation of a possible earliest stage in life's origin on this earth, although a number of the preceding papers in this volume show the need for lesser details to be added.

In microspheres, the basis for replication has been clearly brought out to involve the self-ordering of amino acids (Fox, 1968, 1978, 1980a; Dose and Rauchfuss, 1972; Melius, 1977), a process, not sufficiently understood, that has also been referred to as nonrandomness in the association between those molecules (Dillon, 1978, pp. 50-54). While such a mechanism undoubtedly was sufficient for the needs of prelife particles and then for the very earliest actual protobiont, a point would eventually have been reached where a more precise and perhaps more rapid process of ordering amino acids into polypeptides and proteinoids became mandatory. Thus some sort of coded genetic mechanism was needed, both to maintain populations of the now more complex protobionts and to permit faster and less haphazard evolutionary progress. This requirement for further progress in life's beginnings has certainly not gone unnoticed, so that a number of studies have been made from a diversity of standpoints toward a solution of this problem. An analysis of these varied approaches to this aspect of protobiogenesis provides the basis of the present paper, together with a proposal to stimulate additional progress, in the hope that greater realism can be achieved through the efforts of experimental constructionists.

CURRENT APPROACHES TO THE ORIGINS OF THE CODING MECHANISM

Because the various approaches to the origin of a coding apparatus have been reviewed extensively on earlier occasions (Dillon, 1973, 1978), it is not necessary here to provide a full historic account. Accordingly, attention is given primarily to more recent papers, with occasional discussion of others as clarity demands. Although highly diversified, the approaches that have been employed toward solution of the problem can be grouped into three categories--theoretical, biochemical and constructionistic, and biological.

Theoretical Approaches

A number of completely theoretical proposals have appeared in print in recent years, the more thoroughly documented one of which has received broad acceptance in a diversity of quarters. Since at times it provides a basis for other proposals that have been advanced, it needs attention first.

The hypercycle concept. First published as a series of three articles (Eigen and Schuster, 1977, 1978a,b), this theory was subsequently issued in book form (Eigen and Schuster, 1979). Briefly stated it proposes that the metabolic machinery of organisms became organized very early in life's existence in the form of cyclically arranged series of self-replicative units, each of which was itself cyclic in nature. The linkages between the component parts are viewed as functional in that an intermediate product of one cycle is employed by the next one in the series and so on, an intermediate of the *n*th cycle being employed by the first (Fig. 1). At each step an external supply of "energy rich" building material is conceived to be drawn upon, the molecules of which arose spontaneously in the primitive seas.

In applying the concept to the problem of biogenesis, a series of phylogenetic stages is outlined, beginning with simple self-replicating polynucleotides as stage 1. At stage 2, the original single-stranded GC-rich molecules replicate by way of negative counterparts (replicative intermediates), much as in bacteriophage Qβ. At stage 3, first codon assignments are made, and translation of the plus strands begins, yielding glycine and alanine polypeptides while the minus strands possibly represent primitive adapters (tRNAs?). Hypercyclic fixation of the GC-frame code provides the basis for enlargement of the code catalogue to encode aspartic acid and valine. Further repetitions of these steps then ultimately

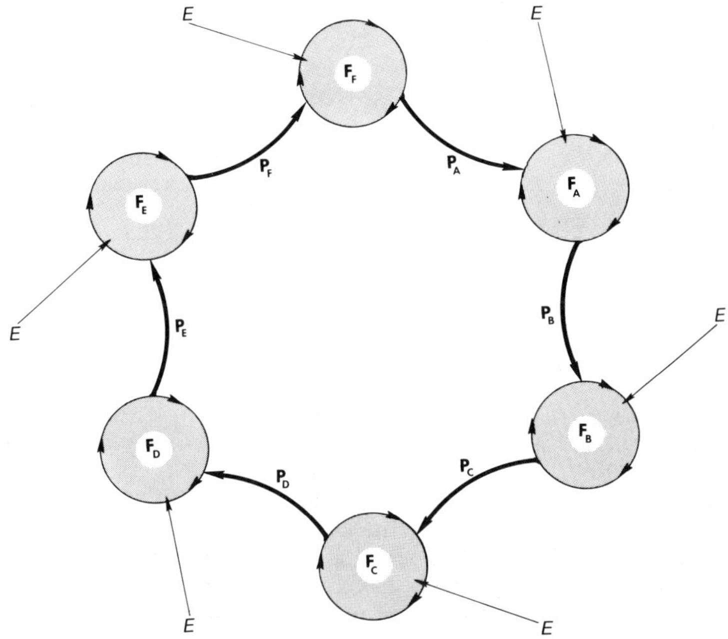

Fig. 1. The hypercycle concept. Here six cycles (F_A- F_F) are diagrammed, each of which is driven by an exogenous energy source (E). The individual cycles are visualized as being bound into a hypercycle through mutual need of an intermediate product (P_A-P_F) resulting from the preceding cycle. No such hypercycles are known to exist in any modern organism.

lead, by means of Darwinian selection, to the full complement of codons and the eventual establishment of the present complex genetic mechanism.

Analysis of the hypercycle concept. Several conclusions reached in the original scholarly presentation of the hypercycle concept are basic principles for any serious study of life's origins. The need for self-replication at the earliest levels of protobiont organization is an obvious truth, for the purpose of "conservation of information", to quote Eigen and Schuster (1979). A second important point is the recognition of glycine and alanine as the earliest amino acids to become encoded, a conclusion reached by the present author in earlier papers (Dillon, 1973, 1978, pp. 217-219). Conclusions in support of this concept to a degree have also been reached through other approaches (Orgel, 1972, 1977). Furthermore, the hypothetical scheme of evolution it advocates from the first macromolecular system stepwise to a cell merits close attention of all persons working in this field.

Unfortunately, although most attractive on the surface, the theory is weakened by an absence of realism. In the first place, it assumes nucleic acids to be self-replicating molecules, an idea that is without substantiating evidence (Fox, 1980b, 1981a,b; Dillon, 1978, p. 121). Nucleic acids have been shown to be extremely difficult to produce under aqueous conditions; even with the use of matrices containing such divalent metals as lead, the best efforts have resulted only in trinucleotides (Lohrman, 1982). Only polypeptides have been demonstrated capable of self-ordering their monomeric constituents and thereby conserving the inherent information of the parent molecule. Secondly, no hypercycles are known to exist in modern organisms, not even in viruses, an observation noted in Eigen's (Eigen and Schuster, 1979) own writings. The single example cited in the original presentation is the replication of the bacteriophage Qβ genome. However, replication of that organism's single-stranded RNA molecule requires the entire protein-producing mechanism of the host, E. coli. Hence, it can scarcely be considered a self-reproductive system for at least 189 proteins are involved in that bacterium in the replication, transcription, and translation of the genetic apparatus, none of the steps of which function in cyclic fashion (Dillon, 1978, pp. 206, 207). Moreover in the concept, the RNA strands, besides encoding the replicase unique to the next member of the hypercycle, are envisioned as serving also as tRNAs (Bresch et al., 1980). However, tRNAs do not react spontaneously with amino acids, but require ligases, each specific for individual species of those protein monomers. Additionally, no proposal is advanced to indicate how the code catalogue and coding reactions may have arisen, for the early codons are described merely as being assigned.

Other theoretical approaches. Because of the number of theories concerning the origins of the genetic code that have been advanced recently, only a few selected examples can be given here, and those necessarily briefly. The symmetrical form of the code catalogue has attracted some analyses, one of which (Findley et al., 1982) attempts to explain the existing codon assignments from a group-theory viewpoint. In the cited study, the more than one hundred alternative codings that have been reported are analyzed mathematically within the framework of what is referred to as the "generalized genetic code" in contrast to the actual existing "standard genetic code". Since the codons are viewed as being arbitrarily assigned to the amino acids, rather than products of chemical or biological interactions, the significance of the results to the present topic is not in evidence.

Another current theory (Dyson, 1982) is frankly stated to be an "empty mathematical framework" within which questions concerning the origin of life can be posed with a degree of precision, and thus merits no further attention here. Still another concept involves a protobiontic model called the "autogen", which is envisioned as consisting of two protogenes that encode two protoenzymes, one of which catalyzes replication, the other translation (White, 1980; White and Raab, 1982). Although the replicatory processes of each gene may be viewed as cyclic, the remaining reactions are not; hence, the proposal is not related to the hypercycle as claimed. Obviously the hypothesis suggests the simultaneous origins of two oligonucleotides, whose products are mutually essential to their perpetuation and respective functioning, a most remote possibility. Moreover, the suggested translation of the oligonucleotide chains into the polypeptide enzymes implies the prior existence of a code, an even more distantly possible condition.

Biochemical and Constructionistic Approaches

Since the last analysis of the literature on life's origins (Dillon, 1978, pp. 1-64, 211-233), the quantity of experimental contributions appearing annually has diminished by at least one order of magnitude, and the vast majority of these have dealt with the prebiotic synthesis and polymerization aspects of the overall problem (for example, Ferris et al., 1978; Oró, 1980; Subbaraman et al., 1980; Bossard and Toupance, 1981; Hawker and Oró, 1981; Miller and Van Trump, 1981; Voet and Schwartz, 1981, 1982; Schwartz and Goverde, 1982; Schwartz et al., 1982). As most of the constructionistic studies pertaining strictly to the origins of the coding mechanism have emanated from the Fox laboratory or are related to those researches, they alone require attention here.

A constructionistic approach. The most pertinent results have been obtained through use of a lysine-rich proteinoid, which was found to catalyze peptide synthesis from amino acids and ATP (Fox and Nakashima, 1980; Nakashima and Fox, 1980). Later, with proteinoid microspheres prepared so as to be both acidic and lysine-rich (Nakashima and Fox, 1981), the ATP was added at intervals during the syntheses, resulting in both homopeptides and heteropeptides (Matsuno, 1982). In the presence of mixtures of glycine and phenylalanine, dipeptides were produced, glycylphenylalanine predominating when poly(A) was added, phenylalanylglycine when either poly(C) or poly(U) was the additive.

These studies are realistic in that ATP is involved here for the synthesis of peptides as in extant organisms, a process imitated at the protobiontic level by the proteinoids serving as a catalyst in place of the modern proteins. Moreover, the ability of the latter to synthesize polynucleotides appears well founded and in keeping with biological phenomena. But the suggestion that such simple reactions support the view that a simultaneous coded nucleic acid-and-protein synthesis arose directly from the self-ordered system that characterizes proteinoid microspheres is contrary to many facts. The implication underlying that proposal, as in the hypercycle, is that polynucleotides of various types code directly for specific amino acids.

The lack of interaction between nucleic acids and amino acids. The fallacy of concepts involving direct particular polynucleotides (including trinucleotides) reacting with specific amino acids has been described elsewhere (Dillon, 1978, pp. 215, 216), as pointed out earlier. But since the mistaken point of view still persists, its lack of realism needs to be enlarged upon.

Conceptual difficulties. The concept of self-replicating polynucleotide chains that produce polypeptides directly is most attractive superficially, for it seems to solve the difficult problem of the origins of the genetic coding mechanism at one blow, in that the codons are viewed as an intrinsic property of trinucleotides themselves. But when a more critical look is taken, other problems arise that diminish the attractiveness of the model. In essence, the concept suggests, in early stages of life, the existence of a single polynucleotide chain that encodes a fairly simple protein. Since polynucleotides are not capable of self-replication, it must be assumed that the polypeptide chain in similar fashion encodes the polynucleotide molecule. However, it must be remembered that the latter consists of sets of trinucleotides, whereas the former is merely a series of single amino acids; in short the polynucleotide molecule is three times as complex as the polypeptide chain it is supposed to encode or is encoded by. Thus if the protoprotein molecule consists of ten amino acids, its nucleic acid generator would contain 30 monomers.

Hence, two problems arise immediately: (1) If during protein synthesis the amino acids lie upon or within such large trinucleotides, the successive monomers necessarily would be held widely apart in any given series, thus a mechanism would be required to bring the amino acids together to permit peptide linkages to form. (2) On the opposite side of the scale, during replication of the RNA, how could each monomer of the polyamino acid chain possibly hold three nucleotides, induce the trio to interconnect in a precise manner, and then bring the combined set into alignment with adjacent ones to permit the trinucleotide units to polymerize into the full chain? These mechanical problems are magnified to even greater degrees if the actual differences in molecular size of the monomeric units are considered, as they must be. The most abundant amino acid in the primitive seas undoubtedly was glycine as stated earlier, whose molecular weight is 75, whereas the trinucleotide GGG that encodes it has a molecular weight of 1,053 (Fig. 2). Even an entire polypeptide consisting of ten glycine residues would not be equal in mass to a single trinucleotide.

But even if these size differences are ignored, as they consistently have been in the literature, two other difficult problems emerge as the model is scrutinized further. (3) If the polypeptide encodes the polynucleotide, what induces the bonds between monomeric nucleotide residues to bind them into a chain? (4) If the polypeptide

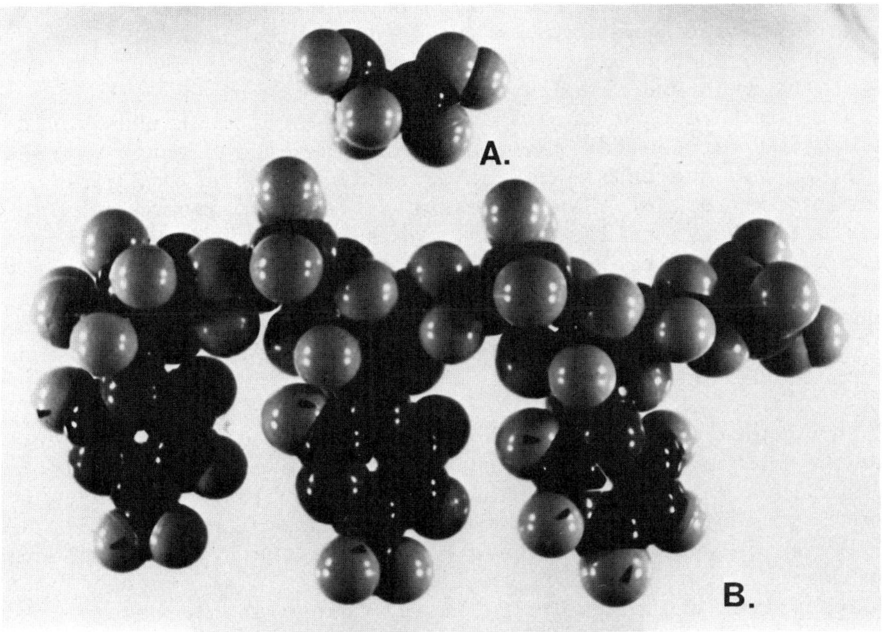

Fig. 2. Space-filling models of a glycine molecule and a trinucleotide. The problems arising from dimensional differences between the glycine molecule (A) and the trinucleotide CCU-(B) immediately become evident from visual comparison of these reproductions of molecular models.

itself is visualized as serving as the necessary polymerase, what enzyme then can possibly be present to bring about peptide formation? To suggest that the same small polypeptide chain performs all three functions by itself is certainly not acceptable.

Physicochemical difficulties. In addition to the molecule size discrepancies just pointed as theoretical problems but which could be considered physicochemical ones as well, there are a number of other known universal conditions in living cells that negate the spontaneous chemical origins of the genetic code. All that is necessary to disclose facts that are contrary to concepts of direct interaction between nucleotide chains and amino acids is to examine the primary and secondary structures of the innumerable macromolecules whose sequences have now been established and verified.

If there were a natural specific interaction between nucleotide polymers and particular amino acids in primitive earth times, those same relationships should be displayed in living cells today. Hence, in modern organisms both prokaryotic and eukaryotic, the numerous types of nucleotide chains present should spontaneously interact with the free amino acids in the cytosol, at least on the exposed loops that characterize all types of the former. Each such loop of the mRNA and precursorial molecules should be found to have interacted with one or more amino acids, a statement that would likewise be true for the several ribosomal classes and their precursors, as well as for the hnRNAs and other small varieties. Further, since the anticodon loop of tRNAs necessarily remains exposed even in the latter's tertiary configurations, the anticodon and other sequences should interact directly with amino acids, thereby effectively preventing reactions with the codons and other aspects of the protein synthesizing mechanism. In short, the genetic coding system could not function in living things were amino acids to react spontaneously with nucleotide chains.

Then there is also the other side of the coin to consider. Since the interaction between the two classes of biochemicals is necessarily a mutual one, the amino acids combined into peptide chains in the cell should react with the nucleotides present, an activity which would occur whether or not either type of molecule is in monomeric or polymeric form. As a result all exposed portions of polypeptides and proteins in living things should be expected to have interacted with nucleotides or nucleic acids, a condition that certainly does not prevail. In other words, every protein in all cells would necessarily be a nucleoprotein, an observation that is obviously unsubstantiated.

A biological difficulty. An additional set of facts that militates against the existence of a genetic code derived from specific and direct trinucleotide-amino acid interactions includes data that are both biochemical and biological in their nature. Were the information system a fundamental characteristic of the molecules themselves, the codons should be expected to be absolutely universal. Although the genetic code catalogue has proven to prevail among organisms in general, it is not completely catholic, mitochondria having provided the most frequent exception. In *Neurospora crassa*, yeast, and human mitochondria, UGA (along with UGG) encodes tryptophan, whereas in viruses, prokaryotes, and eukaryotic cytoplasm it signals termination of peptide synthesis (Macino et al., 1969; Heckman et al., 1980). In bovine mitochondria AUA encodes methionine, not isoleucine as elsewhere, and AGA and AGG, typically encoding arginine, appear to signal termination (Young and Anderson, 1980). To the contrary, in yeast mitochondria AGA is the only codon employed for arginine (Bonitz et al., 1980), and no codon has been found for that amino acid in the human organelle (Barrell et al., 1980). Moreover, in yeast mitochondria CUA and possibly other related codons encode threonine rather than leucine as elsewhere (Li and Tzagoloff, 1979). Since the macromolecules involved are precisely the same in all cases, no support can be offered to the concept of trinucleotide-amino acid interaction.

Additionally, the lack of universality is also indicated by the numerous reported mutant forms of tRNAs that carry the wrong amino acid and insert it into the growing polypeptide chain (Dillon, 1978, pp. 306-312, 336). Furthermore, were direct interactions involved, it is unlikely that the normal tRNAs of prokaryotes could be mischarged by eukaryotic enzymes as they have frequently been reported to do (Dillon, 1978, pp. 243-245).

An additional biological point. If the codons or anticodons react directly as hypothesized, it is extremely difficult to perceive what survival value the present protein synthesizing mechanism would have offered evolving organisms. Had information exchange originally been direct in protobionts, what additional survival value could the gigantic and complex indirect coding system afford, with its requirements for multitudes of tRNAs, tRNA ligases, nucleotidyltransferases, base-modifying enzymes, and all the host of other proteins, not to mention the additional burden of increased genome size to encode all those substances? In what way could these elaborations and the necessary round-about procedures be a bit superior to the immediate precise reaction of codon and amino acid? Thus the hypothesis of nucleotide-amino acid interaction leads ultimately into a blind alley from which no exit can be made on any perceivable evolutionarily sound basis to lead to the realities of living cells.

APPARENT STEPS IN THE ORIGINS OF CODING

Although modern organisms have obviously diverged widely from the primitive forebears in almost all aspects of structure and metabolic activities, the genetic processes nevertheless still retain hints of the primitive events in their establishment and elaboration, as Fox (1981a) points out. And once the details of the coding processes had been fully determined, they were so fundamental to survival that they were not subject to change, as is clearly demonstrated by their near universality. Because of the fatal consequences of deviation from normal, the coding mechanism must have arisen essentially in its present format, even though at a far less complex level of organization than that which now prevails. That is to say, the enzymes, carriers, and other molecules involved in the coded processes undoubtedly consisted each of only a relatively few residues and consequently were simple, as primitive biochemicals should be. But the codons necessarily were trinucleotides and the amino acids they specified were those of proteins today--if other codes or coded amino acids ever existed, they have been lost and are forever beyond knowledge as far as present evidence indicates. The following account of the steps in establishing the coding apparatus, presented in less detail in two former contributions (Dillon, 1973, 1978), is necessarily speculative, as is any other outline on this subject, but it adheres firmly to the principle of so-called backward evolution, in which the essential end product arises first, with the mechanism for producing it evolving in reverse sequence (Goldschmidt, 1940). Moreover, the attempt is made to keep the proposal within what the author considers to be the framework of scientific theory, in which it is acceptable to theorize just one step beyond the known facts. When theories are built upon other theoretical events, they become strictly philosophical in nature, not scientific.

The Earliest Steps

The first assumption that is made is well documented, namely that the earliest protobionts were formed of proteinlike molecules, referred to either as proteinoids or protoproteins. This conclusion is supported by all known organisms, including viruses, consisting in large measure of proteins. Moreover, one primitive type of virus, the scrapie agent, has been shown to consist entirely of proteins (Prusiner et al., 1980a,b, 1981; Diener et al., 1982; Prusiner, 1982), while none, except the viroids, consist totally of nucleic acids (Sänger et al., 1976; Gross and Riesner, 1980). Indeed the viroids are not so exceptional as may appear, for while the infective dispersal stage is apparently entirely RNA, the large genome that comprises it would be meaningless if it did not encode a series of proteins synthesized within its host. Furthermore, and most convincingly, the proteinoid particles of the laboratories of Fox (1980a,b, 1981a,b) and elsewhere (Rosmus and Deyl, 1971; Melius and Sheng, 1975), have proven to be highly acceptable models of protobiontic life, whereas no laboratory anywhere has as yet produced a comparable model constructed of nucleic acids or even protonucleic acids. The latter substances are always replicated by proteinaceous enzymes; only proteins and related molecules show the self-ordering necessary at the earliest levels of primordial life. Finally, according to the principle of reverse evolution, protein molecules had to evolve before the genome and coding mechanism, for they are undeniably the ultimate focal point of all genetic processes.

The first coding event. As attested by the size of the proteinaceous virus just mentioned--hamster scrapie agents have a molecular weight between 64,000 and 150,000 (Diener et al., 1982),--the early proteinoids must have existed for a considerable period of time before nucleic acids were added to the replicatory processes. Thus for an extended period, protobionts replicated merely through self-ordering, depending on the relatively rich seas for the amino acids, possibly including many types no longer employed today. If scrapie is to be believed, this method of reproduction is very slow; the hamster agent has been reported to require several years from time of infection to the first evidence of the resulting disease (Prusiner et al., 1981). Once a sufficient size had been gained, the first step toward accelerating the processes was that of attaching the amino acids to ribonucleic acid chains, for this is the initial action in organisms today. Moreover, such tRNAs are known to conduct amino acids to many points in cells besides ribosomes, participating in such diverse functions as synthesis of cell wall materials and in ribosome-free addition of amino acids to certain proteins (Dillon, 1978, pp. 143, 144).

However, evolution proceeds by small steps at the molecular level as elsewhere, each added ingredient of necessity being functional and having probable survival value. Consequently, it is quite possible that the first interaction of a carrier and an amino acid was preceded by a simpler reaction involving only the amino acid and ATP, perhaps mediated by a simple enzyme, an early tRNA ligase.

At the primordial level, any tRNA would have had to be as simple as possible. It seems most likely that the original forebear of these carrier agents was the trinucleotide -CCA which terminates all known tRNAs and is the only entirely universal structural sequence of those macromolecules. The probable primitive nature of this terminal sector is further indicated by its being added enzymatically to many precursorial tRNA molecules in the absence of template during the processing of the precursors. Since adenosine has been demonstrated to inhibit charging of tRNAs (Lawrence, 1973; Lawrence et al., 1973, 1974), it appears necessary to propose that the entire CCA served as the first carrier, rather than that trinucleotide arising by successive stages leading to the final sequence.

Bound to this simple carrier, the various amino acids would doubtlessly have been more readily transported and then arranged on the protein molecule being replicated, for as pointed out in a preceding section, trinucleotides have masses more than ten times those of the smaller amino acids. Since the CCA-amino acid interaction is nonspecific, all types of the latter substances were probably involved in those reactions, whether or not they actually entered into the self-ordering processes. Once aligned on the protein template, the amino acids would have been freed from the carrier by the same enzyme that bound them, because ligases in extant organisms deacylate the tRNAs in reactions that are completely nonspecific (Bonnet and Ebel, 1972, 1974; Dillon, 1978, p. 143).

The earliest steps toward specificity. First it should be recalled that except under the very special conditions required to produce lysine-rich proteinoids, the majority of experimental prebiotic snytheses have yielded glycine in far greater abundance than any other amino acid (Miller, 1953; Harada and Fox, 1964; Fox and Windsor, 1970; Bar-Nun et al., 1970). The only notable exception resulted with temperatures in excess of 1000°C, 90% of the product proving to be β-alanine (Lawless and Boynton, 1973; Dillon, 1978, pp. 8-14). Consequently, it is not unlikely that the early protein molecules in protobionts consisted largely of glycine, with α-alanine second in relative abundance.

Since glycine probably formed the greater part of the protobionts, a device to distinguish that amino acid from the others would have afforded a means of enhancing protein replication, and thus would have had great survival value. Quite probably the original trinucleotide carrier had become somewhat more elongate before such a distinguishing feature evolved, so that it now may have consisted of five or more nucleotidyl residues, including CCCCA or CCCCCA and many other combinations. The critical step was not so much these chance aberrant types, which at first involved no specificity, but the occurrence of a mutation to the original simple tRNA ligase, resulting in a mutant that reacted solely (or at least predominantly) with glycine on one hand and with the (C)CCCCA variety of tRNA on the other.* These earliest coding events thus involved tRNA-charging activity and resulted in distinguishing the most fundamental amino acid-carrier combination from the remainder in the immediate environment.

A further ingredient. If this early system is viewed more closely, an additional ingredient may be deemed to have been active, too, but there is no evidence as to its actual existence. One feels that distinguishing glycine and its carrier in the manner proposed is not quite sufficient, that some sort of primitive enzyme which could recognize this species of amino-acylated tRNA and emplace it properly on the template would also have been essential to acceleration. But there seems little room for doubt that any such recognizing enzyme (which really would have been an early form of protein polymerase)** would have been advantageous to the organism only if there were something to be distinguished. Hence, the glycine-specific ligase would have had to originate first, with the protein polymerase arising shortly thereafter.

Later Events in Coding Development

It is not proposed here to repeat the complete elaboration of the evolution of the coding mechanism presented in two earlier contributions (Dillon, 1973, 1978)--space limitations prohibit that procedure. Nevertheless those accounts form a necessary portion of the present discussion, because further progress in the evolution of the genetic coding system obviously involved many complexly interwoven activities (Egami, 1981). Briefly stated, the main points established in those references are threefold: (1) In order to establish the genetic code catalogue as it is known today, it first was

*This new polypeptide was replicated, as was its forebear, by the slow processes of superpositioning amino acids upon the molecule to be replicated, in conjunction with the self-ordering properties of the monomers.

**Such protein polymerases are probably numbered among the many enzymes embedded in the ribosomes of modern organisms.

essential that 16 amino acids, one by one in a long series of events, became encoded by sets of four related triplet codons (one such set being GGG, GGC, GGA, GGU). (2) Later, certain of the sixteen were deprived of two codons each (one in several instances) that then became employed to encode other species of amino acids. This need for subsequent sharing of codons was first suggested by Jukes (1966) but has been enlarged upon only in the foregoing references. (3) The third point shows that the genetic code and the amino acids probably evolved together, an observation based on structural identities of all amino acids common in proteins of existing organisms and their relationships to the codons. Every proteinogenic amino acid is of the α type, and has a levorotatory configuration, except glycine, which is symmetrical, and rare noncoded species that occur in limited varieties of proteins in a few advanced taxa of living things. One of the inviolable principles of phylogenetic biology is that structural identity indicates descent from a common ancestry, a principle that is just as applicable to the biochemicals of life as it is to appendages of vertebrates and the parts of a flower. The structural identities among these macromolecules and their monomers and their correlations to the codon catalogue are most readily perceptible in what has been referred to as group I amino acids, those which are encoded by codons having the initial base G (Fig. 3; Dillon, 1973, 1978, pp. 218-222).

The first messenger RNA. Perhaps the origins of the first messenger RNA will always remain a mystery, unless some simple virus should eventually be found that provides pertinent concrete data, for no clear support for its first formation is known to exist today. With the suggested establishment of anticodons on the most abundant amino acid, a chain of protonucleotides may then have become polymerized in association with the protein template (which actually represents a proteinaceous genome), a chain that originally may have been a random mixture of nucleotides. Regardless of its obscure origins, this chain eventually seems to have become comprised primarily of guanylic acid residues.* When such chains were arranged parallel to the template, it may be visualized that interaction between its series of G's with the triplet C's of the primitive glycine-bearing tRNA might have provided greater speed to the former strictly self-ordering proteinogenic processes.

*The ability of the proteinaceous cytosol of cells to synthesize an RNA chain in multiple identical copies in the absence of an RNA template has been forcibly demonstrated in vitro (Sumper and Luce, 1974; Biebricher et al., 1981a,b). This and other manifestations of genome control by the cytosol has been ascribed to a second genetic mechanism in cells (Dillon, 1981, pp. 556-558, 1983, pp. 442-446).

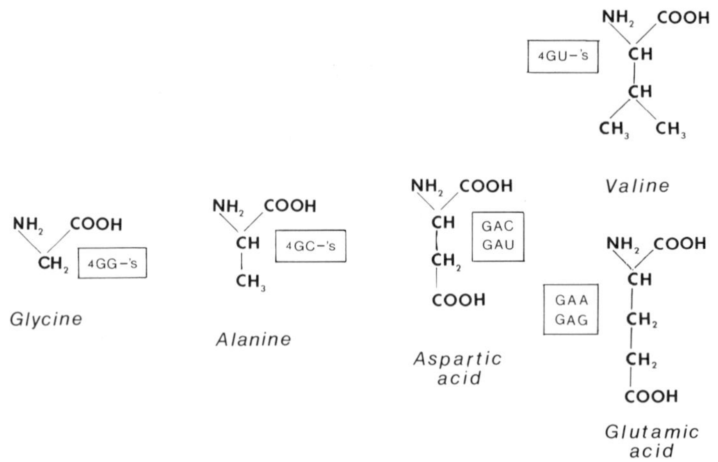

Fig. 3. The amino acids of Group I and their codons. The five members of Group I amino acids arranged as above display apparent successive stages of relations to the first one, glycine, through stepwise additions to the carbon skeleton of the preceding member, shown in stipple. All are encoded by codons with the initial letter G.

Because of the repetitious nature of the protomessenger RNA, many errors probably were made during the early steps, with glycine being inserted into the chain in sites where other amino acids should have been placed. Thus many of the products probably were malfunctional. But with time, mutations in the simple tRNAs that altered the 5' terminal (C)CC- sequence of the glycine carrier to (C)GC- gave rise to a type specific for the second most abundant amino acid, alanine. But this specialization only became operative when a new tRNA ligase had evolved that recognized with reasonable precision both the new carrier and alanine. And this alanyl-tRNA did not become effective in the protein synthesizing system until the protomessenger underwent mutation from $(G)_n$ to chains carrying C's in the correct positions, and, possibly, when the protein polymerase had acquired the capability of distinguishing the new alanyl-tRNA. Once the two most frequent amino acids of early proteins had thus been encoded and distinguished, the replicatory processes of the protobionts were not only accelerated but more efficient, the greater precision that now was provided reducing the production of inviable molecules by a large factor.

The sharing of codons. By and large, the subsequent steps in elaboration of the coding system were basically repetitions of those just outlined, except that new species of amino acids were added one by one as the necessary synthetic steps were gained by processes outlined earlier (Dillon, 1973, 1978, pp. 223-231). With each addition, (1) mutations to the 5' terminal sector of the carrier had to occur, giving rise to new varieties of simple tRNAs. (2) The added tRNA became specific for the recently created type of amino acid only after a tRNA ligase had evolved that specifically used both of those new types of molecules as substrate. (3) None of the foregoing were effective in the protein replicatory steps until appropriate mutations had also come into existence in the protomessenger, and (4) the protein polymerase could recognize the new combinations. These steps continued over geologic time until many amino acids had evolved and had become properly encoded both by the tRNA-ligase complex and the messenger-protein polymerase system. Thus eventually 16 typical proteinogenic amino acids were each encoded by sets of four related codons.

Further elaboration of the coding system involved the sharing of codons between amino acids already alluded to, but those processes were basically similar to those employed in the first 16 types. When an additional amino acid was biosynthesized and incorporated into proteins by self-ordering procedures, a new tRNA ligase could have arisen that recognized the amino acid and bound it to a nonspecific carrier, in each case the new ligase being derived from an earlier one. Thus the ligase for aspartic acid possibly also recognized glutamic acid when the latter substance was first produced. Eventually mutation and natural selection resulted in a mutant ligase that was specific for glutamate and for those former codons of aspartic acid that terminated in purines, whereas the parental ligase still recognized aspartate but subsequently evolved to be specific for the codons terminating in pyrimidines. Thus the sharing of codons between amino acids involves no unusual steps, only the ordinary processes of mutation, natural selection, and geologic time.

Although many steps throughout the entire foregoing discussion are necessarily speculative to a degree, each is nevertheless supported by processes known actually to occur in extant organisms. Undoubtedly some will prove at least partly erroneous as greater knowledge of the most primitive extant organisms is gained. The prime purpose here is that of proposing how the genetic mechanism could have arisen within a framework in keeping with sound biological principles and knowledge, in the hope that it will guide future experimental and constructionist efforts to a firmer realistic approach.

REFERENCES

Bar-Nun, A., Bar-Nun, N., Bauer, S. H., and Sagan, C., 1970, Science, 168:470.

Barrell, B. G., Anderson, S., Bankier, A. T., de Bruijn, M. H. L., Chen, E., Coulson, A. R., Drouin, J., Eperon, I. C., Nierlich, D. P., Roe, B. A., Sanger, F., Schreier, P. H., Smith, A. J. H., Stoden, R., and Young, I. G., 1980, Proc. Nat. Acad. Sci. U.S., 77:3164.
Biebricher, C. K., Eigen, M., and Luce, R., 1981a, J. Mol. Biol., 148:369.
Biebricher, C. K., Eigen, M., and Luce, R., 1981b, J. Mol. Biol., 148:391.
Bonitz, S. G., Berlani, R., Coruzzi, G., Li, M., Macino, G., Nobrega, F. G., Nobrega, M. P., Thalenfeld, B. E., and Tzagoloff, A., 1980, Proc. Nat. Acad. Sci. U.S., 77:3167.
Bonnet, J., and Ebel, J. P., 1972, Eur. J. Biochem., 31:335.
Bonnet, J., and Ebel, J. P., 1974, FEBS Lett., 39:259.
Bossard, A., and Toupance, G., 1981, in: "Origin of Life," Y. Wolman, ed., pp. 93-100, D. Reidel, Dordrecht, Holland.
Bresch, C., Niesert, U., and Harnasch, D., 1980, J. Theor. Biol., 85:399.
Diener, T. O., McKinley, M. P., and Prusiner, S. B., 1982, Proc. Nat. Acad. Sci. U.S., 79:5220.
Dillon, L. S., 1973, Bot. Rev., 39:301.
Dillon, L. S., 1978, "The Genetic Mechanism and the Origin of Life," Plenum, New York.
Dillon, L. S., 1981, "Ultrastructure, Macromolecules, and Evolution," Plenum, New York.
Dillon, L. S., 1983, "The Inconstant Gene," Plenum, New York.
Dose, K., and Rauchfuss, H., 1972, in: "Molecular Evolution: Prebiological and Biological," D. L. Rohlfing, and A. I. Oparin, eds., pp. 199-217, Plenum, New York.
Dyson, F. J., 1982, J. Mol. Evol., 18:344.
Egami, F., 1981, Origins Life, 11:197.
Eigen, M., and Schuster, P., 1977, Naturwiss., 64:541.
Eigen, M., and Schuster, P., 1978a, Naturwiss., 65:7.
Eigen, M., and Schuster, P., 1978b, Naturwiss., 65:341.
Eigen, M., and Schuster, P., 1979, "The Hypercycle: A Principle of Natural Self-Organization," Springer, Berlin.
Ferris, J. P., Joshi, P. C., Edelson, E. H., and Lawless, J. G., 1978, J. Mol. Evol., 11:293.
Findley, A. M., Findley, G. L., and McGlynn, S. P., 1982, J. Theor. Biol., 97:299.
Fox, S. W., 1968, BioSystems, 2:235.
Fox, S. W., 1969, Naturwiss., 56:1.
Fox, S. W., 1978, in: "The Nature of Life," W. H. Heidcamp, ed., pp. 23-92, University Park Press, Baltimore.
Fox, S. W., 1980a, Naturwiss., 67:378.
Fox, S. W., 1980b, Naturwiss., 67:576.
Fox, S. W., 1981a, Int. J. Quantum Chem., QBS8:441.
Fox, S. W., 1981b, J. Chromatog., 215:115.
Fox, S. W., and Nakashima, T., 1980, BioSystems, 12:155.
Fox, S. W., and Windsor, C. R., 1970, Science, 170:984.

Goldschmidt, R. B., 1940, "The Material Basis of Evolution," Yale University Press, New Haven.
Gross, H. J., and Riesner, D., 1980, Angew. Chem., 19:231.
Harada, K., and Fox, S. W., 1964, Nature, 201:335.
Hawker, J. R., and Oró, J., 1981, in: "Origin of Life," Y. Wolman, ed., pp. 225-232, D. Reidel, Dordrecht, Holland.
Hechman, J. E., Sarnoff, J., Alzner-De Weerd, B., Yin, S., and RajBhandary, U. L., 1980, Proc. Nat. Acad. Sci. U.S., 77:3159.
Jukes, T. H., 1966, "Molecules and Evolution," Columbia University Press, New York.
Lawrence, F., 1973, Eur. J. Biochem., 40:493.
Lawrence, F., Blanquet, S., Poiret, M., Robert-Gero, M., and Waller, J. P., 1973, Eur. J. Biochem., 36:234.
Lawrence, F., Shire, D. J., and Waller, J. P., 1974, Eur. J. Biochem., 41:73.
Li, M., and Tzagoloff, A., 1979, Cell, 18:47.
Lohrmann, R., 1982, J. Mol. Evol., 18:185.
Macino, G., Coruzzi, G., Nobrega, F. G., Li, M., and Tzagoloff, A., 1979, Proc. Nat. Acad. Sci. U.S., 76:3784.
Matsuno, K., 1982, BioSystems, 15:1.
Melius, P., 1977, in: "Bioorganic Chemistry," Vol. 3, pp. 123-136, Academic Press, New York.
Melius, P., and Sheng, J. Y. P., 1975, Bioorg. Chem., 4:385.
Miller, S. L., 1953, Science, 117:528.
Miller, S. L., and Van Trump, J. E., 1981, in: "Origin of Life," Y. Wolman, ed., pp. 135-141, D. Reidel, Dordrecht, Holland.
Nakashima, T., and Fox, S. W., 1980, J. Mol. Evol., 15:161.
Nakashima, T., and Fox, S. W., 1982, BioSystems, 14:151.
Orgel, L. E., 1972, Isr. J. Chem., 10:287.
Orgel, L. E., 1977, in: "Organization and Expression of the Eukaryotic Genome," E. M. Bradbury, and K. Javaherian, eds., pp. 499-504, Academic Press, New York.
Oró, J., 1980, MBL Lectures in Biology, Vol. 1, pp. 47-69.
Prusiner, S. B., 1982, Science, 216:136.
Prusiner, S. B., Groth, D. F., Cochran, S. P., Masiarz, F. R., McKinley, M. P., and Martinez, H. M., 1980a, Biochemistry, 19:4883.
Prusiner, S. B., Groth, D. F., Cochran, S. P., McKinley, M. P., and Masiarz, F. R., 1980b, Biochemistry, 19:4892.
Prusiner, S. B., McKinley, M. P., Groth, D. F., Bowman, K. A., Mock, N. I., Cochran, S. P., and Masiarz, F. R., 1981, Proc. Nat. Acad. Sci. U.S., 78:6675.
Rosmus, J., and Deyl, Z., 1971, Chromatogr. Rev., 13:163.
Sänger, H. L., Klotz, G., Riesner, D., Gross, H. J., and Kleinschmidt, A. K., 1976, Proc. Nat. Acad. Sci. U.S., 73:3852.
Schwartz, A. W., and Goverde, M., 1982, J. Mol. Biol., 18:351.
Schwartz, A. W., Joosten, H., and Voet, A. B., 1982, BioSystems, 15:191.
Subbaraman, A. S., Kazi, Z. A., Choughuley, A. S. U., and Chadha, M. S. 1980, Origins Life, 10:343.

Sumper, M., and Luce, R., 1974, Proc. Nat. Acad. Sci. U.S., 72:162.
Voet, A. B., and Schwartz, A. W., 1981, in: "Origin of Life," Y. Wolman, ed., pp. 217-223, D. Reidel, Dordrecht, Holland.
Voet, A. B., and Schwartz, A. W., 1982, Origins Life, 12:45.
White, D. H., 1980, J. Mol. Evol., 16:121.
White, D. H., and Raab, M. S., 1982, J. Mol. Evol., 18:207.
Young, I. G., and Anderson, S., 1980, Gene, 12:257.

EVOLUTION AND ENZYMES

J. Lawrence Fox

Department of Molecular Biology
Abbott Laboratories
North Chicago, Illinois 60064

INTRODUCTION

The molecular weight for a given enzyme from presumed primitive microbes to "simple" eukaryotes and on to higher eukaryotes is seen to increase in a consistent fashion for a variety of enzymes (Dayhoff, 1972; Klotz et al., 1975; Bisswanger and Schmincke-Ott, 1980; Fox and Lynen, 1980). The significance of this progression is not obvious, but can be used to increase our understanding of the evolutionary process (De Ley and Kerstens, 1975; Williams and Fox, 1974, 1980; Fox and Williams, 1978; Schwartz and Dayhoff, 1978). The evolution of organisms is intimately tied to the evolution of its enzymes. We will develop a scenario for the evolution of enzymic systems which defines a general path for organismal evolution in this paper.

Perhaps we should first note that the evidence which has accumulated since the initial enunciation of the "unity of biochemistry" by Kluyver and van Niel (1956) has only served to enforce the concept that all living matter is based on the same fundamental biological chemistry. It is true that a great many variations on the theme are noted in nature, but it is important that we not get distracted by comparing the patterns of branches on one tree with those from another and lose track of the fact that these are all trees in the same forest.

EVOLUTIONARY THEORIES

How much can we depend upon the data obtained from contemporary organisms, the products of 3.5 billion years of evolution to reveal the details of the early stages of evolution? To answer this question, we must present a few possible mechanisms for evolution in which to properly frame the question. We can describe two ends of the continuum with grossly different results. The first is the continual evolution theory; the other is the punctuated equilibrium evolution theory.

The continual (or gradual) theory of evolution (see Olson, 1981, and Gould, 1982, for discussions) indicates that organisms experience continual change. The first obvious conclusion then is that nothing can truly be primitive. However, it is quite apparent among invertebrates, if not even primates, that rates of change are not constant over time. Armadillos, for example, appear to have remained relatively morphologically unchanged for over 100 million years whereas many changes have been noted for higher primates during that same period. We are not in a position to enquire about the details of their evolution at molecular levels, but it is reasonably clear that a series of small molecular changes can be correlated with considerable morphological changes in eukaryotes, at least. Some molecular evolutionists argue for a straight line of mutational change, but in all fairness there is far too little data to do more than define the end points. For higher primates the data suggests a monotonically decreasing rate of change, not a linear rate. A great many more examples exist which serve to suggest that the simpler forms of this theory are inaccurate, and it must be modified.

The quantum or punctuated equilibrium theories of evolution (Gould and Elderedge, 1977) suggest "instantaneous" bursts of change. Instantaneous probably means a number of generations as it defies comprehension to imagine a complex, coordinate series of changes. The simpler model requires a series of individual mutational events. The end result of this burst is the creation of a new species more fit for its environment. This would serve to select the new species over its ancestors. However, since the environment undergoes continual change over long periods of time, the selective pressures do not remain constant. The culmination of a series of mutational events has been to create an irreversible path.

It might also be worth emphasizing at this point that of necessity most of the models we have created in science must lay on the printed page or the blackboard. This imparts a static nature to them which probably affects our thinking about models. Unfortunately, this is quite erroneous; all matter is in constant motion. Our models thus have the initial effect of steering us from reality by their very static nature.

EVOLUTION DRIVING FORCE

What is the driving force for this burst of change? The most acceptable answer would be a single mutational event which has a very profound effect on the organism. Since metabolism is a very complex, intertwined set of reactions, drastic change will lead to instabilities. There would, therefore, be great pressures for other mutational events of lesser magnitude to aid in reestablishing some sort of steady state. This set of events serves to establish yet another set of irreversible alterations that culminate in a new species.

We may ask the question again on a more physical basis. The driving force may be taken to be the optimization of the dissipative flux for an organism (Prigogine et al., 1972). Unfortunately the current state of the development of the formalism for non-classical thermodynamics does not permit us to ask detailed questions about this process, but the point that we are not dealing with a closed, equilibrium system has been strongly made by Prigogine's work. Very few direct measurements for organisms to substantiate this point have been made. We (Williams and Fox, 1980) have argued that some of the difficulty in dealing with this problem experimentally is derived from our reliance on inherently reductionistic modes of conceptualization. Yano and Hasegawa (1974) have calculated that the entropy of proteins increases as examples are taken from species found "up the evolutionary ladder." This observation could not be explained by the authors, but suggests that degree of order may be correlatable with evolution.

RATES OF CHANGE

It should be obvious that different organisms will be subjected to adaptation to environmental changes at different rates. Those organisms which are the most simple would be the most flexible for dealing with changes and will experience less change. The conclusion is that the more primitive organisms will be more stable whereas more complex organisms which are more dependent upon their environmental conditions, and are, hence, more highly evolved, should experience higher rates of mutation. Another corollary of this statement is that we would expect to see the mutational frequency increase as we go up the evolutionary scale.

GEOLOGICAL EVIDENCE

Is it then possible that some organisms are "living fossils"? It will depend on the simplicity and availability of heterotrophic nutrients. The fossil record tells us that a highly significant alteration in the atmosphere of the earth occurred about a billion years after the emergence of the first fossilized life forms (Cloud, 1976). The primitive atmosphere went from a rare, reducing atmosphere to a denser, oxidizing atmosphere. This drastic change must have been devastating to most, if not all, of the primordial organisms. As evolution proceeded, even more adaptive anaerobes must have been able to outcompete their ancestors, ultimately leading to their demise. Thus, it is extremely difficult to accept the notion that any organisms observed on the contemporary earth could be very "primitive."

When did photosynthesis occur? The discrepancies between the proposed date of the origin of living forms and the switch in the earth's atmosphere from reducing to oxidizing as well as simple logic suggested that heterotrophs must have preceded autotrophs. The extant photosynthetic prokaryotes possess proteins with larger polypeptides than the extant heterotrophic anaerobes for a given enzymic function; a premise in line with the hypothesis of an increase in size and complexity correlating with progress up the evolutionary scale. The relatively constant size increase in going from the non-photosynthetic to the photosynthetic prokaryotes suggests a fundamental process being responsible for this change, in particular it emphasizes a stability point at a new level of complexity.

EMERGENCE OF PHOTOSYNTHESIS

We have previously described how a coupled oscillator conceptualization of biological structures can manifest itself in a synthetic manner in new, higher levels of structure (Williams and Fox, 1974, 1980; Fox and Williams, 1978). To make this point clearer, we can resort to an analogy: the comparison of individual amino acids with the alpha helices or beta pleated sheets produced after the linear linkage of amino acids. The constraints on the alpha helix are much greater than the constraints on the individual amino acids, yet the emergence of an alpha helix introduces a new level of complexity with properties not apparent with individual amino acids.

The photosynthetic system requires the emergence of a new series of membraneous boundaries within the cell. These bear a strong resemblance to the plasma membrane. We could imagine that primitive systems contained a number of electron transfer components and probably existed for non-photosynthetic purposes first. Efficiency would increase if these redox components had lost a degree of freedom by incorporation into a membraneous structure.

Only later did the evolution of light harvesting chromophores yield a photosynthetic system. Since most electron transferring proteins have chromophoric properties, it is possible to imagine that primitive electron transport proteins evolved into photosynthetic sensitizing substances. The first photosynthesizers may be considered to have been nonaqueous photooxidizers. Reactive oxygen species would have been highly deleterious to such primitive systems.

EMERGENCE OF INTERNAL MEMBRANES

The efficiency of these primitive photosynthesizing systems would be enhanced by the formation of the concentric propagation of membraneous structures now obvious in the cyanobacteria. The emergence of these semi-autonomous systems would permit the separation of the photolysis of water and its oxygen by-products from the unstable products of contemporary photosynthesis: ATP and NADPH, or their equivalents.

The emergence of an oxidizing atmosphere from the primitive, reducing atmosphere presented a very difficult environmental challenge to most organisms. Our scenario now suggests that the primitive cells learned to cope with this by breaking up their internal photosynthetic membraneous sacks into smaller pieces. In order for this to occur, the cells needed to be able to produce genomic copies for each new membraneous sack. All this required was the loss of binary fission of the plasma membrane, but the continued fission of the internal membraneous sacks. One solution to the mechanism for this process is to envisage that the increasing oxygen concentrations externally had deleterious effects on plasma membrane division and yielded this result. This is analogous to the maternal fission theory of chloroplast and mitochondrial lineage for contemporary organisms. Another possibility is the result of the development of the extrachromosomal genomic element, the plasmid. If plasmids did arise early in evolution, they could have served as the prototype for subsequent linear chromosome development.

Once the cell had multiple versions of these membraneous sacks, a new set of specialization events can be envisaged. One set continued on its path towards the contemporary chloroplast. A second set began to forget how to photosynthesize, but continued with its evolutionary development of its electron transporting components. This evolved into contemporary mitochondria. The last set forgot about all electron transfer reactions and evolved into what we now know as nuclei. This occurred from the duplication of the primitive single prokaryotic chromosome to cell which possessed several copies of the chromosome. Since the normal function of only one set of gene products was essential for survival, the second set of genes could undergo mutation with little negative effect. This could help to explain the observations that related polypeptide chains, such as the alpha and beta chains of hemoglobin, are found on separate chromosomes, but experience coordinate regulation.

This model then provides some accommodation for the remnants of prokaryotic-like replication, transcription and translation mechanisms still found in chloroplasts and mitochondria. It, however, provides for a simpler and more easily understood pathway for the emergence of eukaryotic cells than the symbiotic theories that are currently in discussion. The remaining problem still is the emergence of the eukaryotic type nuclear organization.

Broda (1978) has presented some of the arguments lately for contrasting theories of the evolution of the mitochondria and the chloroplasts. The endosymbionic origin theory (Margulis, 1970) has attracted attention for a century. Margulis proposes that mitochondria evolved first, then mitosis appeared and then the chloroplasts appeared. This argument is quite difficult to swallow as there seems little need for true mitochondria prior to the emergence of an oxygen rich atmosphere. Stanier (1970) has advanced counterarguments showing the evolution of chloroplasts into mitochondria.

Klein (1970) has argued that the special relationships between blue-green algal (cyanobacteria) and red algal photosynthetic accessory pigments and membraneous structures places the algae in a special position in the prokaryote → eukaryote transition. This argues that the evolution of eukaryotes occurred via photosynthetic organisms and that the animal kingdom then branched off from the primitive eukaryotes. This would support the argument we are presenting in this paper. Stanier has further pointed out that prokaryotes are not known to harbor any endosymbionts as are known in eukaryotes. It is, perhaps, again our simplistic thinking which casts this in such a problematic manner. Broda and Peschek (1979) argue "prerespiration" may have preceded photosynthesis, but that only once oxygen evolved into the atmosphere did contemporary style respiration have a chance to evolve. We are adopting this line of argument here.

NUCLEAR ORIGINS (Eukaryotic Evolution)

We should return to the question raised above about origin of the eukaryotic nucleus. As this eunuclear system began to sort out with the chloroplast and mitochondria structures the partitioning of various synthetic activities, the eunuclear genomic content began to increase. A circular organization simply became too complex, so the DNA was broken into linear pieces. Once this development had begun, a whole new series of propagative mechanism began to emerge, forcing the separation of these genetic mechanisms continually farther and farther from each other. To provide finer levels of control over synthetic reactions, the less identical the two sets of systems were, the more adaptive the organism would become.

Once the true eukaryotic cell had evolved, the basic elements of regulation had developed. This then permitted the emergence of multicellular organizations. A couple of steps still remained. The multicellular organism had to reach a critical mass before the specialization of determined cell types occurred, defining tissue types. From tissue types, there remained the evolution of organ systems. These latter topics are too far removed from the molecular level of understanding to be discussed in any greater detail at this time by this author.

The thread currently holding this argument together is the effect of the alteration of the genetic mechanism on enzymic structure. It is only too easy in such a discussion to overlook the fact that selection is made at the level of activity of the protein. There are now several cases in the literature which suggest that this conservation occurs in three dimensional terms (actually functional terms) that are often obscured in proteins with low levels of sequence conservation (Rossmann et al., 1974; Richardson, 1981). This point is clearer as more protein three-dimensional structures are obtained. The data base is still small, but would be greatly extended if theoretical structure prediction methods become reliable. The expense and difficulties in obtaining crystal structure solutions will severely limit this data base otherwise.

EPILOG

The development of recombinant DNA technology and new protein sequencing methodology has simplified the difficulties of protein sequence determinations. However, it is the structure-function relationships which will yield the proper understanding of evolutionary relationships. Sequence data is already clearly at too fine a level of detail in a two-dimensional sense for us to observe many of the more important criteria for evolutionary comparison. However, we currently lack the ability to translate this information into three-dimensional relationships, except for short stretches of sequence. New methods for simple identification of macromolecular structure are required. Advancements in theoretical structure prediction may be invaluable in these studies, but ultimately these predictions must be testable by experimental confirmation. Perhaps Raman resonance or multiquantum spectroscopy can also provide this data.

REFERENCES

Bisswanger, H., and Schmincke-Ott, E., eds., 1980, "Multi-functional Proteins," Wiley-Interscience, New York.

Broda, E., 1978, "The Evolution of the Bioenergetic Process," revised edition, Pergamon Press, Oxford.

Broda, E., and Peschek, G. A., 1979, J. Theor. Biol., 81:201.

Cloud, P., 1976, Paleobiology, 2:351.

Dayhoff, M. O., 1972, "Atlas of Protein Sequence and Structure," Volume V, National Biomedical Research Foun-ation, Washington, D.C.

De Ley, J., and Kerstens, K., 1975, Compreh. Biochem., 29B:1.

Doolittle, R. F., 1979, in: "The Proteins," Volume IV, 3rd edition, H. Neurath, and R. L. Hill, eds., pp. 1-118, Academic Press, New York.

Fox, J. L., 1972, in: "Molecular Evolution: Prebiological and Biological," D. L. Rohfling, and A. I. Oparin, eds., pp. 23-34, Plenum, New York.

Fox, J. L., and Lynen, F., 1980, Eur. J. Biochem., 109:417.

Fox, J. L., and Williams, M. D., 1978, in: "Origin of Life," H. Noda, ed., pp. 441-446, Japan Scientific Societies Press, Tokyo.

Gould, S. J., 1982, Science, 216:380.

Gould, S. J., and Elderedge, N., 1977, Paleobiology, 3:115.

Klein, R. M., 1970, Ann. NY Acad. Sci., 175:623.

Kluyver, A. J., and van Niel, C. B., 1956, "The Microbe's Contribution to Biology," Harvard University Press, Cambridge, Massachusetts.

Margulis, L., 1970, "Origin of Eukaryotic Cells," Yale University Press, New Haven.

Olson, E. C., 1981, Quart. Rev. Biol., 56:405.

Prigogine, I., Nicolis, G., and Babloyantz, A., 1972, Physics Today, 25:23; 38.

Richardson, J. S., 1981, Adv. Prot. Chem., 34:168.

Rossmann, M. G., Moras, D., and Olsen, K. W., 1974, Nature, 250:194.

Schwartz, R. M., and Dayhoff, M. O., 1978, Science, 199:395.

Stanier, R. Y., 1970, in: "Organization and Control in Prokaryotic and Eukaryotic Cells," H. P. Charles, and B. C. Knight, eds., Symp. Soc. Gen. Microbiol., 20:1.

Williams, M. D., and Fox, J. L., 1974, in: "The Origin of Life and Evolutionary Biology," K. Dose, S. W. Fox, G. A. Deborin, and T. E. Pavlovskaya, eds., pp. 461-468, Plenum, New York.

Williams, M. D., and Fox, J. L., 1980, in: "Chemical Recognition in Biology," F. Chapeville, and A.-L. Haenni, eds., pp. 86-96, Springer-Verlag, Berlin.

Yano, T., and Hasegawa, M., 1974, J. Mol. Evol., 4:179.

PHYLOGENIC STATUS OF ARCHAEBACTERIA: CONSIDERATIONS BASED ON mRNA

Tairo Oshima

Mitsubishi-Kasei Institute of Life Sciences
Machida, Tokyo
Japan

INTRODUCTION

Based on the sequence homology of ribosomal 16SRNA, Fox et al. (1977) have proposed a new taxonomic kingdom called "archaebacteria." They proposed that organisms can be divided into three major groups, eubacteria (cyanobacteria and bacteria except archaebacteris), archaebacteria, and eukaryotes. They speculate that archaebacteria may have been the dominant form of life in early Archaen times (Woese and Fox, 1977). At the present, four types of unusual microorganisms are classified as archaebacteria: methane producing anaerobic bacteria, extreme halophiles, extreme acid-thermophiles, and extremely thermophilic anaerobes (Woese et al., 1978; Zillig et al., 1981).

In contrast to the idea that archaebacteria might be the direct descendants of primitive life, the hypothesis that archaebacteria are closely related to eukaryotes has also been proposed since certain biochemical properties of archaebacteria, including the sequence homology of biopolymers, are similar to those of eukaryotes (Searcy et al., 1978; Hori and Osawa, 1979; Van Valen and Mariorana, 1980).

In any case, archaebacteria can be regarded as living fossils which provide clues to early biological evolution. According to paleontological records, eukaryote cells evolved more than 1.5 billion years ago (Schopf and Oehler, 1976). Biochemical studies on archaebacteria may give some insight into the evolutionary events that occurred more than 1.5 billion years ago.

BIOCHEMICAL DIFFERENCES IN EUBACTERIA AND EUKARYOTES

Many of the biochemical properties of eukaryotes differ from those of prokaryotes (Margulis, 1981). In a sense, there is a gap between prokaryotes and eukaryotes. For instance, eubacteria (except Mycoplasma) are surrounded by peptide-glycan cell walls, whereas eukaryotes have no such walls. Usually animal cells have no cell walls. Likewise, the structure of eukaryote flagella is quite different from that of eubacteria.

However, the most remarkable difference is in the genetic information machinery. Eukaryotes differ from eubacteria in that they possess nuclei, histones and nucleosomes. DNA dependent RNA polymerase is different with respect to the number of subunits. The ribosomes of eukaryotes consist of 40S and 60S subunits, whereas those of eubacteria, of 30S and 50S subunits. Ribosomal RNAs in eukaryotic ribosomes are larger than the corresponding prokaryotic rRNAs. Polypeptide elongation factors and tRNAs are other examples of definite differences in eukaryotes and eubacteria. Protein synthesis in eubacteria can be initiated only with formylated methionyl-tRNA, whereas in eukaryotic cells, formylation is not required for the initiation.

These differences in chemical structure are reflected by varied sensitivities toward antibiotics and other toxins. Rifampisin inhibits RNA synthesis in eubacteria, but has no effect on eukaryotic RNA polymerases. Chloramphenicol and streptomycin are specific inhibitors for eubacterial protein synthesis. Eukaryotic translation can be specifically depressed by anisomycin or cycloheximide.

PHYLOGENIC STATUS OF ARCHAEBACTERIA

In considerations the differences stated above, archaebacteria may be regarded as unique organisms. Their cells do not possess nuclei, but histone-like proteins and nucleosome-like particles have been detected (Searcy et al., 1978; Ohba and Oshima, 1980). DNA dependent RNA polymerase from archaebacteria is also unique in its subunit structure, which is similar to one of the eukaryotic enzymes. Like eukaryotic polymerase, the archaebacterial enzyme is resistant to rifampisin (Prangishville et al., 1982).

Studies on the sequence homology of ribosomal 5S RNA suggest that halophilic archaebacteria evolved from an ancestor to eukaryotic organisms (Hori and Osawa, 1979). The sequence homology of ribosomal A protein (Matheson et al., 1980) and a tRNA (Kuchino et al., 1982) also suggests a closer relationship of archaebacteria with eukaryotes, although sedimentation coefficients of archaebacterial ribosomes, ribosomal subunits, and rRNAs are similar to those of eubacteria counterparts. Homology studies of 16S RNA base sequence indicate that archaebacteria are equally distant from eubacteria and eukaryotes (Woese and Fox, 1977).

A definite difference in eukaryotic and eubacterial polypeptide elongation factors can be seen in the sensitivity toward diphteria toxin. The factor EF-2 from eukaryotes is sensitive to the toxin in the presence of NAD. The toxin catalyses ADP-ribosylation of EF-2 and the factor loses its activity. In contrast, the corresponding factor from eubacteria, EF-G is quite resistant to diphitheria toxin. Surprisingly, it has been reported that archaebacterial factors are ADP-ribosylated by the toxin (Kessel and Klink, 1980).

Based on these observations, the hypothesis that archaebacteria may be the direct descendants of primitive eukaryotes has been proposed (Searcy et al., 1978; Van Valen and Mariorana, 1980; Hori and Osawi, 1979). In particular, Thermoplasma, an acidothermophilic archaebacterium, is considered to be a possible candidate in the symbiotic theory (Margulis, 1982), as a host cell, which had been infected by an aerobic bacterium that was eventually converted into mitochondria.

To come to a definite conclusion in regard to the phylogenic status of archaebacteria, it would be necessary to compare every detail of the replication, transcription, and translation processes of archaebacteria with those of eubacteria and eukaryotes. In this regard, the author and his colleagues have directed their attentions to the mRNA of archaebacteria.

POLY(A)RNA OF ARCHAEBACTERIA

A gap can be seen in the chemical structure of eukaryotic mRNA and that of eubacteria. The mRNA of eukaryotes has a "cap" structure at the 5'-end, and a poly(A) tail at the 3'-end. The messenger is monocistronic, whereas eubacterial mRNA, in general, possesses none of these properties.

We found that poly(A)-cluster containing RNA is present in archaebacterial cells. Poly(A) containing RNA can easily be isolated by affinity chromatography using poly(U)- or poly(dT)-cellulose. When the total RNA extracted from an acidothermophilic archaebacteria, Sulfolobus acidocaldarius, was applied on a poly(dT)-cellulose column, 0.23% of it was retained on the column (the data will be reported elsewhere). This value is close to the ratio of poly(A) containing RNA/total RNA extracted from eukaryote cells, and is much larger than that reported for eubacterial RNA (Nakazato et al., 1975).

Chemical analysis using ribonucleases and a kinase showed the poly(A) cluster to be located at the 3'-terminal of the isolated poly(A) containing RNA from S. acidocaldarius. The size of this cluster was estimated to be 4.5S or 120 nucleotides, which is also close to that reported for the poly(A) tail of eukaryotic mRNA.

The presence of poly(A)RNA in eubacteria has also been reported (Gropalakrishna and Sarkar, 1982). However, the size of the poly(A) tail seems to be shorter than those of eukaryotic messengers. The poly(A) cluster found in three types of archaebacteria are similar to those of eukaryotic with respect to size.

Poly(A) containing RNA was found in other archaebacteria, Thermoplasma and Halobacterium halobium; 0.05% of total RNA extracted from Thermoplasma and 0.09% of total RNA of H. halobrium was retained on the poly(dT)-cellulose column (Ohba and Oshima, 1983).

The isolated poly(A) containing RNA of S. acidocaldarius performed the role of mRNA in a cell-free extract of retinculocytes for in vitro protein synthesis. However, protein synthesis directed by the archaebacterial poly(A)RNA was more active in a cell-free extract of S. acidocaldarius than in a reticulocyte extract.

The proteins produced were analyzed by SDS-polyacrylamide gel electrophoresis. The major product was protein(s) of a molecular weight of 23 K. This protein has not yet been characterized. When the total RNA was added to the reaction mixture instead of the poly(A)RNA, other polypeptides were synthesized. This suggests that only specific protein(s) is translated by the poly(A)RNA as the messenger. The product(s) might be of special importance for the growth of acidothermophile and a poly(A) tail may possibly stabilize the messenger for the protein.

CAP STRUCTURE

The presence of a cap at the 5'-terminal of a messenger RNA can be easily examined by studying the effect of 7-methylguanosine, a structural analog of the cap, on protein synthesis (Asselbergs et al., 1978). The modified nucleoside inhibits protein synthesis in vitro directed by the capped mRNA. Cell-free protein synthesis in the presence of the poly(A)RNA of archaebacteria was not inhibited by 7-methylguanosine (Ohba and Oshima, 1983). These results suggest the absence of a cap structure in archaebacterial mRNA.

Our studies on the poly(A)RNA of archaebacteria have not provided conclusive evidence for the phylogenic status of archaebacteria. The presence of a poly(A) tail and absence of a cap structure suggest that the organisms are phylogenically distant from either eubacteria or eukaryotes, although the presence of a poly(A) tail favors the idea that archaebacteria are more closely related to eukaryotes than to eubacteria.

Our study suggests the presence of mRNAs both with and without poly(A) tails in archaebacteria. The polyadenylation of mRNA seems to have evolved at an early stage of evolution before the divergence of eubacteria, archaebacteria and eukaryotes. Archaebacteria and eukaryotes share the common ability to make long poly(A) tails consisting of more than 100 nucleotides. Eubacteria lack this ability. In contrast, modification at the 5'-end of mRNA seems to have evolved only in eukaryotes after the divergence of eukaryotes and archaebacteria. Such changes in the translation apparatus might have induced changes in other parts of the machinery such as ribosomes, tRNA, and elongation factors. Table 1 summarizes the comparison of the biochemical properties of the translational machinery of eubacteria, archaebacteria, and eukaryotes.

Table 1. Biochemical properties of the translation machinery of archaebacteris

	Properties
Ribosomes	
Subunits	eubacterial type
5S RNA	homology to eukaryote
16S RNA	independent
A protein	homology to eukaryote
EF-G	similar to eukaryote
Initiation	similar to eukaryote(?)
tRNA sequence	partially similar to eukaryote
mRNA	
Cistron	?
Cap	may be absent
Poly(A) tail	present

REFERENCES

Asselbergs, F. A. M., Peters, W., Van Venrooij, W. J., and Cabooche, M., 1978, Eur. J. Biochem., 88:483.

Fox, G. E., Magrum, L. J., Balch, W. E., Wolfe, R. S., and Woese, C. R., 1977, Proc. Nat. Acad. Sci. U.S., 74:4537.

Gropalakrishna, Y., and Sarkar, N., 1982, Biochemistry, 21:2724.

Hori, H., and Osawa, S., 1979, Proc. Nat. Acad. Sci. U.S., 76:381.

Kessel, M., and Klink, F., 1980, Nature, 287:250.

Kuchino, Y., Ihara, M., Yabusaki, Y., and Nishimura, S., 1982, Nature, 298:684.

Margulis, L., 1981, "Symbiosis in Cell Evolution," W. H. Freeman, San Francisco.

Matheson, A. T., Moller, W., Amous, R., and Yaguchi, Y., 1980, in: "Ribosomes: Structure, Function and Genetics," G. Chamblis et al., eds., pp. 297-332, University Park Press, Baltimore.

Nakazato, H., Venkatesan, S., and Edomonds, M., 1975, Nature, 256: 144.
Ohba, M., and Oshima, T., 1981, in "Origin of Life," Y. Wolman, ed., pp. 543-549, D. Reidel, Dordrecht, Boston and London.
Ohba, M., and Oshima, T., 1983, Origins Life, in press.
Prangishvilli, D., Zillig, W., Gierl, A., Biesert, L., and Holz, I., 1982, Eur. J. Biochem., 122:471.
Schopf, J. W., and Oehler, D. Z., 1976, Science, 193:47.
Searcy, D. G., Stein, D. B., and Green, G. R., 1978, BioSystems, 10:19.
Van Valen, L. M., and Mariorana, V. C., 1980, Nature, 287:248.
Woese, C. R., and Fox, G. E., 1977, Proc. Nat. Acad. Sci. U.S., 74:5088.
Woese, C. R., Magrum, L. J., and Fox, G. E., 1978, J. Mol. Evol., 11:245.
Zillig, E., Tu, J., and Holz, I., 1981, Nature, 293:85.

THE ADAPTATIONIST PROGRAMME IN MOLECULAR EVOLUTION: THE ORIGINS OF GENETIC CODES

John R. Jungck

Department of Biology
Beloit College
Beloit, Wisconsin 53511

Salthe (1982) has recently argued that a few notable contemporary evolutionary spokespersons are "guilty of excluding the origin of life from evolutionary theory." This Anglo-American position has had strong currents ever since Darwin and has been criticized since then, especially by scholars from a Germanic-monistic tradition (Farley, 1972):

> British scientists, whether followers of Bacon, Mill, or Whewell, were in agreement that any discussion of the origin of life was beyond the scope of science . . . a reluctance to discuss the origin of life was [due to] the strong background of Natural Theology. To dismiss God as the designer of organic life was one thing, but to dismiss him as the creator of life was quite another.

On the other hand, Lewin (1982a) has stated "molecular biologists still don't understand evolution." I personally take umbrage at such a statement since I believe that the quality of many articles in The Journal of Molecular Evolution, Origins of Life, and BioSystems speak for themselves. Many of these authors have had formal educational background in population genetics, systematics, molecular genetics, and prebiotic chemistry. Nonetheless, many of the fundamental problems in molecular evolution have been obfuscated by a number of unwarranted assumptions (Jungck, 1973, 1977, 1983). Hence, the recent challenge of Gould and Lewontin (1979), Gould and Vrba (1982), and Lewin (1982b) (with replies by Mayr (1983) and Brown (1982)) on the "adaptationist programme" in evolutionary thinking warrants our serious attention.

Gould and Lewontin (1979) describe six salient features of the "adaptationist programme":

1. An organism is atomized into "traits" and these traits are explained as structures optimally designed by natural selection for their functions.
2. After the failure of part-by-part optimization, interaction is acknowledged via the dictum that an organism cannot optimize each part without imposing expenses on others. The notion of a 'trade off' is introduced, and organisms are interpreted as best compromises among competing demands.
3. If one adaptive argument fails, try another.
4. If one adaptive argument fails, assume that another must exist; a weaker version of the first argument.
5. In the absence of a good adaptive argument in the first place, attribute failure to imperfect understanding of where an organism lives and what it does.
6. Emphasize immediate utility and exclude other attributes of form.

I believe that 'the origins of genetic codes', as a problem in molecular evolution, brings these philosophical challenges well into focus. The two fundamentally new concepts of Gould and Vrba (1982) for alleviating the language problems of the adaptationist programme, namely aptation and exaptation, are shown here to be useful for examining *de novo* aspects of the origins of life.

I believe that 'the origins of genetic codes', as a problem in molecular evolution, brings these philosophical challenges well into focus. The two fundamentally new concepts of Gould and Vrba (1982) for alleviating the language problems of the adaptationist programme, namely aptation and exaptation, are shown here to be useful for examining de *novo* of the origins of life.

Exaptation refers to "features that now enhance fitness but were not built by natural selection for their current role" (Gould and Vrba, 1982). Obviously, since *de novo* materials are not the product of natural selection, the features of molecules in the earliest stages of chemical evolution should be understood more in terms of structural and thermodynamic constraints and not in terms of selective constraints. On the other hand, just because selection may have acted on a particular characteristic does not mean that it will not have adaptive consequences. Such features can have both "dispositional" and "propensity" aspects to them. For example, a dispositional property of sodium chloride is its solubility in water and a propensity of NaCl is salinity. Thus, features of primordial matter will have well defined dispositional properties, but the probability of expressing propensities are far more dependent on the context within and between selected systems.

Elsewhere I (Jungck, 1983) have argued that philosophers of science should be interested in three other evolutionary theory debates over (1) units of selection, (2) levels of selection, and (3) limits of selection. Here I will instantiate these assertions with examples from the models for the origins of genetic codes.

ORIGINS OF GENETIC CODES

The "just-so" stories in molecular evolutionary "adaptationist programmes" usually are grounded in one of two extreme interpretations of randomness. Wimsatt (1980) has extensively explored "Randomness and Perceived-Randomness in Evolutionary Biology" with particular attention to different ways of defining the degree of randomness of a sequence. Miller and Orgel (1974), for one example among many, illustrate one extreme. They presume that amino acids polymerize randomly in the absence of a template or an ordering mechanism such as the genetic code. They believe: "The production of a well-defined and fairly long polypeptide sequence, except by residue-by-residue replication, requires a new 'enzyme' for each kind of link." This argument has two adaptationist flaws: (1) an inherent assumption that evolution of the contemporary system is a given constraint, and (2) that of processes 'require' a particular path to their products (counterexamples to residue-by-residue replication can be given from formal language theory where sequence growth is specified by some previous elements in the growing chain). Crick's (1968) "frozen accident" which assumes a lack of stereochemical interaction between amino acids and nucleotides is also representative of this kind of thinking.

These perspectives ignore the "aptations" of amino acids. Amino acid side chains come in various grades from small to bulky, hydrophobic to hydrophilic, charged (basic or acidic) to uncharged, aromatic to nonaromatic, chromatic to nonchromatic, etc. The specific nature of the side chain of an amino acid can have tremendous adaptive significance in the selective history of a protein such as cytochrome C where the residues which bond to the prosthetic group heme are highly conserved or one such as trypsin with its cascade-like charge transfer complex. Yet these properties of the amino acid pre-existed their utilization in such proteins and cannot be said to have emerged because of the selective history. In other words, in molecular evolution of sequences, the selection is on composition, sequence, length (quantity) and interaction between units, but not on the atomic constituents of the units themselves. While the notion is still anthropomorphic, Jacob (1977) and Gould (1978) analogize evolution to a "tinkerer" rather than a grand designer.

Tinkering is done with spare parts; make do with what you have on hand. Similarly, evolution is opportunistic to the degree to which it makes use of the features of ubiquitous materials in selecting for contraptions which have worked well enough to survive. Intriguingly, while Gould uses the "Panda's thumb" as his example, Jacob's example was a molecular one, namely the complement system in immunology. And while others have avoided the origins of life in evolutionary thinking, he addresses it head on: "But once life had started in the form of some primitive self-reproducing organism, further evolution had to proceed mainly through alterations of already existing compounds." Elsewhere, I (Jungck, in preparation) have examined the question: "Genetic Code: Contraption or Contrivance?" and have used Gould's and Jacob's metaphorical analysis.

Adaptationist models are consistent with the slogan: "order out of chaos." They neglect two important kinds of data in the molecular evolutionary literature. First, in the absence of a template, amino acids do not polymerize randomly (Nakashima et al., 1977; Melius, 1982; Steinman, 1967; Calvin, 1969; Fox et al., 1982). In other words, selective tacticities between amino acids affect the polymerization process so as to yield nonrandom sequences. I (Jungck, 1971) inconclusively searched contemporary sequences for possible vestiges which might be remnant of these primordial constraints, but these should not be excluded on _a priori_ grounds because given the combinatorial complexities of potential sequences, selection may have simply operated on which of these constrained sequences preferentially survived (with some novelty arising by mutation and breakage-reunion). Second, they neglect the selective tacticities of amino acids or proteins with supposed template molecules (nucleotides or nucleic acids). Strong correlations between the properties of amino acids and anticodonic dinucleotides exist (Weber and Lacey, 1978; Jungck, 1978). Thus, both the assumptions of "random polymerization" and "lack of stereochemical interaction" are empirically false.

These assumptions persist presumably because of the attractive logic of a just-so adaptationist story which probably goes something like this: sequences in contemporary living systems are highly nonrandom, and are ordered by genetic coding mechanisms, therefore, before the existence of a proto-gene and coding mechanism, no selective tacticities of any consequence existed. I believe that this is within the _deus ex machina_ tradition: If the god of design is unwarranted, then nonrandomness is due to natural selection as a design agent. On psychological grounds, it seems that creationist thinking can breed well when "adaptationist programmes" fail; i.e., because the products of evolution are so nonrandom, they look for an external source. This group even includes some scientists: Crick (1982), Hoyle and Wickramasinghe (1978), Yockey (1981), and others.

Another just-so adaptationist story is inherent in the logic of the vast majority of "vocabulary expansion" models, so named by Woese (1969) for origins of genetic codes. Jukes (1965, 1967), Dillon (1978), Hartman (1975, 1978), Wong (1975, 1976, 1981), Hoffman (1975), Crick et al. (1976), Shepherd (1981), and Clark (1982) have all described models of primitive codes which began with a few amino acids and/or a few nucleotides and later added more units of each. While these models have been criticized for such problems as how to change codon size from one nucleotide to two to three without destroying the translation mechanism, I do not know that they have been criticized as a just so story. As with the "order out of chaos" slogan of the previous examples, these models have an inherent slogan: "simple to complex." Again, there is an underlying anthropomorphic logic here: if you were to design something tremendously complex, you would probably begin with a simple model. Furthermore, there is evidence that selection has operated simultaneously on two, three, and four nucleotide sequences in coding of amino acids (Lagerkvist, 1980; Cedergren et al., 1980) and that selection moves sometimes from complexity to simplicity such as in the evolutionary history of the mitochondrial genome.

By ignoring exadaptive properties of amino acids and nucleotides, these models fail to begin with empirical data. For example, the nucleation size of nucleotide sequences which provides thermodynamically and kinetically metastable complexes (i.e., with significant association constants) is three. Thus, the specificity of codon:anticodon triplet pairs may be reflective of such a physicochemical limitation rather than as the result of selection <u>for</u> trimers. Sober (1982) has eloquently differentiated the selection <u>for</u> and selection <u>of</u> properties by using a child's sorting toy in which when shook all the blue marbles fall to the lowest level and other colored marbles reside on upper levels. The selection for was for smaller size which allowed blue marbles to fall through the holes in the roof of the bottommost level and selection of blue color was a mere hitchhiker effect. If we simply look at the products of selection (such as small, blue marbles in one section of the toy), we may incorrectly infer the nature of the selection. Thus, the question should seemingly remain open: Was the triplet size of codons selection <u>for</u> or selection <u>of</u>? The scientific norm of parsimony may operate in the preference of one just so story over another but it may not always lead us to reliable conclusions about what actually happened.

A third just-so aspect of design-type thinking is an underlying premise of linear causation. The genes-first model probably stems from at least two different sorts of reductionist thinking. First, the central dogma of molecular genetics: DNA → RNA → protein was linear and reified DNA as the "secret of life." Secondly, our culture is replete with examples of a management-control ideology. Both of these have been roundly attacked by investigators with a more cybernetic view (Commoner, 1968) and a more leftist view (Hubbard, 1982). As one example of feedback, DNA polymerases and enzymes which catalyze the formation of monomers are involved in DNA replication. Recently, Eigen's and Schuster's (1979) hypercycle and White's and Raab's (1982) autogen are developed models for the evolution of coding mechanisms which contain explicit dynamical equations to describe the nature of cybernetic interaction. While the *de novo* pre-existence of proteins or nucleic acids could have historical consequences on the way in which the second one evolved (Fox and Dose, 1977), it seems to me that the physicochemical nature of their *interaction* with one another is paramount (Jungck, 1978; Root-Bernstein, 1982a).

The opposite extreme of just-so stories is inherent in the orthogenetic, deterministic school of molecular evolution. Kenyon and Steinman (1969) and Fox and Dose (1977) have argued for a "biochemical predestination." They feel that the selection tacticities of molecules are so strong as to lead unambiguously to our current form of life, especially the genetic code. Fox (1983) states:

> The self-ordering, i.e., self-limiting, nature of these processes would then have resulted in products that would have been much alike from one event of spontaneous generation to another. This would be a literal manifestation of molecular determinism; molecular determinism would have been the evolutionary precursor of genetic determinism. . . . This kind of internal directedness has much in common with concepts of orthogenesis.

The power of these models compared to the previous ones discussed is that they are eminently falsifiable; i.e., they are not only based upon experimental support, but they are further testable via explicit experiments. Weber and Miller (1981), Balasubramanian (1982), Balasubramanian, Seetharamulu, and Raghunathan (1980), Sukhodoletz (1980), and Shimizu (1982) are similarly deterministic. The latter states: "The lock and key relation can hold only for the twenty kinds of protein amino acids and for no other non-protein amino acids."

These genetic code models are an extreme example of what Woese (1969) called the "direct interaction" hypothesis. Bertman and Jungck (1978, 1979) and Jungck (in press) have showed that, even if strict combinatorial restrictions are placed on the number of codons per amino acid, there still exist more possible alternative codes than could have been selected *for* in all of the evolutionary history of our universe. For example, we calculated that even if the genetic codes were limited to arise from the group of codes with the distribution of code words to 21 categories in the same frequency as our present code, then there would still be 10 to the 70th power possible codes to select among. I believe that this philosophy of determinism was a psychological (perhaps religious) overreaction to the empirically false assertions of the "orders out of chaos" and "simple to complex" schools of thought.

When adaptationism thinking goes awry, some authors have sought nonselective scenarios which do not depend simply upon the properties of the molecules themselves. Medawar (1978) roundly criticizes one such example:

> I lost confidence in the threesome of Lacey, Weber and Pruitt writing on the 'Edge of Evolution' when I read the opening paragraph, which draws a parallel between organic evolution and learning. . . . The authors go on bravely to try to envisage molecular mechanism by which a cell might learn from its environment as we do when we learn to adapt to it. It is, of course, the very difficulty of this enterprise that has caused instructive theories of evolution to be supplanted by selective theories.

Thus, there appear to be four kinds of argument which might be rejected by the previous series of analyses: (1) hyper-adaptationist just-so stories, (2) creationist or supernatural or extraterrestrial explanations, (3) orthogenetic determinism scenarios, and (4) instructive theories.

Each deterministic school of thought ignores the other half of Gould's and Vrba's (1982) new vocabulary; namely, aptation. Functional properties of contemporary biological systems can be either adaptations or exaptations. An exaptation is a:

> structure that by chance performs an essential current role . . . because such a structure, having been co-opted, is fit (aptus) by reason of (ex) its form, or ex aptus. This contrasts with a structure that has been pushed toward fitness, ad aptus, by natural selection, and this is an adaptation . . . the term adaptation is currently used to describe the static phenomena of being fit. Gould and Vrba suggest that a more appropriate word would be simply, aptation, the modes of which would be adaptation and exaptation, depending on the historical circumstances. (Lewin, 1982a)

Part of the difference lies again in selection for and selection of thinking. The predestination model presumes that properties of the materials preclude alternative avenues; this is naive from the standpoint of biotic and geophysical environmental affects on a biotic system in question. Elsewhere, I (Jungck, 1972) have shown how a thermodynamic perspective which focuses on order within a system to the exclusion of its environment leads to faulty conclusions about the interaction of entropy and evolution.

Molecular evolution is largely a stochastic process, not an equi-probable, arbitrarily random process nor a strictly determined orthogenetic process. Stochastic processes are historically understandable; however, the next event is related to but not a function of the previous event or events (see Gatlin (1972) on the methematical difference between relations and functions). Since 1964, Kimura (1982) has developed robust stochastic models of molecular evolution. His and others' neutralist models provide excellent null hypotheses which are alternative to either kind of adaptationist models. Such null hypotheses introduce another sort of scientifically acceptable norm of parsimony into molecular evolution. While Hoffman (1975) presumes a vocabulary expansion model, he does an admirable job of applying stochastic modeling to the origins of genetic codes problem. Since we are now aware of mitochondrial genetic codes in addition to "the" genetic code, the belief in a direct interaction model for the origins of genetic codes which would have resulted in only one contemporary code is negated.

What alternative models for the origins of genetic codes exist? The mutational damage minimization model of Sonneborn (1965) and the translational ambiguity reduction model of Fitch (1966, 1973) are two alternatives. Goodman and Moore (1977) demonstrated that the underlying premise of Sonneborn's model was false; namely, that the genetic code is structured so that mutations minimally affect the three dimensional structure of proteins is false, based on actual comparisons of proteins whose three dimensional structures and phylogenetic histories are well agreed upon in the literature. Furthermore, we could have rejected it _a priori_ as a model on the grounds that it is teleological; it presumes a purposive role of the goal state (3D structure) in the evolutionary constraints on an evolving code.

On the other hand, Fitch's model is consistent with both stochastic models of evolutionary processes and _ab initio_ and persistent criteria of some effect of direct stereochemical interactions among the nonfeatureless components of molecular evolutionary systems. The power of the latter, the direct interaction hypothesis, will persist because (1) it is eminently testable in the laboratory and (2) it is coordinated with the search for a "recognition code" (Carauthers, 1980) to explain the interaction of repressors with operators, polymerases with templates, topoisomerases and restriction endonucleases with specific sequences, and rRNA with ribosomal proteins, just to name a few. Some binding data on amino acids and nucleotides has already been reported (Reuben and Polk, 1980; Woese et al., 1966). These data are consistent with the correlations between amino acids and anticodons previously reported (Jungck, 1978; see updated version in Table 1). If we look at the contemporary code, we see that 19 out of 20 of the amino acids can be specified by codons assigned to them which would only be specified by the composition of the triplet, not the sequence (see Table 2), as suggested originally by Gamow (1954).

Table 1. Intercorrelations between the properties of amino acids and nucleotides

	All 16 Dinucleoside Monophosphates						Mononucleotide (AMP)	
	Weber-Lacey R_f		Barzilay et al. R_f		Garel et al. R_f		Reuben & Polk	
Amino Acid Properties	Anti-Codon	Codon	Anti-codon	Codon	Anti-codon	Codon	log K_D	$1/K_D$
Grantham Polarity	+0.90	-0.59	+0.78	-0.40	-0.77	+0.59	+0.269	-0.245
Woese Polarity	+0.89	-0.58	+0.75	-0.63	-0.70	+0.72*	+0.214	-0.211
Levitt Hydrophobicity	-0.83	+0.55	-0.68	+0.61	+0.67	-0.65	-0.253	+0.267
Zimmerman Polarity	+0.64	-0.42	+0.56	-0.44	-0.48	+0.51*	-0.156	+0.087
Bull & Breese Hydrophobicity	+0.76	-0.42	+0.68	-.043	-0.61	+0.50	+0.487	-0.437
Jones Hydrophobicity	-0.66	+0.49	-0.55	+0.53	+0.60	-0.44	-0.638	+0.577
Bulkiness	-0.67	+0.41	-0.63	+0.37	+0.58	-0.31	-0.572	+0.469
Specific Volume	-0.62	+0.36	-0.66	+0.03	+0.55	-0.25	-0.268	+0.148
Weber-Lacey Hydrophilicity	+0.50	-0.37	+0.30	-0.49*	-0.52	+0.40	+0.759	-0.802
Hydration Potential	-0.65	+0.46	-0.65	+0.40	+0.54	-0.68*	+0.360	-0.369
Surrounding Hydrophobicity	-0.71	+0.17	-0.68	+0.30	+0.62	-0.55	-0.198	+0.111
Cloister Energy	+0.71	-0.46	+0.73	-0.37	-0.65	+0.68*	-0.061	+0.095
Beta Sheet Conformational Parameter	-0.69	-0.01	-0.61	+0.21	+0.49	-0.45	-0.261	+0.210
CPK Volume	-0.21	+0.12	-0.10	+0.21*	+0.27	+0.00	-0.905	+0.866

Legend: Rows 1-9 (Columns 1-6) are from Jungck (1978). All other correlations are newly presented here. Log K_D and $1/K_D$ values are from Reuben and Polk (1980). Hydration potential values are from Wolfenden, Cullis, and Southgate (1979). Surrounding hydrophobicity values are from Manavalan and Ponnuswamy (1978). Cloister energy and CPK volumes values are from Swanson (1980). Beta Sheet Conformational Parameter values are from Goodman and Moore (1977). Underlined values are significant at the 0.001 level. Values without "*" have anticodon correlations with amino acids greater than respective codon correlations (36 out of 42 pairs). All signs are in favor of the anticodon hypothesis.

Table 2. Combination code embedded within contemporary code

Nucleotides	Codon Sequence	Amino Acid
A_3	AAA	LYS
A_2C	CAA	GLN
A_2G	AGA	ARG
A_2U	AAU	ASN
C_3	CCC	PRO
C_2A	CAC	HIS
C_2G	CCG (ACG)	THR
C_2U	CUC	LEU
G_3	GGG	GLY
G_2A	GAG	GLU
G_2C	GCG	ALA
G_2U	UGG	TRP
U_3	UUU	PHE
U_2A	UAU	TYR
U_2C	UCU	SER
U_2G	UGU	CYS
AUG	AUG	MET
AUC	AUC	ILE
GUC	GUC	VAL
GAC	GAC	ASP

Legend: Gamow (1954) listed the twenty unique nucleotide compositions of triplets. The center column takes a specific codon which has a composition corresponding to that in the first column and the third column indicates which amino acid that codon specifies. Only one of the twenty amino acids is excluded from this set and it, threonine, is coded for by a codon containing 2 of the 3 nucleotides in the left over composition. Thus, one could say that the combination code accounts for 19-2/3 out of 20 possible correlations. Note that the assignment of 20 amino acids to 20 triplets could be achieved in 20! or 3×10^{17} different ways. Four of the assignments, namely, AAA, UUU, GGG, and CCC, are non-ambiguous in my selection process. Some permutations code for similar amino acids. For example, AUU = ILE, UUA = LEU, and UAU = TYR, these are all hydrophobic amino acids. The most arbitrary selections on my part are from codons composed of three different nucleotides; three out of four of these combinations code for six different amino acids and the fourth one includes the initiator and two termination codons (AUG, UGA, and UAG). Thus, while I had considerable latitude in selection of the above set, it was limited to 16×3^{16} which is 2.3×10^{-7} per cent of 20!; therefore, this result is still highly significant.

The possibility of actually building chemical translation systems will also tremendously affect the growth of this field. Krampitz and Fox (1969), Fox et al. (1982), Paecht-Horowitz, Berger and Katchalsky (1970) and others have used templates to affect the sequence specifity of polymerization of amino acid adenylates. White, Lacey, and Weber (1973) have used imidazole activated amino acids to study the preferential loading of RNA by amino acids. Usher (1980) would like to build proto-tRNA's out of short oligonucleotides which would be capable of catalyzing directed protein synthesis.

SELECTIONIST MODELS ON ORGINS OF GENETIC CODES

If selection does not operate on the actual properties of amino acids and nucleotides, then where has selection played a part in the evolution of genetic codes? I believe that there are a variety of properties which are easily amenable to selectionist analysis: (1) the rate of translation; (2) the accuracy of translation; (3) the dependence of translation on environmental conditions; and (4) the constraints on coding due to selection operating on other components in the translation machinery. In fact, Gatlin (1972) has asserted that most of the positions represented in the earlier portions of this paper are primarily concerned with the mapping function of genetic codes:

> I shall present the heresy that the genetic code is a <u>relation</u> rather than a <u>function</u>. Since I view the genetic code as a small part of a master program which structures the information processing system as an ascending functional hierarchy of controls, ambiguity is no specter in the closet in such a scheme but rather a useful member of the household. Game theoretic principles govern the emergence of higher levels of control and organization, whereas programming functions oversee the routine operation of each level. The mapping function is at the lowest level of this functional hierarchy.

The rate of transmission is amenable to selection in two well known ways. There may also be many others of which I am simply unaware. First, the length of code words or the number of synonymous code words are characteristics amenable to selection. I believe that the minimal length of code words is somewhat constrained, as mentioned before, by the nucleation size of nucleotide sequences, which is three, for readily stable complexes.

The most intriguing proposal I have seen is by Mackay (1967). In Table 3, I present a revised version of Mackay's Huffman code analysis of the genetic code which employs a larger sample and

Table 3. A comparison of Huffman coding and genetic coding of amino acids: Is the genetic code optimized for transmission rate?

Amino Acid	Relative Frequency (P_i)	Huffman Coding	No. of Bits (N_i)	No. of Codons	No. of Bits in Codons ($N_\circ$)
GLY	0.0868	111	3	4	4
ALA	0.0841	0111	4	4	4
SER	0.0728	0101	4	6	3.4
LEU	0.0713	0011	4	6	3.4
LYS	0.0706	0010	4	2	5
VAL	0.0679	0001	4	4	4
THR	0.0602	1101	4	4	4
GLU	0.0524	1011	4	2	5
ASP	0.0491	1010	4	2	5
ASN	0.0487	1001	4	2	5
ILN	0.0484	1000	4	3	4.4
ARG	0.0469	01101	5	6	3.4
PRO	0.0449	01100	5	4	4
GLN	0.0384	01001	5	2	5
PHE	0.0372	01000	5	2	5
TYR	0.0359	00001	5	2	5
CYS	0.0296	11001	5	2	5
HIS	0.0190	000001	6	2	5
MET	0.0162	000000	6	1	6
TRP	0.0122	110001	6	1	6
TERM.	0.0074	110000	6	3	4.4
21	1.0000	21	97	64	96

$E = -\sum_{i=1}^{21} P_i \log_2 P_i = 4.000$ $\quad \sum_{i=1}^{21} N_i P_i = 4.256$ $\quad \sum_{i=1}^{21} N_\circ P_i = 4.346$

Legend: Mackay (1967) originally prepared such a table, but he did not employ a heterologous, phylogenetically and functionally diverse, and large sample. The amino acid frequencies employed in calculating the data in this table were from a larger set defined by the principles in Jungck (1971). The genetic code value of average word length (4.346 bits) is remarkably close to electrical engineer's design limit (4.256 bits). A Morce code like model of the genetic code came out with a value above 7; thus, it seems valid to infer that the genetic code is partially the result of selection on transmission rate.

greater variety of protein sequences. Nonetheless, the inference which Mackay drew, and which I do as well, is that the genetic code is close to the limit on codes illustrated by Shannon's first theorem of information theory. Instead of stating the difference between the value computed for average word length in the genetic code of the distribution of code words among amino acids as due to selection operating on more characters than just this one, I believe that a valid interpretation of this measurement is that selection operates to some degree on the rate of transmission. Obviously, this is different from stating that it is the only or the major selectionist force operating on the evolution of genetic codes.

Second, there is a fundamental limit upon the rate of translation imposed by diffusion rates of the component parts to one another if reaction is to take place. No selection on the sequence of amino acids in enzymes involved in translation or on the sequence of nucleotides of tRNAs or rRNAs can conceivably overcome this diffusional constraint. There could also be mechanical constraints on rate because selection is operating on large, bulky molecules.

Accuracy in coding is usually subdivided into two separate parts: encoding and decoding. For encoding, early attention was given to comma free codes by Crick, Griffith and Orgel (1957). In comma free codes, only one possible reading frame is understandable and all other reading frames are nonsense. They work well in a variety of contexts such as in the seven bit code used in the NASA Viking missions which were susceptible to very noisy channels. Crick, Brenner, Klug, and Pieczenik (1976) have even given a model for how such a comma free code could have given rise to the present code. Shepherd (1981) and Clarke (1982) infer that the RNY pattern which is statistically significant in contemporary nucleic acid sequences is a vestige of a primordial comma free code. Note however, that in contrast to the compositional code, the correspondence of comma-free codes to the contemporary code is 13 out of 21 (see Table 4).

On the other hand, Andrews and Boss (1970), Findley and McGlynn (1979), and Bertman and Jungck (1979) have looked at the decoding aspects; namely, error detection and error correction. These invoke Shannon's second theorem of information theory:

> The incredible thing about the second theorem is that it promises error-free transmission in noisy channels without reduction in transmission rate; and the upper limit to this error-free transmission is the channel capacity. The second theorem tells us that all this is possible if and only if we can find precisely the right code. It does not tell how to find this code. This theorem, like so many theorems of higher mathematics, simply tells us that' such an entity exists.

Table 4. Comma-free code embedded within contemporary code

Comma-free Codon	Amino Acid
UUC	PHE
CUC	LEU
UGU	CYS
UGC	CYS
CGU	ARG
CGC	ARG
GGU	GLY
GGC	GLY
AAG	LYS
AAC	ASN
AAU	ASN
GAG	GLU
GAC	ASP
GAU	ASP
CAG	GLN
CAC	HIS
CAU	HIS
UAG	AMBER (stop)
UAC	TYR
UAU	TYR

Legend: The twenty comma-free codons described by Crick, Griffith, and Orgel (1957) are listed with their twenty translations. This set of twenty comma-free words accounted for more amino acids than any other comma-free set which I checked. Note that only 12 out of the 20 amino acids are encoded in this set. Compare this with the results of Table 2.

> The search for general methods of constructing such codes has long been the Holy Grail of information theory . . . , and it is just possible that the living organism in its quest for evolutionary advancement has the same goal. (Gatlin, 1972)

Group theoretic symmetries are used in code design by electrical engineers to enhance error detection and error correction possibilities. We (Bertman and Jungck, 1979) were able to extend the work of Danckwerts and Neubert (1977) in illustrating that completely degenerate codons, i.e., codons which are completely specified by the first two nucleotides, constitute an octet within the genetic code which has group theoretic symmetries. Furthermore, we employed the notion of a Hamming distance to record the actual errors in coding made by amino acyl tRNA synthetases and our group graph of the genetic code. The correspondence seems more meaningful than simply discussing the number of nucleotide changes from one codon to another. Again, as with the rate argument above, we do not argue that error detection and error correction were the fundamental constraints on code evolution, but since we had no a priori reason to believe that group theoretic properties would be present in the genetic code, the selectionist reasoning was helpful in our exploration of properties of the code. Cullmann and Labouygues (1983) have also employed Hamming's distance to investigate "noise immunity of the genetic code" and conclude that it operates as an optimizing principle in the selection of codes. Labouygues and Gourgand (1981) are convinced that there was an active selection in this regard against the code containing too many termination codons.

The effects of environmental conditions on coding can also be studied through the use of group theoretic approaches. Instead of using the Klein-four group which we employed, Findley, Findley, and McGlynn (1982) have employed cyclic groups. It should be noted that much earlier Gurel (1969) had suggested the use of cyclic permutation groups for understanding the structure of the genetic code. They conclude that all of the so-called mistakes in the translation of the genetic code (including both missense suppression and high salt effects on ambiguity) and the alternative mitochondrial codes are explainable within their group theoretic model. Furthermore, they even go on to predict other coding possibilities consistent with our current knowledge of the code which might be found in the future. Laughrea (1981) has reported empirical evidence that both speed and accuracy can increase under some environmental conditions; thus, Shannon's first and second laws could both be operating to select similar coding constraints.

The fourth and final constraint listed above is the selection on an evolving code due to interaction with other components in the protein synthesizing apparatus. Hopfield (1978) has explored the effects of secondary structure of transfer RNAs on the loading specificity of amino acids. He concludes that: "Kinetic proofreading through diffusion can amplify a low level of intrinsic selectivity of tRNA for its amino acid." Grosjean, de Henau, and Crothers (1978) also explore kinetic proofreading models for the origin of the genetic code and even feel that selection operated on the "folding [of] the mRNA on the ribosome into a loop similar to the anticodon loop." Lestienne (1978) invokes the tRNA-tRNA interaction as the important one in determining the high fidelity of amino acylation. Finally, Wetzel (1978) has explored the amino acid sequences and three dimensional structures of amino acyl tRNA synthetase families and their relationship to the genetic code. While I do not accept his conclusion of a vocabulary expansionist model, I do believe that he is correct in assuming that the origins of the code lie immersed in the framework of the coevolution of tRNAs and these enzymes.

There are many oher places that we could look for selection operating on the evolution of genetic codes. Swanson (1980) has employed Gray codes to examine the minimization of the effects of mutation in translation. I (unpublished) have looked at memory wheels (de Bruijn sequences, Eulerian washing machines, circular concatenated codes) to examine the constraints on codes susceptible to frameshifts and selection on overlapping reading frames. While I personally will continue to look at more coding theoretic models in my attempts to understand the emergence of genetic codes, I share Berlinski's (1978) caveat with you: "Still there are things that can be learned from code theory. . . . But a sense of algebraic rapture shifts the biologist's attention to the medium much at the expense of the message; deflated, he will require concepts that code theory cannot express." While I could cavilierly reply with Marshall McLuhan's canard: _The Medium is the Message_, I only point out that this position was taken in response to a variety of models which abused the "adaptationist programme" by arguing from grounds of necessity rather than trying to investigate modes of selection on coding.

Hofstadter (1982) has recently argued that the genetic code is an arbitrary one; other codes would work as well. From the inferences of the discussion above, I would react that, based on physicochemical interactions and intermolecular specificities, the genetic code is surely not an arbitrary one. Second, while other codes may have been _selected for_ which would have worked as well, the genetic code does function well on a variety of measurable variables. The simplistic changes which Hofstadler suggests probably would not work as well because he has not taken these constraints into consideration. Hence, he believes that we must "content ourselves with a sense of wonder and awe, rather than with an answer. And perhaps experiencing that sense of wonder and awe is more satisfying than having an answer--at least for a while" (Hofstadler, 1979). I believe that we can go beyond this simple sense of awe by applying the tools discussed above and by a nonabusive application of selectionist thinking which avoids teleological arguments of necessity, perfect optimization, absolute determinism, and anthropomorphic design.

REFERENCES

Andrews, D. H., and Boss, M. L., 1971, _Yale Scientific Magazine_, 45:2-4, 6-7, 9-11.

Balasubramanian, R., 1982, _Trends in Biochemical Sciences_, 7:9.

Balasubramanian, R., Seetharamulu, P., and Raghunathan, G., 1980, _Origins of Life_, 10:15.

Berlinski, D., 1978, "On Systems Analysis: An Essay Concerning the Limitations of Some Mathematical Methods in the Social, Political, & Biological Sciences," MIT Press, Cambridge, Mass.

Bertman, M. O., and Jungck, J. R., 1978, Notices Am. Math. Soc., 25:A-174.
Bertman, M. O., and Jungck, J. R., 1979, J. of Heredity, 70:379.
Brown, J., 1982, "Chemical Evolution: Molecular Evolution Towards the Origin of Living Systems on the Earth and Elsewhere," Oxford Univerity Press, New York.
Carauthers, M. H., 1980, Accounts Chemical Research, 13:155.
Cedergren, R. H., LaRue, B., Sankoff, D., Lapalme, G., and Grosjean, H., 1980, Proc. Nat. Acad. Sci. U.S., 77:2791.
Clarke, C. H., 1982, J. Theor. Biol., 99:397.
Commoner, B., 1964, Amer. Sci., 52-365.
Crick, F. H. C., 1968, J. Mol. Biol., 38:367.
Crick, F. H. C., 1982, "Life Itself: Its Origin and Nature," Simon and Schuster, New York.
Crick, F. H. C., Brenner, S., Klug, A., and Pieczenik, G., 1974, Origins Life, 7:389.
Crick, F. H. C., Griffith, J. S., and Orgel, L. E., 1957, Proc. Nat. Sci. U.S., 43:416.
Cullman, G., and Labouygues, J. M., 1983, BioSystems, 16:9.
Danckwerts, H. J., and Neubert, D., 1975, J. Mol. Evol., 5:327.
Dillon, L. S., 1978, "The Genetic Mechanism and the Origin of Life," Plenum Press, New York.
Eigen, M., and Schuster, P., 1979, "The Hypercycle: A Principle of Natural Self-Organization," Springer-Verlag, New York.
Farley, J., 1972, J. History Biol., 5:285.
Findley, A. M., Findley, G. L., and McGlynn, S. P., 1982, J. Theor. Biol., 97:299.
Findley, G. L., and McGlynn, S. P., 1979, Int. J. Quantum Chem., QBS6:313.
Fitch, W., 1966, J. Mol. Biol., 16:1.
Fitch, W., 1973, Ann. Rev. Genetics, 7:343.
Fox, S. W., 1974, Mol. Cell. Biochem., 3:129.
Fox, S. W., in: "Evolution and Creationism," A. Montagu, ed., Oxford University Press, New York, in press.
Fox, S. W., and Dose, K., 1977, "Molecular Evolution and the Origin of Life," 2nd edn., Marcel Dekker, New York.
Fox, S. W., Nakashima, T. Przybyski, A., and Spren, R. M., 1982, Int. J. Quantum Chem., QBS9:195.
Gamow, G., 1954, Nature, 173:318.
Gamow, G., 1964, "Applied Combinatorial Mathematics," Wiley, New York.
Gatlin, L. L., 1972, "Information Theory and Living Systems," Columbia University Press, New York.
Goodman, M., and Moore, G. W., 1977, J. Mol. Evol., 10:7.
Gould, S. J., 1978, New Scientist, 80:700.
Gould, S. J., and Lewontin, R. C., 1979, Proc. Roy. Soc. Lond., B205:581.
Gould, S. H., and Vrba, E. S., 1982, Paleobiology, 8:4.
Grosjean, J. J., de Henau, S., Crothers, D. M., 1978, Proc. Nat. Acad. Sci. U.S., 75:610.

Gurel, O., 1969, Biophys. Soc. Abstracts, 9:A-254.
Hartman, H., 1975, Origins Life, 6:423.
Hartman, H., 1978, Origins Life, 9:133.
Hoffmann, G. W., 1975, Ann. Rev. Phys. Chem., 26:123.
Hofstadter, D. R., 1979, "Godel, Escher, Bach: An Eternal Golden Braid," Basic Books, New York.
Hofstadter, D. R., 1982, Sci. Amer., 246:18, 23-26, 28-29.
Hopfield, J. H., 1978, Proc. Nat. Acad. Sci. U.S., 75:4334.
Hoyle, F., and Wickramasinghe, C., 1978, "Life Cloud, the Origin of Life in the Universe," Harper and Row, New York.
Hubbard, R., 1982, "Towards a Liberatory Biology: The Dialectics of Biology Group," pp. 62-78, S. Rose, ed., Schocken Books, Inc., New York.
Jacobs, F., 1977, Science, 196:1161.
Jukes, T. H., 1965, Biochem. Biophys. Res. Commun., 19:391.
Jukes, T. H., 1967, Biochem. Biophys. Res. Commun., 27-573.
Jungck, J. R., 1971, BioSystems, 3:309.
Jungck, J. R., 1972, in: "Molecular Evolution: Prebiological and Biological," pp. 101-109, G. L. Rohlfing, and A. I. Oparin, eds., Plenum Press, New York.
Jungck, J. R., 1973, "Some unwarranted assumptions in molecular evolution," Abstracts of the First International Congress Systematic and Evolutionary Biology, Boulder, CO.
Jungck, J. R., 1977, Amer. Biol. Teacher, 39:124-137, 157.
Jungck, J. R., 1978, J. Mol. Evol., 11:211.
Jungck, J. R., 1983, "Is the neo-Darwinian synthesis robust enough to withstand the challenge of recent discoveries in molecular biology and molecular evolution?", in PSA 1982, Volume 2, P. Asquith, and T. Nickles, eds., Philosophy of Science Association, East Lansing, Michigan.
Jungck, J. R., "A circular concatenated genetic code: Selective constraints on overlapping codes," BioSystems, accepted for publication.
Jungck, J. R., ed., "On the Origins of Genetic Codes," Dowden, Hutchinson and Ross (distributed by Academic Press, New York), in preparation.
Kenyon, D. H., and Steinman, G., 1969, "Biochemical Predestination," McGraw-Hill, New York.
Kimura, M., 1982, "Molecular Evolution, Protein Polymorphism, and the Neutral Theory," Springer-Verlag, New York.
Krampitz, G., and Fox, S. W., 1969, Proc. Nat. Acad. Sci. U.S., 62:399.
Labouygues, J. M., and Gourgand, J. M., 1981, C. R. Acad. Sc. Paris, 292:1105.
Lacey, Jr., L. C., and Weber, A. L., 1977, Precamb. Res., 5:1.
Legerkvist, U., 1980, Amer. Sci., 68:192.
Laughrea, M., 1981, Biochimie, 63:145.
Lestienne, P., 1978, J. Theor. Biol., 73:159.
Lewin, R., 1982a, Science, 216:1212.
Lewin, R., 1982b, Science, 216:1091.

Mackay, A. L., 1967, Nature, 216:159.
Manavalan, P., and Ponnuswamy, P. K., 1978, Nature, 275:673.
Mayr, E., 1983, Amer. Nat., 121:324.
McLuhan, M., and Fiore, Q., 1967, "The Medium is the Massage: An Inventory of Effects," Bantam Books, New York.
Medawar, P. B., 1978, Nature, 272:772.
Melius, P., 1982, BioSystems, 15:275.
Miller, S. L., and Orgel, L. E., 1974, "The Origins of Life on the Earth," Prentice-Hall, Englewood Cliffs, New Jersey.
Nakashima, T., Jungck, J. R., Fox, S. W., Lederer, E., and Das, B. C., 1977, Int. J. Quantum Chem., QBS4:65.
Paecht-Horowitz, M., Berger, M. J., and Katchalsky, A., 1970, Nature, 228-636.
Reuben, J., and Polk, F. E., 1980, J. Mol. Evol., 15:103.
Root-Bernstein, R. S., 1982a, J. Theor. Biol., 94:885.
Root-Bernstein, R. S., 1982b, J. Theor. Biol., 94:895.
Salthe, S. N., 1982, Nature, 295:452.
Shepherd, J. C. W., 1981, J. Mol. Evol., 17:94.
Sober, E., 1982, "Selection of and selection for," Comments on papers by G. C. Williams and M. Wade, University of Chicago Conference on Persistent Controversies in Evolutionary Theory, March 3-5, 1982, transcript from author.
Sonneborn, T. M., 1965, in: "Evolving Genes and Proteins," V. Bryson and H. J. Vogel, eds., pp. 377-397, Academic Press, New York.
Steinman, G., 1967, Archiv. Biochem. Biophys., 121:533, 119:76.
Sukhodoletz, V. V., 1980, Genetika, XVI:759.
Swanson, R., 1980, "The genetic code: Almost a gray code," Abstracts of the Tenth Northeast Regional American Chemical Society Meeting: A-296.
Usher, D. A., 1980, "Some constraints on new and old suggestions for the origin of protein synthesis," Abstracts of the Tenth Northeast Regional American Chemical Society Meeting: A-299.
Weber, A. L., and Lacey, Jr., J. C., 1978, J. Mol. Evol., 11:199.
Weber, A. L., and Miller, S. L., 1981, J. Mol. Evol., 17:273.
Wetzel, R., 1978, Origins Life, 9:39.
White, D. H., and Raab, M. S., 1982, J. Mol. Evol., 18:207.
White, Jr., W. B., Lacey, Jr., J. C., and Weber, A. L., 1973, Biochem. Biophys. Res. Commun., 51:283.
Wimsatt, W. C., 1980, Synthese, 43:287.
Woese, C. R., 1969, J. Mol. Biol., 43:235.
Woese, C. R., Dugre, D. H., Saxinger, W. C., and Dugre, S. A., 1966, Proc. Nat. Acad. Sci. U.S., 55:966.
Wolfenden, R., Andersson, L., Cullis, P. M., and Southgate, C. C., 1981, Biochemistry, 20:849.
Wolfenden, R., Cullis, P. M., and Southgate, C. C., 1979, Science, 206:575.
Wong, J. T. F., 1975, Proc. Nat. Acad. Sci. U.S., 72:1909.
Wong, J. T. F., 1976, Proc. Nat. Acad. Sci. U.S., 73:2336.
Wong, J. T. F., 1981, Trends in Biochem. Sci., 6:33.
Yockey, H. P., 1981, J. Theor. Biol., 91:13.

ON THE PROBLEM OF THE ORIGIN OF BIOLOGICAL INFORMATION

N. M. Chernavskaya and D. S. Chernavskii

P. N. Lebedev Physical Institute
Moscow State University
Moscow, U.S.S.R.

INTRODUCTION

The works by Oparin (1965), Fox (Fox and Dose, 1977; Fox, 1981) and their further developments (Egami, 1981; Thomas, 1970) represent a whole epoch in the study of protobiological evolution. They formulated the problem of self-formation of the material basis of life, i.e., of polypeptides, polynucleotides and their spatial organization in concervates and microspheres. The contribution made by Fox enables one to make the next step and to formulate the question of self-formation of the informational basis of the biosphere. Quastler (1964) was the first to formulate this question which then has been discussed by many investigators, including Eigen (Eigen, 1971; Eigen and Schuster, 1979), Kuhn (1972), Ebeling (Ebeling and Feistel, 1977; Feistel, 1977; Feistel et al., 1980), and the present authors (Chernavskaya and Chernavskii, 1975; Romanovskii et al., 1975; Chernavskii, 1980). The problem of the origin of biological information is to a considerable extent a theoretical one, and therefore among other methods mathematical approaches should be used. The problem is not yet solved, at least a unique and conventional view is absent. There exist only some approaches and alternative hypotheses. There are two reasons for this. First, many papers don't formulate this question clearly and, second, the proposed theoretical solutions are often inaccessible for experimental verification.

This paper represents an attempt to give an explicit formulation of the problem (more precisely, several problems), to discuss the existing ways of solving the problems, and to suggest consequent experimental possibilities.

THE ORIGIN OF BIOLOGICAL INFORMATION

The appearance of new information, according to Quastler (1964), consists in the choice of one (out of very many) a'priori equivalent version and its storage. The system, able to provide this process, must possess the following properties (Volkenstein and Chernavskii, 1975, 1979):

(a) It must be multistationary, otherwise the choice is impossible.
(b) It must be dissipative, otherwise the storage is impossible.
(c) The initial state (prior to the choice of a stable state) must be unstable, otherwise the choice is predetermined and new information is absent.

Systems possessing these properties will be referred to as informational systems (I.S.). Note that not every system can be an informational one. Thus, thermodynamically equilibrium systems are not I.S. since in this case there is only one equilibrium state. The Hamiltonian systems are not informational also because of the absence of dissipation. Thus, information can arise only in thermodynamically non-equilibrium and dissipative systems. An example of I.S. may be Chinese billiards--a plate with holes, one of which accepts the ball.

The very act of the choice of any of the possible states is not improbable. Moreover, it is inevitably realized (i.e., the probability is equal to unity). The choice of some version "appropriate" (or valuable) for achieving a certain goal is, on the contrary, improbable. Its probability is equal to $W = N^{-1} = 2^{-I_m}$, where N is the number of possible versions, $I_m = \log_2 N$ is the maximum information possible for a given system (i.e., the informative capacity).

As a rule, the number of versions N is sufficiently large (as we show further, in our case $N \simeq 10^{60}$); correspondingly, the quantity I_m too is large, and therefore the quantity W is very small. This is just the essence of the problem.

If the choice is repeated, the probability to choose an "appropriate" version in each attempt increases to become

$$P = JW(1 + JW)^{-1}$$

where J is the number of attempts.

Now let us discuss these general points in application to biology.

Biological information is understood as nucleotide sequences in DNA, which determine (code) the initial sequences of proteins and thus their properties. Information transmission in the DNA-RNA-protein chain is realized by the transcription and translation apparatus which contains transport RNA, ribosomes and proteins-amino-acylsynthetases. Note that the latter are of particular importance in the information aspect since they contain information on the code (i.e., on the correspondence of three nucleotides (anticodon) and amino acid). For brevity we shall refer to them as adaptors (Chernavskaya and Chervanskii, 1975). Note that the presence of adaptors providing the correspondence between codons and amino acids is necessary in any translation mechanism, but their construction (the same as that of the translation apparatus itself) may in principle be essentially simpler, and their number smaller than that in the contemporary biosphere.

The choice of the version in biology implicates a certain sequence of nucleotides in DNA or amino acids in protein. The number of versions in DNA containing about 600 nucleotides is equal to $N^{(DNA)} = 4^{600} \simeq 10^{360}$; (correspondingly $I_m^{(DNA)} = 1200$); the number of versions of a protein sequence containing 200 amino acids $N^{(P)} = 20^{200} \simeq 10^{260}$ ($I_m^{(P)} = 860$). Here the numbers 200 amino acids and 600 nucleotides are chosen as typical of enzyme-proteins and their coding genomes. The difference between $I_m^{(DNA)}$ and $I_m^{(P)}$ is caused by code degeneracy. The estimates demonstrate that the numbers of combinations (and information capacities) in biology are rather large.

Information storage in biology implicates not only fixing of the chosen sequence but also its ancestral transmission. The latter suggests genome replication. The modern replication system contains several enzymes, including protein-replicase comprised of six subunits. Its main function is initiation of the replication. The primary replication mechanism could be essentially simpler, but the presence of at least one protein-replicase is necessary.

Thus, the biological information system must contain polynucleotide and protein promoting the replication of the former.

In the contemporary biosphere such a system itself contains much information and can be constructed only in the presence of such information. In other words, a spontaneous, random occurrence of all the contemporary translation and replication systems is quite improbable, according to the estimation of its probability in one attempt as $W_m \simeq 10^{-1000}$ and less. The large number of attempts ($J \simeq 10^{30}$), (Quastler, 1964; Chernavskii, 1980) does not improve the situation. This estimate of J is rather approximate, but from the comparison of the values of J and W_m, it is clear that inaccuracies in the estimates are not very important.

The initial information system could contain essentially less information. We shall present its estimates.

The simplest system capable of replication must contain a polynucleotide and a protein-replicase (according to some hypotheses, a set of adaptors is not necessary at first stages of evolution).

Let us estimate the information necessary for creating protein with replicase properties or, in other words, the probability of the random synthesis of such a protein. First, such a protein must have the form complementary to a sufficiently large polynucleotide region. Second, it must be capable of breaking hydrogen bonds at the moment of initiation. Third, it must prevent the polynucleotide from being hydrolyzed.

It is clear that the initial replicase is a protein of a large size and a certain shape which is capable of corresponding conformational changes.

Apparently, the protein-replicase sequence was the same or even larger than that for protein enzymes (i.e., contained about 200 residues). The assumption that the setting up of the function requires the fixation of all the residues leads to the above mentioned estimates: $I_m^{(P)} \simeq 860$ and $W_m^{(P)} \simeq 10^{-260}$. This is evidently an underestimation of the probability since it is known that many amino acids can be replaced by others without loss of the function. A realistic estimate may be obtained by the assumption from Csanyi (1980) that only one-fourth of the residues is fixed. Then the probability of a random appearance $W_\tau \simeq 10^{-65}$ and, correspondingly, the necessary (valuable) information $I_\tau \simeq 215$ bits. Note that the quantity W_τ is, however, absurdly small since the product $JW_\tau \simeq 10^{-35}$ remains very small. In other words, the information contained in a biological I.S. cannot arise spontaneously. This is the first problem:

1. In what way could the biological I.S., i.e., the replicating protein-nucleotide complex, arise?

We are dealing with a closed circle typical for many self-organization problems: information occurs within the information system, while the latter cannot be formed without information.

Before discussing possible ways of solving the problem, it is appropriate here to make some remarks:

1.1. There is an opinion that such a small probability is not an obstacle since the life originated only once, and this is possible at an arbitrarily small probability. Many investigators, including us, do not agree with this opinion, although it is difficult to disprove it within a formal approach.

1.2. It can be supposed that the amount of information necessary for setting up the specific protein function (including the replicase one) is not large. The supposition is based on the fact that the estimates for I_τ and W_τ are not yet sufficiently accurate. For example, it may be suggested that the setting up of the specific function is satisfied with the fixation of the amino acid residues in an active center only (its amount being approximately four). Then the necessary information is equal to $4 \log_2 20 \simeq 20$ and, correspondingly, the probability $W_{max} \simeq 10^{-6}$. In this case the product $JW_{max} \simeq 10^{24}$, and the problem disappears. We do not treat this view as being realistic (although it is difficult to argue against it), and we do not discuss it.

1.3. In the general case, reception of information from the I.S. and its subsequent use must be provided. In biology it is realized by the protein translation and biosynthesis mechanism. In the entire biosphere (with rare exceptions (Hall, 1979; Barrel et al., 1979)) a unique code is used; it is degenerate (i.e., to one amino acid there correspond several codons), but an opposite degeneracy (i.e., one codon vs. several amino acids) is absent. We point out that the presence of the code and the absence of inverse degeneracy are necessary conditions for the reception of genetic information. However, genrally speaking, the code need not necessarily be unique when we deal with one individual or a uniform population. A unique code becomes necessary for development of the biosphere as a whole, including interactions of different populations.

The information contained in the apparatus of translation in protein synthesis and in the code itself is extensive (its value is much higher than the above-mentioned rather $I_\tau = 215$ bits), but we disregarded it. The point is that the appearance of I.S., the arising of the information reception apparatus and then the formation of a unique code are different acts of evolution. They could (and must) have been separated in time, their sequence being obvious. Note that in a number of papers these acts are not separated, which leads to some difficulties. Thus, another two independent problems can be formulated.

2. What is the original mechanism of the translation apparatus (i.e., the set of adaptors) while in the presence of autoreproduction protein-nucleotide complex (i.e., a biological I.S.)?

3. Where did the code unique for the entire biosphere originate from (already, in the presence of I.S. and translation)?

4. Besides the indicated problems of evolution, there exist others also associated with the appearance of new information. They include: species formation (already in the presence of the translation mechanism and the unique code), and the appearance of essentially new functions.

It is of importance to emphasize the following: First, these processes proceeded after the one under consideration. Second, the formation of species and of new functions are divergent processes, whereas the formation of the translation mechanism and the unique code is a convergent process. Divergence and convergence in evolution are complementary processes, i.e., conditions favorable for one are unfavorable for the other. Therefore, it is necessary to distinguish between them, to clearly formulate the process under consideration. This analysis is necessary and specifies the conditions for discussing physicochemical mechanisms and for mathematical modeling of this or that process. Either mixing of problems or the unclear formulation of problems leads to misunderstanding.

WAYS OF SOLVING THE PROBLEMS

We try here to give a clear formulation of the problem and the possible ways of its solution discussed in the literature.

1. The problem of a spontaneous appearance of I.S., i.e., a protein-nucleotide complex capable of replication. This problem arises if one assumes (as is widely accepted now) that the information for the protein-replicase (i.e., a polynucleotide with a certain sequence) was first to appear, then being realized in the protein. This chain of events can be illustrated by the following scheme

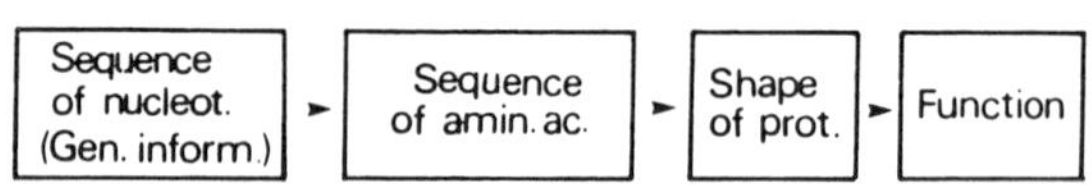

(1)

However, the problem is out of consideration if one assumes that the initial I.S. were organized according to another principle, namely, spontaneous synthesis of protein-replicase in the presence and with the participation (as a heterogenous catalyzer) of polynucleotides. This possibility is discussed in more detail in Chernavskaya and Chernavskii (1975) and Chernavskii (1980). In this case the method of synthesis determines first of all the shape of polypeptide, which becomes an envelope covering the polynucleotide. The shape of the protein is responsible for its function. Indeed, in such a protein even small conformational changes (e.g., caused by periodic changes in the environment) are able to initiate polynucleotide replication. On the other hand, with such a way of synthesis only those sequences are selected which are compatible with its form. This can be illustrated as follows:

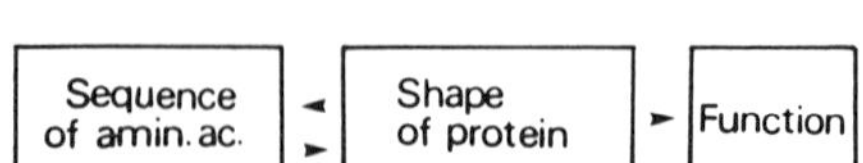

(2)

Double arrows implicate here that there is feedback between the form of the protein and its sequence. This scheme does not include yet the sequence of nucleotides, and indeed, the synthesis of protein-envelopes may proceed on a polynucleotide with any sequence. Owing to this, the probability of such a process is not small.

Generalizing, one can say that I.S. may arise spontaneously and no a'priori information is required for that. Let us clarify this statement by the example of Chinese billards. In order to create it one needs the plan, i.e., information on the position of the holes. However, a pattern of a billard which fulfills the function of I.S. can appear spontaneously--its role can be played by an rough surface with holes.

2. The problem of formation of the translation apparatus, i.e., a set of adaptors, was discussed in a number of papers. Two hypotheses are proposed.

According to the first hypothesis (Mekler, 1980; Altstein and Koverin, 1980), there exists a direct correspondence (strong interaction) between a codon and a coded amino acid such that on a polynucleotide of a given sequence a corresponding chain of amino acids is arranged. (Here we abstract ourselves from the details, namely, the original code could be not ternary but rather binary, a correspondance could exist between amino acid and anticodon (Mekler, 1980). The problems of the origin of adaptors, of moment of appearance of the translation apparatus, and the probability values of these events are not discussed in the framework of this hypothesis.

The problem of code uniqueness is solved simply here: One of the versions is considered to be a'priori clearly pronounced as compared with the others; in other words, information on the code already pre-exists in the interaction of amino acids and codons. At the present time this idea is disputable, and in our opinion there are no physicochemical arguments for such a definite statement.

According to the second hypothesis (Chernavskaya and Chernavskii, 1975; Chernavskii, 1980), adaptors were synthesized spontaneously, following the principle of heterogenous catalysis. In this case a random connection between codons and amino acids arose,* i.e., there appeared an individual "code" for each population. The process can be illustrated by the following scheme.

*The possibility of a preferable interaction of codons with certain amino acids is not excluded, but the role of this factor is not treated to be of primary importance.

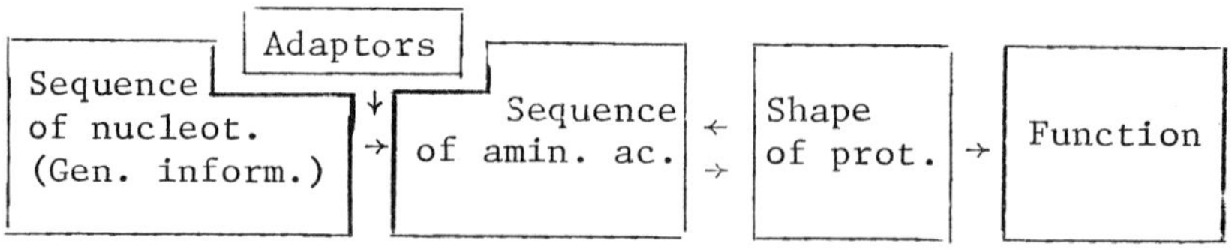

(3)

Double arrows implicate feedback of sequences. A set of adaptors (even an incomplete one) is advantageous for evolution, and therefore, at subsequent stages of evolution the process proceeds from the left to the right, i.e., as in scheme (1).

3. The problem of uniqueness of the biological code is urgent if one assumes (as was done in the second hypothesis) the possibility of a spontaneous appearance of several different versions.

Two ideas of solving the problem are proposed. The first one suggests that a unique code appeared as a result of the choice of the best version (Eigen, 1971; Eigen and Schuster, 1979). The second hypothesis suggests the choice of one of the possible versions (not necessarily the best one) (Chernavskaya and Chernavskii, 1975; Chernavskii, 1980). The latter is possible if the coexistence of populations using different codes is unstable.

In both cases the formation of a unique code results from interaction of populations. In this interaction two factors are essential, namely, a competition for a general substrate (or the "tightness effect"), and antagonistic interaction of individuals which use different versions of the code. The latter means that when encountering, two individuals with different codes die (this may happen due to mixing of adaptors, which leads to inverse degeneracy (Chernavskaya and Chernavskii, 1975; Romanovskii et al., 1975)). Realization of both hypotheses requires a mathematical approach, i.e., construction and investigation of models of code selection (or choice). This was done by Eigen (1971), Eigen and Schuster (1979), Ebeling and Feistel (1977), and Chernavskaya and Chernavskii (1975). Here we present the results:

3.1. In the presence of antagonistic interaction, a unique code has to be chosen, i.e., there survives and develops a population that uses one version of the code (not necessarily "the best one"). This process is unsensitive to distortions, i.e., the result remains the same also in the presence of code-breaking mutations (Chernavskii, 1980). The situation is the same when both antagonistic and competing interactions are taken into account.

3.2. In the presence of a competing interaction alone, the process of the choice of a unique code is sensitive to the distortions, particularly if the "advantages" of the version are not large.

Thus, in the framework of this approach a unique code was created as a result of a long interaction of "objects already capable of reduplication and possessing the translation apparatus.

Now let us discuss the informational aspect of this process. The amount of information contained in the code itself is sufficiently large. This information is valuable both for the individual and for the entire biosphere, since there "survive" only those using the same code as all the others. The probability that this valuable information appeared at a time (in one act or in one attempt) in all the individuals is absurdly small. However, the probability of the above-mentioned process is not small. The point is that the probability of a random choice of one (arbitrary) of the possible code versions (equivalent or even unequivalent) is not small. At the beginning, before the choice is over, the information is not valuable. In the course of selection it becomes increasingly valuable. Thus, we deal here with a gradual generation of valuable information: new information appears in one act of a random choice, but its price is not high; it increases gradually as a result of object interactions. The probability of such a process is not low. Note that similar phenomena take place also in other self-organization processes: evolution of language, alphabet, higher nervous system, etc.

We have discussed here three problems: The origin of the biological informational system, of the apparatus of information reception, and of unique codes (i.e., the alphabet with which the information is written). We have not touched upon problems of further accumulation of new information; they require special discussion and must be clearly distinguished from the above-mentioned ones.

We would like to stress our conclusion, that the problems under consideration can be solved theoretically without any mystic elements or without appealing to improbable random events. However, in biology a theoretical solution is not final; experimental verification is necessary. The most relevant is a spontaneous formation of I.S., i.e., a complex of polynucleotide with protein-replicase. The general scheme of the experiments is the following. The synthesis of polypeptides under abiotic conditions becomes available now. It is possible also to provide this synthesis in the presence, as well as in the absence of the polynucleotides (for example, DNA). According to the above discussed theory, the synthesized polypeptides should have in the first case higher affinity for the polynucleotides than that in the second case.

The affinity of the protein for the polynucleotides should lead to the formation of the protein-nucleotide complex. The initiation of the polynucleotide replication in this complex is possible under certain, periodic, external conditions.

These predictions can be proven experimentally. Forming the details of the experimental conditions requires special discussion. We believe that such an experiment is quite realistic now, owing to the contributions of Professor Fox and his collaborators. The methods developed and the results obtained by them are the basis for solving the fundamental theorectical information problems.

REFERENCES

Altstein, A. D., and Koverin, N. V., 1980, Jour. Vsesoyusnogo Chemicheskogo Obchestva (in Russ.), 25:383.

Barrel, B. G., Banier, A. T., and Drouin, I. A., 1979, Nature, 282:189.

Chernavskaya, N. M., and Chernavskii, D. S., 1975, J. Theor. Biol., 53:13.

Chernavskii, D. S., 1980, Jour. Vesosyushogo Chemicheskogo Obchestva (in Russ.), 25:404.

Csanyi, V., 1980, Acta Biol. Acad. Sci. Hung., 31:409.

Ebeling, W., and Feistel, R., 1977, Ann. Phys. (Leipzig), 34:81.

Egami, F., 1981, "New Horisons in Biological Chemistry," M. Koike, T. Nagatsu, J. Okuda, and T. Ozawa, eds., pp. 1-7, Japan Scientific Societies Press, Tokyo.

Eigen, M., 1971, Natuzwiss., 58:465.

Eigen, M., and Schuster, P., 1979, "The Hypercycle," Springer-Verlag, Berlin.

Feistel, R., Romanovskii, Y. M., and Vasiliev, V. A., 1980, Biophysica (in Russ.), 25:882.

Fox, S. W., 1981, Int. J. Quantum Chem., QBS8:441.

Fox, S. W., and Dose, K., 1977, "Molecular Evolution and the Origin of Life," 2nd edn., Marcel Dekker, New York.

Hall, B. D., 1979, Nature, 282:129.

Kuhn, H., 1974, Angew. Chemie, 84:838.

Mekler, L. B., 1980, Journ. Vsesoyusnogo Chemicheskogo Obchestva (in Russ.), 25:473.

Oparin, A. I., 1965, "The Origin of Life on the Earth," Academic Press, New York.

Quastler, H., 1964, "The Emergence of Biological Information," Yale University Press, New Haven, Connecticut.

Romanovskii, Y. M., Stepanova, N. V., and Chernavskii, D. S., 1975, "The Mathematical Modeling in Biophysics," (in Russ.), Nauka, Moscow.

Thomas, B. R., 1970, Bioch. Biophys. Res. Comm., 40:1289.

Volkenstein, M. V., and Chernavskii, D. S., 1979, J. Soc. Biol. Struct., 1:95.

Volkenstein, M. V., and Chernavskii, D. S., 1979, Isv. Acad. Nauk. Biol., 4:531.

DEDUCTIVE PREBIOLOGY

Otto E. Rössler

Institute for Physical and Theoretical Chemistry
University of Tübingen
7400 Tübingen, W. Germany

INTRODUCTION

Darwin's (1859) approach to biology is basically a self-consistent mathematical formalism. Its connection to reality is of secondary importance. What is assumed mathematically is a single-variable, autonomously growing and self-maintaining (at an upper stable steady state) dynamical system that is subjected to a slow parametric change. The simplest version assumes that more than one such system is possible. If one is already there, others of the same type are assumed to arise sufficiently frequently to make sure that when the first system loses its upper attracting regime under the progressive parametric change, at least one of the more recently emerged systems does not.

Little is assumed initially about the internal structure of the single-variable autonomously growing and self-maintaining systems presupposed. The possibility of a more detailed description in which more variables appear, each contributing to the overall behavior ('survival') of the single main variable ('biomass') assumed, is not excluded. Such auxiliary (internal) variables are helpful when 'mechanisms' explaining the emergence of the next system in the sequence are to be formulated.

An alternative specification instead focuses on the global properties of the overall system. Here the whole system, with all self-maintaining subsystems included, is modelled in terms of a single essential variable. The lumped system then again has the property of preserving its autonomous self-maintenance (that is its upper stable steady state) despite continual external parametric change. The whole system, therefore, possesses a dynamical property that could be called 'ultrastable self-maintenance.' Ashby (1952) described the first 'ultrastable' system--a system that after external disequilibration (tampering) always found a way to go on with its former business (which in Ashby's case amounted to doing nothing, in an equilibrium that was a stable rest point). Ashby's system shared an essential property with the 'Hydra' of Greek mythology: it could not be perturbed indefinitely. (The Hydra could not be killed.) Life too, when taken as a single system, is more or less resistant against being extinguished, becoming ever more so as time progresses. This might be life's ultimate property. A fitting name is 'ultra silience.' This term generalizes the dynamical notion of resilience (Holling, 1973; Grümm, 1976) in the same way as the dynamical notion of stability was generalized by Ashby's notion.

Is it possible to observe the spontaneous emergence of the property of ultra silience in a relatively simple class of dynamical systems already? This is the question to be addressed in the following.

THE PRIVILEDGED ZERO

Living systems viewed as dynamical systems possess a strange property: they cease to exist when the main variable (biomass) reaches zero. This feature is shared by many chemical reaction systems. A watery soup in which the concentrations of all other reactants are zero is pure water. By contrast, an electronic set in which all state variables (voltages) are zero is not an empty case--even when the dynamical equations are exactly the same as before. An example is provided by an electronic analogue to the well-known oscillatory reaction system of Belousov and Zhabotinsky (see Field et al., 1972). At least 20 capacitors (plus resistors and amplifiers, etc.) are needed for the electronic implementation. They are still needed if one particular initial condition: "all concentrations zero," or (to be less radical): "all concentrations except one (or two or three) zero," is to be simulated. The same strange identity between state variables and supportive hardware is typical also for systems made up from nuclear reactions. Economic and ecological state variables also frequently observe the same rule, whereas biophysical state variables (like membrane potentials) do not. A practical criterion that frequently permits one to distinguish between the two cases is whether the state variables of the system in question are necessarily nonnegative. If so, the system may belong to the privileged zero class. There are, however, chemical reaction systems that despite their non-negativity behave just like the electronic class (compare the 'chemical Eccles-Jordan trigger'; Rössler, 1974a).

Physical systems obeying the privileged zero principle are very convenient in cost/efficiency terms. Building an electronic computer with (say) 10^{10000} state variables is not just forbiddingly expensive but even in principle impossible because the known cosmos contains much fewer particles (some 10^{80} baryons). Creating a chemical system with the same huge number of state variables is, in contrast, very easy. Just purchase some ten substances at a pharmacist's: a liter of distilled water, a gram of ammonium chloride, and some grams of sodium carbonate plus some other salts. This mixture, when shaken continually at 40°C, is bound to implement an equilibrium between some 10^{10000} substances. With infinite observation time granted, the expectation values of all these state variables become non-zero.

Mathematically, this whole machine exists the very moment the ten substances have been poured together. The dynamics of the machine appears rather dull, of course. But this is not the point.

The "of course" of the last paragraph actually was not justified. If one of the ten reactants was 'energy rich' initially, a very complicated sequence of 'transients' could be expected to develop. This sequence is not qualitatively different from the one that applies when the energy-rich substance is assumed present as a so-called pool substance, that is, with a back-up mechanism supplied that makes sure that its concentration remains constant in the presence of consumption. This pool-type assumption can be extended to all ten variables. In this way, a second prototypic example of an (initially) 'invisible machine' is formed.

A WELL-STIRRED EVOLUTIONARY CHEMICAL SYSTEM

Organic chemistry, based on carbon, is capable of generating a virtually unlimited number of substances. With other solvents given than water, and in other temperature ranges, further types of atoms can likewise form 'backbones' of virtually unlimited length, whereby the backbone need not consist of one type of atom only (-C-C-C-), but may as well alternate (as in the -N-B-N-B- chain; cf. for example, Lipscomb, 1963). As to nuclear reactions, a similar diversity appears possible, although at the time being the gap in atomic weights between the known chemical nuclei (some 10^{2}) and the nuclei making up neutron stars (some 10^{60}) has yet to be bridged--possibly in the interior of neutron stars.

The Miller-Urey experiment (cf. Miller and Urey, 1959), in which organic substances were created under energetically open conditions, provides one implementation of a virtually infinitely large dynamical system of dissipative type. It belongs to the nonnegative variety discussed in the preceding section. 'Hot' photons whose concentration is kept constant on the average, and a cool primordial ocean, provide another implementation of such an open system. For simplicity, this system may again be assumed well-stirred even though this is hard to do with an ocean.

The reaction system then in either case is the one considered in the preceding section: Some 10^{10000} state variables are immediately generated, along with an equally huge, more or less random network of reactions in between. The right-hand sides of all the 10^{10000} ordinary differential equations are filled-in from the beginning. Use of deterministic differential equations implies that expectation values stand for concentrations. This is misleading in the case of very small concentrations since the associated time lags are not represented. A more realistic deterministic description assumes that all near-zero influx and outflux terms on the right-hand sides are replaced by threshold-type terms which become non-zero only after the precursor substance in question has reached a certain minimum concentration (like one molecule per ocean, for example). The resulting dynamical system then is something more nonlinear than when only mass-action type (quadratic) nonlinearities were incorporated (cf. for example, Clarke, 1980).

To model this system of equations, one would need a very large computer indeed. The qualitative behavior of this class of systems can nevertheless be indicated (Rössler, 1971): The huge random network contains both very many 'regenerative cycles' and very many 'over-regenerative cycles.' The first are generalized catalytic reactions, the latter are generalized autocatalytic reactions. These cyclic reaction sequences (formed if one product somewhere down the line is identical to an earlier reactant of the chain) may be very long and complicated. The potentially autocatalytic (autonomously growing, toward an upper stable steady state of autonomous self-maintenance) ones are of special interest. They become autocatalytic only after all the necessary substrates (on which the cycle 'feeds') have acquired sufficiently large concentrations, while at the same time all the potential side reactants (whose presence 'poisons' the cycle as far as its capability of autonomous growth is concerned) have sufficiently small concentrations. Most of the over-regenerative cycles contained are, therefore, 'dormant' (that is, non-triggerable) for most of the time--especially so at the beginning.

Initially, only those potentially autocatalytic subsystems are non-dormant (or 'primed,' that is, triggerable) whose substrates happen to comprise only members of the set of initially non-zero substances (including their first mutual products). Whenever a member of this set of primed (that is, potentially growing) cycles is now provided with a single molecule of any one of the substances involved in the over-regenerative reaction chain, this substance (as well as all the other substances of the cycle) will start growing autocatalytically The triggered cycle then approaches its upper stable steady state of autonomous self-maintenance in finite time. Thereafter, all the substances contained in this cycle have become effective 'pool' substances themselves. That is, the whole system is now suddenly equivalent to a newly set up analogous system of 10^{10000} state variables in which initially not 10 but (say) 20 variables are exogeneously maintained at constant concentrations.

This means two things: First, the set of primed (triggerable) autocatalytic cycles is much larger this time; second, the number of potentially triggering reaction paths (yielding at least one molecule after a while) is also much larger. Therefore, the probability for another cycle to be 'set' is considerably greater this time. And so forth.

Thus, a branching domino process can be set in motion in the present class of systems by providing 'sufficiently many' (but still extremely few) initially non-zero variables. Once the process has actively set in, it is bound to go on 'burning' for a long time.

With this autonomous process, evidently a member of the axiomatic class of systems considered by Darwin has formed. As more and more autocatalytic systems are being triggered into 'positivity' (visibility), in a chain reaction of triggering-type bifurcations, the probability for an arbitrary environmental change to extinguish all of them is bound to decrease.

A POLLUTION PROBLEM

As more and more autocatalytic systems are 'set' (triggered), the overall dynamics becomes more and more complicated. For one thing, many of the newly generated (pseudo) pool substances are 'weak,' that is, of low carrying power. This weakness increases in proportion to the distance (in terms of intermediary reaction steps) to the primary (genuine) pool substances. Only those newly set cycles which happen to be close to some of the primary sources can support long 'food chains.' Another problem is generated by the fact that the larger the number of substances already produced, the larger is the number of potential 'pollutants.' This increased linkage leads to 'cross inhibitions' among already existing cycles. An even stronger effect is exerted on not yet existing cycles, however. Many of the potentially over-regenerative cycles are poisoned (unprimed) before having had a chance to be triggered. Even more are prevented from being primed in the first place.

All of this means that one cannot be sure that as the growing network of reactions becomes more and more strongly coupled (and more and more strongly and irregularly oscillating), the paradisiac growth conditions of the beginning continue to exist. The growing 'fringe' of the expanding web is still bound to reach individual subsystems from time to time which are functionally unaffected by most of the ongoing concentration movements to a sufficient degree for the notion of 'autonomous self-maintenance' still to be applicable. But this special subset may soon become insignificant quantitatively.

What is needed therefore, if one is bent on focusing on Darwinian subsystems only, is an appropriate mechanism for their 'insulation.' Conversely, if some insulating mechanism ensures the further existence of pure 'domino chains,' further consideration of this subset is justified.

Functional insulation between reaction paths is not easy to obtain in a well-stirred mixture. The idealizing assumption of well-stirredness, made throughout so far, was certainly unrealistic. Dropping it altogether, however, unfortuntely means that our system of ordinary differential equations has to be replaced by a (certainly unmanageable) equally large system of partial differential equations. In this situation, the mechanisms of spatial aggregation in solutions of organic molecules investigated by Oparin (1957) and Fox (1965)--cf. also Matsuno (1981)--offer themselves as a way out. Ready formation of droplet-like spatial domains, as described in these papers, virtually guarantees that functionally insulated subsystems are easily possible under conditions of incomplete well-stirredness. While it is very hard to take all the underlying physicochemical mechanisms into explicit dynamical account, a 'hybrid' description offers itself as a first approximation.

To adopt this strategy, it suffices to axiomatically acknowledge the existence of 'appropriate insulating mechanisms' and go on with the analysis of the qualitative behavior of the well-stirred system.

FURTHER CHARACTERIZATION OF THE DOMINO PROCESS

So far, the triggering of ever new subsystems, among an ever faster-growing set of primed subsystems, could be described without reference to the particular chemical and organizational properties of the subsystem in question. The overall system could, in fact, be considered a randomly wired 'growing automaton' in the sense of Burks (1961), although one of especially simple type.

Specifically, every subautomaton was a bistable switch ('Darwinian switch') that could be either white (non-set) or black (set). New white switches were added to the automaton by the action of 'priming arrows' emanating (at random) from the black switches only. The second type of 'triggering arrows' likewise originated (at random) from black switches only. Unlike most of the triggering arrows, many of the priming arrows required several switches to be black simultaneously in order to become effective.

Knowledge of the chemical underpinnings of this automaton now suggests the following additional axiom: Different switches may be 'related.' Related means here: similar (and partly identical) in their pattern of connections with other switches. This randomness-defying similarity is to have no upper bound since the chemical subatances involved in very many switches (autonomously growing and self-maintaining systems) may be closely related and even identical in part.

With this axiom added, the behavior of the growing automaton becomes more interesting. Specifically, it is now possible to discern a 'trend' in the branching pattern of the growing tree of domino chains. At first, most of the activated branches were mere bifurcations or trifurcations. After a while, however, most of the new branches become multifurcations, so to speak. One has the impression of a 'selection' acting on the overall system such that the branch points become richer and richer. This is a consequence of the possibility that successor systems (daughter nodes) may be chemically related to their predecessors, even if such cases are relatively rare. 'Rich' branches necessarily have more successors, including rich ones. Rich branches with 'related' successors possess an even greater number of rich successor branches, especially so on the level after next, and so on. This amounts to a selection for visibility (that is, quantitative preponderance) of rich subtrees.

Using biological terms, what is thus observed is an (apparent) selection for 'mutability.' Soon the 'growth zone' of the tree of dominos belongs almost exclusively to a subtree with greater and greater branching power at every successive node.

The increase in mutability has an unexpected side effect. Due to the simultaneously increasing chemical similarity between adjacent nodes, a cross inhibition (due to substrate competition) is bound to develop between them. As a consequence, a smaller and smaller proportion of the richest sets of daughter nodes become actually black (set). Therefore, as far as quantitative preponderance in the growth zone is concerned (selection for visibility), the richest branches tend to recede into the background again. At the same time, however, the decision which of the many daughter nodes is actually activated at a rich branching becomes more and more 'fine tuned' to the set of all environmental conditions that happen to apply at the moment. This means that a new type of selection--'differential environmental selection'--has become effective in the rich subset.

Not all rich branchings are equally affected. In principle, two extremal types of richness (diversity among daughter nodes) can be distinguished from now on: 'neutral' (merely chemical) diversity, and 'effective' (differentially selective) diversity. Only the former type is counterproductive as far as visibility is concerned. Rich branchings with a high proportion of effective mutants continue to dominate the growth zone of the tree.

The principle of differential environmental selection now at work favors a maximal average effectiveness of mutational change more than a maximal increase in the number of such changes. This follows already from the fact that only a moderate number of daughter nodes can be 'prepared for triggering' (so that the environment may decide) at any one time. This is because every potential daughter node requires a non-zero influx, and hence a finite fraction of the synthetic activity of the parent node.

Thus, a property of 'unrestricted mutability' (into a virtually unlimited number of potential daughter nodes) will arise only very late in the tree. This contrasts sharply with one's biological intuition. In biology, the property of unrestricted mutability--found in even the most primitive organisms, and based on simple-appearing molecular mechanisms--imposes as a staple feature that calls for a 'direct' evolutionary explanation. The conditions under which such an incipient organization is no longer selectively disadvantageous have yet to be formulated both on the automata-theoretic level and in terms of possible physicochemical implementations.

QUALITATIVE DYNAMICS OF THE EVOLUTION DESCRIBED

What is the attractor in the qualitative dynamical process described above? It certainly is not a point (zero-dimensional) attractor; nor is it a (one-dimensional) periodic attractor; similarly, it is also most probably not an ordinary (two- or more-dimensional) chaotic attractor (cf. Rössler, 1979). With this list, the presently known types of both simple and 'robust' attractors are exhausted, however. The best thing that can be attempted at the time being, therefore, is a more or less verbal description of the new attractor found.

The asymptotic dynamics of the present class of systems is somehow contained in the following verbal statement: Darwinian dynamical systems internally select (uncover into visibility) a sequence of subsystems in which each generates a viable successor even if the environment changes more and more strongly during the process. The property ultimately approached hereby can be called 'ultraresilience.'

Unfortunately, neither this property nor the way it is approached fits the usual repertory of dynamical notions. An attractor is always something that applies to a dynamical system as a whole. The property of ultraresilience applies only to subsystems and is asymptotic only in the sense of an ever more closely observed functional, taken over an indefinitely long sequence of subsystems. How to reconcile both these features with qualitative dynamics is an open problem.

DISCUSSION

A large dynamical system of 'chemical' (or 'non-negative,' or 'privileged zero,' respectively) type was considered. The system belongs to the class of singular perturbation-type ordinary differential equations (cf. Rössler, 1974b). In the present case, attention was focused not on the whole system, which has some 10^{10000} state variables, but on a particular subset only. All the subsystems considered belonged to the triggerable variety discovered by Darwin: There is an unstable near-zero steady state generating autonomous growth toward a stable upper steady state of autonomous self-maintenance ('existence flip-flops'). Only insulated (more or less non-interacting) subsystems were considered. Within this subclass, an 'explosion' was found to apply: The number of new subsystems recruited and fired can itself grow autonomously. During this growth process, an internal 'selection' for greater and greater 'effective mutability' occurs. The subsystems involved in the sequence at the same time show a greater and greater resilience in the sense of Holling.

Several points remain open. First of all, it is not clear whether chemistry (including organic chemistry) indeed provides both the necessary richness and the necessary gradedness of relations that was axiomatically supposed to apply. Here computer-guided approaches like those of Dugundji et al. (1976) might prove helpful. Second, if organic chemistry does provide one candidate system (and other, similarly rich subclasses of chemistry provide others), one would still like to know whether non-chemical physical systems with analogous properties might not exist too. Third, there is the general mathematical question of how to characterize an attractor that can so far be specified only in terms of the dynamical properties of a sequence of subsystems.

Apart from these open theoretical questions, there are also practical problems to be solved. So far, no long-term kinetic studies of well-stirred Miller-Urey type reaction systems have been performed (cf. also Decker et al., 1982). If life is indeed a mathematical (reaction-kinetic dynamical) process in the first place, the discovery under laboratory conditions of the right dynamics (a spontaneous cascade of autocatalytic subsystems) may be as important as the discovery of the right substances. As it happens, the second major type of experimental questions suggested by the present approach (How likely is the spontaneous formation and reproduction of spatially confining chemical structures?) has already found a detailed positive answer--for the special case of carbon-based chemistry (Fox, 1965).

SUMMARY

A class of chemically implementable dynamical systems is described which go through a (for all practical purposes) never-ending sequence of transients. A partial process (generation of ever new autonomously growing subsystems) is singled out and interpreted as a primitive evolutionary process. In this process, a property of 'unrestricted effective mutability' is bound to be displayed eventually by one of the generated subsystems. For the asymptotic functional property in the background--realized more and more closely by successive members in the sequence--, the term 'ultraresilence' is proposed. The attractor organizing such a sequence of subsystems has yet to be characterized mathematically. Whether or not actual prebiology falls under the present equivalence class of dynamical systems is open. Long-term kinetic studies on well-stirred Miller-Urey type experiments will help answer this question.

ACKNOWLEDGMENTS

I thank Michael Conrad, Norman Packard, and Rob Shaw for discussions.

REFERENCES

Ashby, W. R., 1952, "Design for a Brain," Chapman and Hall, London.

Burks, A. W., 1961, Behav. Sci., 6:5.

Clarke, B. L., 1980, Advan. Chem. Phys., 63:1.

Darwin, C., 1859, "On the Origin of Species by Means of Natural Selection," John Murray, London.

Decker, P., Schweer, H., and Pohlmann, R., 1982, Bioids X. Identification of formose sugars, presumable prebiotic metabolites, using capillary GC/GC-MS of n-butoxime trifluoroacetates on OV-225. J. Chromatogr. (in press).

Dugundji, J., Gillespie, P., Marquarding, D., and Ugi, I., 1976, in: "Chemical Appliations of Graph Theory," A. T. Balaban, ed., pp. 108-174, Academic Press, New York.

Field, R. J., Körös, E., Noyes, R. M., 1972, J. Amer. Chem. Soc., 94:8649.

Fox, S. W., 1965, Nature, 205:328.

Grümm, H. R., 1976, "Definitions of Resilience," IIASA Research Report, RR-76-5, International Institute for Applied Systems Analysis, Laxenburg, Austria.

Holling, C. S., 1973, "Resilience and Stability of Ecological Systems," Annual Rev. Ecol. and Systematics 4, reprinted as IIASA Research Report, RR-73-3, International Institute for Applied Systems Analysis, Laxenburg, Austria.

Lipscomb, W. N., 1963, "Boron Hydrids," Benjamin, New York.

Matsuno, K., 1981, BioSystems, 14:163.
Miller, S. L., and Urey, H. C., 1959, Science, 130:245.
Oparin, A. I., (1957), "The Origin of Life on Earth," Oliver and Boyd, London.
Rössler, O. E.,1971, Z. Naturforsch (in German), 26b:741.
Rössler, O. E., 1974a, Springer Lect. Notes Biomath., 4:546.
Rössler, O. E., 1974b, Springer Lect. Notes Biomath., 4:399.
Rössler, O. E., 1979, in: "Structural Stability in Physics," W. Güttinger, and H. Eikemeier, eds., pp. 290-309, Springer-Verlag, Berlin, New York.

EVOLUTION EVOLVING

Thomas O. Fox

Department of Neuropathology
Harvard Medical School
Boston, Massachusetts

INTRODUCTION

A significant achievement in this century is that evolution itself has evolved. What began as a specialized theory intended to explain the ancestry of different biological species and geological change has become a pervasive term in the language of science. Along with the usual questions accompanying observations, descriptions, definitions, and mechanisms, questions about origins and evolution have become integral aspects of virtually all fields. Several interpretations of Darwinian evolution state that selected genes mediate evolution. Progress has been made in formulating, describing, and defining the concept of the gene. Many people have tried to fit genetic concepts into their understanding of biological processes and their evolutionary intuition.

This essay attempts to define mechanisms of inheritance, and in so doing, explore the relationships between genes and other mediators of evolution. While the term "heredity" usually refers specifically to biological genetics, the word "inheritance" is used here because it has a broader meaning which includes both non-genetic and genetic aspects. These include many feedback mechanisms which reinforce previous characteristics, thereby maintaining traits between succeeding generations. A large set of feedback mechanisms may operate at many levels of organization to determine what each generation inherits.

Evolution, furthermore, is now considered a component of personal interactions, social institutions, history, and politics. Popular descriptions of evolution, particularly in connection with social or moral matters, often give rise to uneasiness caused by a sense that biological or genetic evolution does not quite apply, or is in fact irrelevant. This uneasiness is appropriate if one of two theses is correct. According to the first theory, evolution is a special process which applies directly only to biology; other forms of evolution with different premises and rules apply to other disciplines. According to the second theory, the fundamental features and processes of evolution might be appropriate to help us comprehend all that we sense and interact with in this world; we are uneasy because our widely held concepts of evolution are still quite imperfect.

One expression of Darwinian evolution that has been stated often in this century is that natural selection acts upon chance mutations. This is taken to imply that Darwin assumed the existence of variation and set out to understand how it resulted in the differentiation of species. It is possible to conclude, however, that Darwin believed almost the opposite to be true. In Chapter V (Laws of Variation) of _The Origin of Species_, Darwin (1950) writes as follows: "I have hitherto sometimes spoken as if the variations--so common and multiform in organic beings under domestication, and in a lesser degree in those in a state of nature--had been due to chance. This, of course, is a wholly incorrect expression, but it serves to acknowledge plainly our ignorance of the cause of each particular variation." I take these explanations and the subjects of much of Darwin's attention to indicate that Darwin's intellectual goal was initially and primarily to understand variation. Darwin was fascinated by the enormous extent of variation that he was fortunate and perceptive enough to examine. In the Introduction, Darwin (1950) says that "It is . . . of the highest importance to gain a clear insight into the means of modification and co-adaptation." He concludes the Introduction with " . . . I am convinced that natural selection has been the main but not exclusive means of modification." Darwin did not believe that variation was a chance occurrence and that his central task was to explain how descendents were selected from among the varieties. Rather, he realized that if he ignored the source and causes of variation until they could be explained, he could partially explain evolution by recognizing and substantiating a second process, selection. Thus his great contribution was, while in admitted ignorance of one principle, to perceive a second principle, that some amongst the varieties perish while others survive.

LEVELS OF EVOLUTION

Biological evolution occurs at many levels of organization. It applies to sub-cellular molecules and entities, to cells, to organ systems, to organisms, to groups of organisms, and to societal products of organisms. For any system, the highest level of organization can be identified, and lower levels of organization can be described. Each level is comprised of a system of components which are derived from the next lower level. Different tables of hierarchical organization have been described. As we learn more these tables are modified, but they serve as examples for examining the relationships between adjacent levels (Fox, 1972).

At each level natural selection has the potential to act upon a set of major properties which selectively change in increments. These foci of selection (Fox, 1972) are the major source of variation and mediator of inheritance for each level. While in single living cells the focus of selection is genes, in the mental level it is memory, and in social levels of organization, foci of selection are literature, bulk forms of information storage, and computer databases. Additional relationships can be identified between any two adjacent levels. In every case, the higher, or emergent, level expresses properties that are not shown by the constituent parts or by the level to which they correspond. This is easy to see since it is usually the basis for identifying different levels of complexity.

In one such hierarchy, a pride of lions is a social system, the highest level of organization, derived from mental and neural levels. The neural level is a multicellular system, below which is a genetic level, below which are various subcellular levels. The foci of selection are, respectively, communication among lions; thinking patterns of individual lions; dendritic associations of cells (multicellular interactions); developmental events leading to these structures; and genes (at the cellular level). For different systems, such as groups of lions, humans, or insects, not all the same preceding levels are represented.*

The components of an emergent system are restricted in the expression of their inherent characteristics. For example, not all the states of matter that would characterize a given molecule are compatible with living systems. So the potential for chemical interactions of constituent molecules is restricted when molecules

---

*In this essay a series of organizational levels is used to illustrate several relationships between adjacent levels. It is beyond the scope of this chapter to consider several alternative lists; for example, differences must exist between the series of levels leading to societies of insects versus societies of mammals. Molecular genetics is revealing the complexity of gene and chromosome structure and dynamics and the evolutionary relationships of different genetic mechanisms. For biological systems one can design lists which include several different genetic levels, which may or may not be sequential. For example, developmental levels of evolution may follow the lowest genetic level, which is used in examples in this essay, but might precede levels composed of genetic rearrangements, transposable elements, or maternally transmitted viruses.

are constituents of higher levels. Put another way, the properties of an emergent system necessarily influence the properties of the constituents. This phenomenon has been observed in the development of literature by a critic who describes the paradox of a modern writer's influence on his ancient predecessors (Borges, 1964). Historians are aware of how current thinking affects our perceptions of history, which change in time as words and concepts assume new meanings. As another example, human freedom is restricted by social structures, because societal rules of interaction dictate that individuals not express certain impulses and traits.

For biological or social levels of evolution, we can partially describe how new levels of organization emerge and evolve. A new energy input into an existent system can lead to changes in that system or to an emergent system with new properties and a new focus of selection. In social levels of evolution we have observed the effects of new inputs such as the railroad, the telephone, and the automobile. Each of these profoundly affected individual activity; generated new economies, new social systems, and new methods of war; and changed existing social structures such as governments and religions. With dramatic and severe alterations in the availability and distribution of oil in recent years, we can vividly see these principles at work. Sudden fears of oil shortages or gluts, as well as predictions that long-term supplies will decrease, have led to searches for alternate energy sources, shifts in diplomatic relations among nations, and changes in the classes of individuals whose political power and influence are recognized and valued. Energy inputs of various kinds are important at all levels of evolution. The effects of energy inputs upon the dynamics of productivity are as important to an economist as the effects of molecular signals responsible for differentiation of cells are to a developmental biologist.

In order to persist, each emergent level of evolution must be able to evolve more quickly than the lower level from which its constituents are derived. If it does not, then the entities of which it is built may change in a way which alters their potential for cooperating with one another and with the essential energy inputs. Cells, such as certain bacteria, which evolved genetically in ways that allowed survival, might have very little potential to form multicellular organisms. If, due to subsequent mutation and selection, the cells in a simple multicellular colony evolve to allow complete autonomy, the organization at the multicellular level will cease. Since we do not known all the natural experiments that have occurred historically it is quite possible that some of the single cellular organisms we know today descended from cells which at one time were part of a multicellular level which perished. Although these cells have not again cooperated to form a multicellular colony, they may contain in their genomes vestiges of a history that will otherwise be difficult to understand.

A lower level can evolve to become incompatible with higher levels, either by gaining newer or by losing older properties. Assume the existence of an organization comprising six levels, of which A is the lowest and F is the highest. If level A evolves faster than B, not only will B disappear but so will C, D, E, and F, which are built successively upon each preceding level. This can occur whether A evolves to something new and viable or to something less viable than it had been. As an example, a severe and persistent deficiency in an essential nutritional factor such as a vitamin or particular amino acid (level A) could cause errors in metabolic feedback mechanisms (level B) that are involved in differentiation. This could result in poorly formed neural circuits in new generations (level C). As a consequence, mental activity (level D) within a specific locale and culture eventually would be impaired, and over a long time a decline in communications among individuals (level E) could lead to the dissolution--or rather, evolution--of social structures (level F). Likewise, a severe increase in mutational rate, due to biological causes or radiation, could lead to cellular changes that would jeopardize the integrity of mammalian organisms as we know them. This could lead to increases in susceptibility to infection, abnormal tissue development, or inadequate biological repair.

The requirement that emergent levels evolve faster than their generative lower levels can be viewed existentially. I have said that if the components of a given system evolve to something incompatible with the system, then the system will disappear. What is actually required, however, is that on the average the higher levels evolve no more slowly than the constituents. The consequence of this requirement is that evolution of each level influences the evolution of the immediately lower level. This and the other relationships among levels can be illustrated by the following example.

A small animal, accustomed to collecting grasses and seeds from among certain rocks, wanders beyond her usual habitat. She finds and eats some berries of a kind she had not seen before. From the nutrients they contain, certain of which are new to her diet, she acquires greater stamina and skill at foraging. The potential existed before, but now is expressed for the first time as a consequence of the new food intake. The animal seeks out these berries regularly and is able to maintain a better nutritional state. Her offspring share the advantage, being better developed during gestation and while nursing. Because the mother nurtures her offspring more effectively and teaches them to find and eat the same berries, the offspring in this and in several subsequent generations continue to express this heightened level of activity and ability to thrive even though the older source of seeds and grasses slowly disappears.

Succeeding generations express characteristics and habits that result from the discovery and use of the berries; they have inherited a distinctive set of geographical, physical, neural, and behavioral habits and instincts. This inheritance can be attributed to the large set of feedback mechanisms which reinforce the effect of the new food. If the reinforcement lasts for more than a generation, then any changes in these particular organisms will also be maintained. The first feedback loop involves the mother's ability to acquire the new berries. Once she ate them, she enhanced her health in various ways and therefore was able to continue (at an increasing rate) to include this food in her diet. Since this trait was passed on to her offspring and to their offspring it represents inheritance. Such an inheritance gives the offspring a selective advantage. Selection at a behavioral level also affects metabolic feedback. Ultimately, changes at the metabolic level and at lower levels-- even at the genetic level--will be retained if they are compatible with the foraging behavior.

This passing of selective pressure to lower levels can be referred to as contraction, since it represents a transfer of information that is later expressed in ontogeny. It is all the better if changes occurring independently at each lower level also reinforce the potential for superior ability. Time will bring changes in the animal's environment, physiological parameters, and genetic makeup. All of these changes affect inheritance, which is expressed during the development of each new generation and which, therefore, contributes to evolution.

In the above example the genotype of the mother permitted a range of phenotypes. At each level, a range of potential types is possible, such as variations in the degree of gene expression, enzyme activities, or electrical activity of nerve cells. The particular nutrients available or the experiences of the individual affect the outcome of development and learning. A new situation, locale, or stimulus could induce emergent behavior, as illustrated above. No specific new mutation is required, since the set of genes in each individual allows many possible phenotypes, including ones not yet observed. The many feedback mechanisms that operate at various levels limit the possibilities during ontogeny, while preserving the behavior that has emerged.

Over succeeding generations, further changes occur; behavior becomes modified and selected, while genes continue to mutate and chromosomes rearrange. Eventually, a shifted genotype exists which is compatible with a different set of possible phenotypes. Now, this newer set of genes is essential for expression of current behaviors. What had been an unessential emergent behavior may now be necessary for survival in the current locale and time. Accordingly, while at the time of emergence specific new genes were not required for that event, genetic changes that have occurred over a much longer period of time have been selected, since they are necessary for expression of the animals' repertoire of behaviors. What is critical is the difference in timing. Specific genetic changes do not have to occur to allow a new behavior, though generations later specific sets of genes will have been selected that correlate with specified traits.*

These relationships between levels and time can be illustrated further by considering the evolution of video games. The focus of selection at the lowest level considered is the computer chip used in making the games, analogous to genes in the animals above; the focus in the highest behavioral level is the popularity of the games among customers. Consider that two companies use identical chips to make two different games, although the properties of this chip might allow many other variations. With time one of the two games becomes more popular. Its manufacturer experiments with variations of the games, of which some become even more popular and successful. This leads to a more rapid depletion of the inventory of chips, and more profit. Now this manufacturer goes out to purchase a new supply of chips. Meanwhile newer chips have been manufactured which have more memory and, therefore, a wider range of applications. Because the more successful video game producer can afford a large supply of these chips, it uses them to make more of its popular games, and continues to experiment. They find that the new chips allow the addition of previously unavailable special effects. Therefore they develop and now sell yet newer games, which are even more popular than the previous games. When the next marketing season opens they only sell the new game and have discontinued manufacture of the original video sets.

The company above was successful, although its success did not require that the other company perish; they simply evolved in different ways, and this difference was not focused on the chip used, since it was the same for both. But the success of one company through its series of games facilitated the utilization of a newer type of chip, which eventually led to newer types of games not originally possible. Selection at the level of popularity was contracted over time through levels of sales, promotions, production, and development and eventually had an impact on the chips used to make the video sets. Thus selection at a higher level later influenced selection at a lower level.

In biological systems, when a change in behavior is selected at a level higher than the genetic level, the set of compatible

*Many geneticists believe that there is DNA which is not essential or which can undergo mutation with no detriment to the organism at any point in time. By contraction this may eventually be selected to reinforce a higher level activity which had been selected in earlier generations. This DNA may appear to constitute neutral genes with neutral mutations in the intervening time.

individuals that will be preserved will have a large degree of genetic heterogeneity. This heterogeneity will be preserved. At a later time when both the phenotype and the genotype of the species have shifted, correlations will exist between specific traits and sets of genes. For evolutionary levels remote from the genetic level, the emergent, or modified, traits will increasingly be correlated with large numbers, or sets, of genes, while emergent levels close to the genetic level generally will be correlated with relatively small sets of genes.* For complex systems with many levels above the genetic, gene functions will be assigned most directly to intermediate structures or activities. As further evolution occurs, some of these genes may come to influence additional new traits, such that pleiotropic effects--multiple effects of single genes--become more prominent.

Pleiotropic effects are seen in other levels of evolution for which correlations with genetics are infinitesimal. For example, consider multiple consequences of an amendment to the United States Constitution. This is illustrated by the 22nd Amendment, which stipulates that a president can only serve two elected terms. The reasons for this restriction, within the philosophy and history of the United States, are quite obvious. However, this change had a consequence that was not anticipated, and accordingly no compensatory provision was made to accommodate it. The unintended effect was to partially eliminate one of the mechanisms for checks and balances on the performance of the Executive Branch; namely, a president's fear that he will not be re-elected. While not written into the Constitution along with other checks on the President, such as congressional voting authority and the process of impeachment, this is a part of the political system which, along with party politics and opinion polls, have evolved to be integral components of American government. A twice-elected president would not be permitted by the 22nd Amendment to seek a third term, and his actions, and concerns about the electorate, might be much less moderate than if he were eligible for another term. The consequences could be extremely grave. It can be argued that the viability of the system would be enhanced by a further change to compensate for this effect.

As the number of evolutionary levels increases, the effect of selection that is passed to the lowest levels by this process of contraction diminishes. Thus on the one hand it is inevitable that

*While behavior of mammals represents a level significantly removed from the genetic, the correlations of behavior with genetics should be much more direct in simple organisms. In animals, such as garden slugs, which are used as models systems for studying neurobiology and behavior, the distance between genes and specified behaviors may be relatively short.

some traits acquired in social, emotional, and other behavorial and mental levels eventually will be contracted to the genetic level, and will result in genetic changes which were influenced or initiated by behavior in preceding generations. On the other hand, this happens only if the initial feedback at the higher levels is sufficiently strong and long-lasting and if the effect on genomic changes necessarily affects many genes simultaneously, such that correlations between genes and behavior become increasingly indirect. The question is not whether the process of contraction occurs, but rather how significant it is in relation to the very large set of processes and feedback which affect development and behavior.

CONCLUSION

This is an expansion of an earlier essay (Fox, 1972) which considers how levels of evolution evolve. I believe that most of the individual ideas which have been assembled here are faily widely discussed within different disciplines, both by scientists and non-scientists. Evolution is not the exclusive realm of biologists, geologists, and cosmologists, and neither is the game of philosophizing about it. Since the essay above, many aspects of evolution have been debated widely, with much attention focused on sociobiology. It should be clear from this essay that, in my opinion, much of that debate was not as productive as it could have been. Indeed, the term sociobiology seems to imply that evolution is essentially a biological phenomenon and that its application to behavior and to sociology can be made in the same genetic terms and with the same apparent aims to propagate the most successful germ lines. But evolution can be viewed in more general terms than those derived from biology. Feedback mechanisms at many levels result in inheritance--if it is defined as preservation between generations. Necessarily, genetic heredity is one such mechanism--and a particularly important one--but it is only one of several mechanisms which, like the others, not only contributes to, but also is influenced by, other foci of variation and selection.

Discussion and debate about evolution, the origins of life, the origins of intelligence, and the relationships between nature and nurture recur in a cyclic fashion. If evolution is simply a way of classifying the complexity of things and the processes occurring among them, then we can explain the timing of these cycles. They crest with methodological and conceptual breakthroughs is science. Each of these events enables us to ask new questions about the mechanisms--and the fundamental significance--of what makes us what we are and how we are. In the past century, such milestones include the founding of the modern study of development, genetics, and bacterial genetics which led to our very basic (though certainly still incomplete) understanding of the molecular principles of genetics. In physics the milestones reached in understanding matter and energy have been at least as numerous and influential in our thinking and our technology as have been the discoveries of biology and chemistry.

Perhaps the attraction of evolution is that all intellectual pursuits aid our comprehension of it. Therefore, in a sense it is an intellectual synthesis, producing concepts which are not perceived within a single discipline or in any solely descriptive field. Each emergent level of understanding produced by this synthesis feeds back into the individual discipline of its origin, continuing the process of evolution evolving.

REFERENCES

Borges, J. L., 1964, "Other Inquisitions," p. 108, Simon & Schuster, New York.

Darwin, C., 1950, "On the Origin of Species by Means of Natural Selection," a reprint of the First Edition, Watts and Co., London.

Fox, T. O., 1972, in "Molecular Evolution: Prebiological and Biological," D. L. Rohlfing, and A. I. Oparin, eds., pp. 35-42, Plenum, New York.

CONCEPTS OF PROTOBIOLOGICAL EVOLUTION: THEIR IMPLICATIONS ON NATURAL SELECTION AND TIME-COURSE OF EVOLUTION

Laura H-N. L. Hsu

Department of Biology
Rice University
Houston, Texas 77001

This paper is addressed to the development of certain concepts in pre- and protobiological* evolution. Such elucidation sheds light not only on the origin of life but also brings new insight to general principles of evolution such as the nature of natural selection and the time-course of evolutionary change.

THE PROBLEM

The origin of life occupied that evolutionary intermediary period through which the first recognizable biological forms arose from simpler, non-living precursors. By definition, those precursors must not have had the combination of features that characterizes contemporary organisms. Yet, at the same time, one would expect these precursors, whatever they were, to have possessed sufficient complexity to have been capable of evolving into the first life forms.

Most modern scientists have little intellectual difficulty in recognizing the logic of the evolutionary origin of life. It is in consideration of the materials and mechanisms of protobiological evolution where diverse opinions arise. The major stumbling block in search for a definite answer to the problem of the origin of life lies in the fact that there is simply no direct evidence preserved from that period of our evolutionary history from which we can

*Prefix "proto" denotes states immediately or closely before;
"pre" indicates any stage leading to.

derive unequiocal proofs. The most promising method for understanding prebiological evolution is through experimentation. The choice of experimental approach is thus of crucial importance in this line of study. Scientists have turned to a scientific approach in the form of constructive experimentation for elucidation of problems associated with the origin of life.

EXPERIMENTAL APPROACH TO PREBIOLOGICAL EVOLUTION

Constructive procedures start with the very simple and work towards greater complexity. The entire scenario of molecular evolution is based on the constructionistic principle; all molecular species in the universe arise by combinations of simpler precursors (Oparin, 1957). The evolution of the universe is intimately tied to the evolution of matter. Numerous kinds of inorganic and organic compounds have been discovered and more are being identified in the cosmos, scattered in interstellar spaces or in association with celestial systems (Morrison, 1973). Life's raw materials could and can arise via molecular evolution whenever suitable reactants meet under appropriate conditions.

The evolution of the universe and everything within, including life, can be seen as an immense constructionistic exercise. The wise constructionistic experimentor is careful to adhere to relevant experimental conditions, then sits back to allow himself/herself to be taught about what comes naturally. Laboratory simulations of primordial Earth conditions have shown that most of the building blocks of life can form spontaneously and abiologically (Calvin, 1962; Keosian, 1974). Further experimentation has shown amino acids to join in orderly fashions forming polymers of amino acids or proteinoids (Fox and Harada, 1958). Proteinoids possess not only essential protein-like properties, including numerous enzymic functions (Table 1), but also some undergo structural orientation to form cell-like microspheres (Table 2).

Microspheres show dynamic behaviors which are mechanically analogous to inheritance, reproduction and production of inherently heterogeneous progenies (Fox et al., 1967; Hsu and Fox, 1976). The conditions for the operation of natural selection and for evolutionary progression are integrated in the microsphere system (Hsu, 1974).

EVOLUTIONARY PLAUSIBILITY

The plausibility of proteinoid-microsphere systems as models of pecursors of life has been extensively examined and reviewed from the points of view of geological logic, cosmic relevance, thermodynamic soundness, and adherence to principles of physics, chemistry, and biochemistry (Fox and Dose, 1972).

Table 1. Enzymic and Related Activities in Thermal Polyamino Acids (Fox and Dose, 1972)

Reaction, Function, or Substrate	Authors and Year
Hydrolysis:	
p-Nitrophenyl acetate	Fox, Harada, and Rohlfing (1962)
ATP	Fox and Joseph (1965)
p-Nitrophenyl phosphate	Oshima (1968)
Decarboxylation:	
Glucuronic acid	Fox and Krampitz (1964)
Pyruvic acid	Hardebeck, Krampitz, and Wulf (1968)
Oxaloacetic acid	Rohlfing (1967)
Amination:	
α-Ketoglutaric acid	Krampitz, Haas, and Baars-Diehl (1968)
Oxidoreductions:	
H_2O_2 (catalase reaction)	Dose and Zaki (1971)
H_2O_2 and hydrogen donors (peroxidase reaction)	Dose and Zaki (1971)
Photo-activated decarboxylation:	
Glyoxylic acid	
Glucuronic acid	
Pyruvic acid	Wood and Hardebeck (1972)
Synthesis with ATP (in proteinoid microparticles):	
Internucleotide bonds formed	Jungck and Fox (1973)
Peptide bonds formed	Fox and Nakashima (1974)
Hormonal Activity:	
MSH	Fox and Wang (1968) Bagnara and Hadley (1970)

The entire sequence of events in the formation of proteinoid-microspheres from inorganic precursors (Fig. 1) has been done under simulated primitive Earth conditions. Although there are some discrepant opinions among scientists regarding the relative reducing qualities of primoridal Earth's atmosphere (Oparin, 1938; Kerr, 1980) building blocks of biomacromolecules such as amino acids and nucleic acid bases have consistently been recovered from experiments using various gas mixtures ranging from strongly reducing to mildly reducing (Kerr, 1980).

Polymerization of amino acids takes place abiologically by simple anhydrous thermal condensation (Fox, 1969). Polymerization of amino acids can also be facilitated by inorganic catalysts imputed

Table 2. Properties of Acidic Proteinoid Microspheres in Common with Those of Contemporary Cells (Fox and Dose, 1972)

Stability (to standing, centrifugation, sectioning)
Microscopic size
Variability in shape
Uniformity in characteristic sizes
Numerousness
Stainability
Producibility as gram-positive or gram-negative
Osmotic types of sensitivity to changes in tonicity
Ultrastructure (electron microscope)
Double-layered boundary
Selective passage of molecules through boundary
Catalytic activities
Varying and characteristic patterns of association
Propagation by budding and fission
Growth, by accretion
Motility
Selective inclusion of polynucleotide with basic proteinoids (particles are composed of nucleoproteinoid, not proteinoid)
Membrane and electrical activities[a]

[a]Ishima, Y., Przybylski, A. T., and Fox, S. W., 1981, Electrical membrane phenomena in spherules from proteinoid and lecithin, BioSystems, 13:243-251.

to primordial Earth (Fox and Dose, 1972). Certain proteinoid polymers spontaneously aggregate into cell-like microstructures in the presence of water and moderate changes in temperature (Fox et al., 1959). Simulation experimentation demonstrates that many varieties of biomolecular precursors are formed simultaneously, and the overall sequence of reactions from amino acid precursors to evolvable, cell-like microspheres (Fig. 2) takes place with ease, generally in high yield, and in relatively rapidly (minutes and hours).

In view of their abiological origins and their evolutionary potentials, proteinoids and microspheres are plausible life precursors. With proteinoid microspheres as models of protocells, we can begin to study and comprehend the nature of the protobiological evolutionary process, the link between molecular evolution and organic evolution. Comparative studies on molecular, prebiological and biological evolution can bring new insight to our understanding of evolution in general.

Molecular Evolution	Prebiological Evolution	Protobiological Evolution	Biological Evolution	Species in Phylogenetic Tree
→	→	→	→	
(amino acids and precursors of nucleic acids)	(proteinoids)	(microspheres)	(cells)	

Fig. 1. General evolutionary stages from inorganic chemicals to modern biological species. Steps to the left of the vertical line are results from simulated experimentation.

ORDER WITHOUT DNA

When the theory of the DNA basis of inheritance was elucidated in the 1950s, its elegant, precision, and the sheer pervasiveness of its application elevated it to the status of the Central Dogma (Watson and Crick, 1953). It was dogma that all biological order

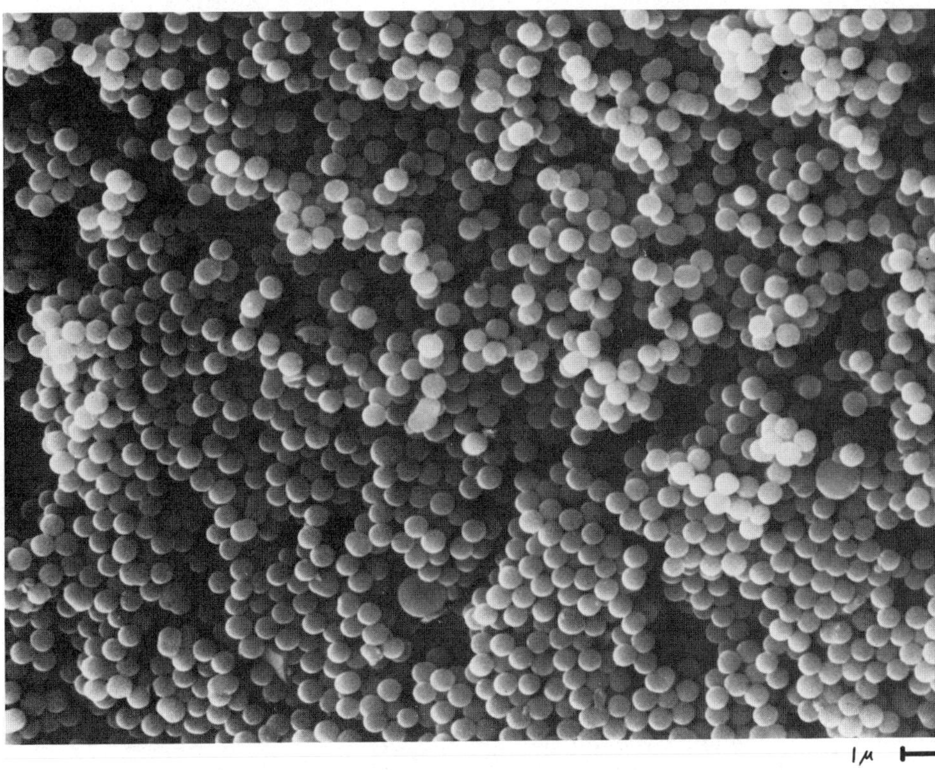

Fig. 2. Electromicrograph of a population of proteinoid microspheres. (Photo by Robert M. Syren)

is ultimately DNA-based. Conversely, no biological order is conceivable without nucleic acid coding.

This kind of mental disposition has led to the postulation that DNA had to come first in the origin of life. However, experimentation in this area shows no plausible possibility for the ab initio origin of biologically meaningful DNA on the prebiotic Earth (Fakhrai et al., 1981). Furthermore, DNA is a very inactive molecule. No evidence has ever been offered to show its ability (unassisted by enzymes) to either replicate itself or to synthesize a protein.

Propensity for existence is a first prerequisite for survival which, in turn, is a primary criterion for evolutionary success. In terms of origins, the emergence of the first life form is the result of the propensities of each of the chemical reactions which led to its appearance. If abiological formation of meaningful polynucleotides were unlikely, what other source of order could have provided the matrix for protobiological evolution?

In considering simple chemical reactions, what theoretically can happen is essentially governed by the stereochemical state of the reactants and the thermodynamics of the system. Under a particular set of circumstances, however, only one of any number of theoretical possibilities is allowed to happen. If we examine molecular changes as a type of evolutionary process, it is possible to distinguish evolutionary components in simple chemical reactions. The laws of physics and chemistry provide the guidelines (order) for a limited variety of possible interactions (limited variability). The environment at large then selects which particular reaction will actually take place. Natural selection does not determine what reactions are potentially possible, but dictates the direction of the change. When simple reactions are sequentially connected into pathways, patterns of molecular evolution become evident.

As we understand molecular evolution and the laws of physics and chemistry, it can be said that orderliness is inherent in all matter. From this point of view, certain degrees of orderliness in polymers of amino acids and the resultant microspheres can be logically assumed. That this level of orderliness can be meaningful to the origin of life has been indicated by experimental evidence.

As S. W. Fox has extensively written, information inherent in amino acid monomers dictates their orderly polymerization into macromolecules of biological significances (Fox and Dose, 1972; Fox, 1969; Table 1). Similarly, L. L. Hsu's experiments with microspheres as model protocells showed highly ordered organization including reproductive mechanisms that produce progenies of controlled variability (Hsu and Fox, 1976; Hsu, 1974). Each level of organization expresses salient patterns of order which are characteristic of its structural complexity and, at the same time, reflect the history of its derivation and the collective properties of its components.

COMPONENTS OF PROTOBIOLOGICAL EVOLVING SYSTEMS

Order in the proteinoid microsphere can be seen as a measure of its resemblance to the biological cell. A summary of such properties is listed in Table 2. Those phenomena which contribute to an understanding of their evolutionary potential are selected for discussion below.

Acidic proteinoid microspheres* have been observed to undergo several types of reproductive cycles:

1. Budlike structures often develop spontaneously, attached to mature-sized microspheres (3-7 mμ). Buds can be separated from the larger microspheres and serve as nuclei for accretive growth in the presence of new proteinoid to form subsequent generations of microspheres (Fox et al., 1967, Fig. 3).

2. Another structure that develop spontaneously in some microspheres aged three weeks or longer are the so-called "endoparticles". These are small packets of proteinoid formed inside microspheres. Some are released into the medium, accumulate fresh proteinoid molecules and form new microspheres resembling those of the previous generation (Hsu and Fox, 1976; Hsu, 1974).

3. Endoparticles from different microspheres can meet after their exit into the medium and behave as multiple nuclei for accretive growth into a heterogeneous second generation (Hus, 1972, Fig. 4). Microspheres with up to four nuclei have been observed. It should be noted that the coexistence of many varieties of microspheres in the same milieu is the norm rather than the exception.

4. Endoparticles from one microsphere can pass through junctions to enter an adjoining microsphere. The two endoparticles from two microspheres can form a couplet inside the recipient microsphere and exit into the medium. After accretive growth, a new "recombinant" microsphere is formed.

The movement of proteinoid polymers across conjugal junctions (Hsu et al., 1971) is a mechanism for controlled variation whereby information can be pooled or exchanged. Proteinoid polymers contain

---

*Acidic proteinoid is made with aspartic acid, glutamic acid and an equimolar mixture of 15 amino acids in the proportion of 2:2:1 by weight.

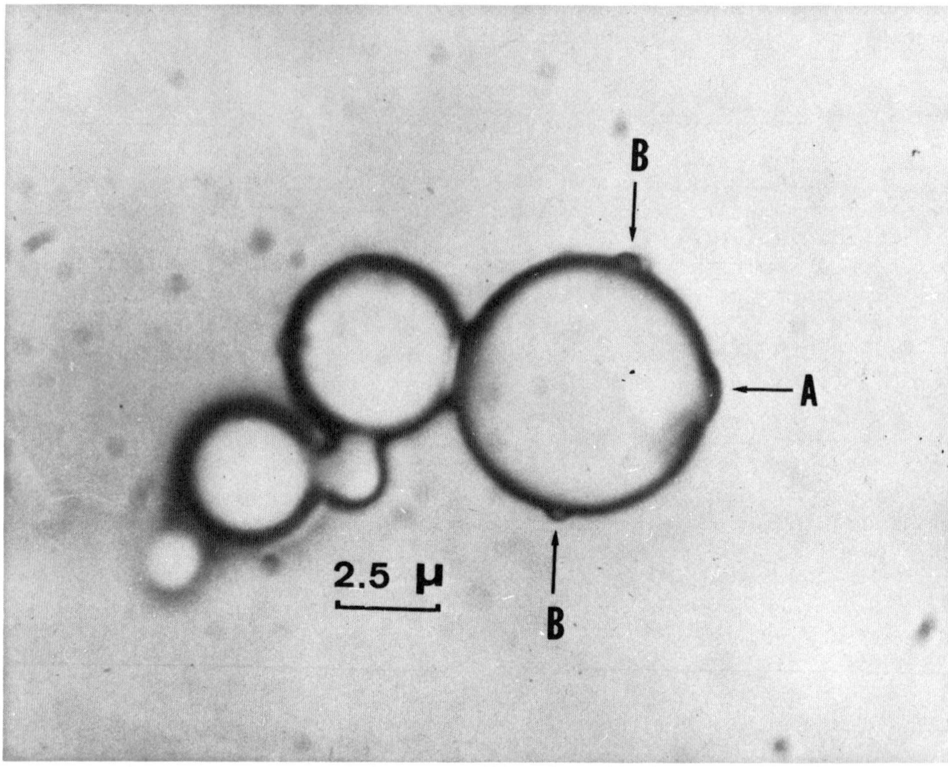

Fig. 3. Budlike structures attached to larger microspheres: type A buds have an apparent intrinsic origin whereas type B may have either an intrinsic or extrinsic origin. Buds separated from larger microspheres serve as "seeds" which can grow by accretion to form new generations of mature-sized microspheres (Hsu, 1974).

information in the sense they are capable of self-ordered assembly (Fox, 1969), have a roster of enzymic qualities (Table 1), and exhibit degrees of specificity in interactions with homopolynucleotides (Fox et al., 1972; Fox, 1974).

When buds and endoparticles from one generation of microspheres are passed on to subsequent generations, the physical basis for non-nucleic acid inheritance is established.

As summarized in Fig. 5, prototypes of asexual and sexual modes of reproduction are modeled, and diversity among members of populations is easily attained. Growth by accreting preformed macromolecules may be seen as a forerunner of heterotrophism. Capacity for

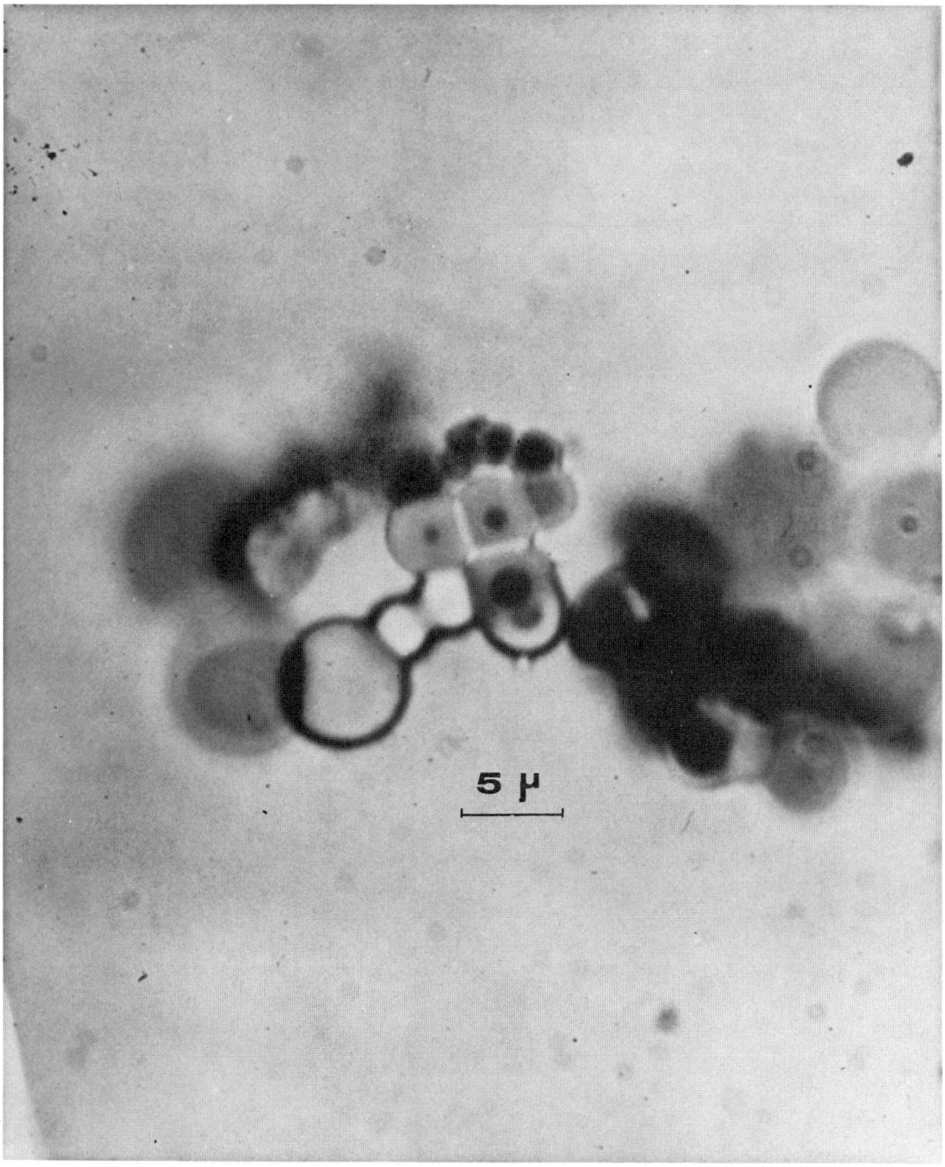

Fig. 4. One or more kinds of endoparticles can be incorporated into new microspheres to give rise to a heterogenous population. Hybrid microspheres containing different proteinoids show expanded enzymic functions either additively or synergetically (Hsu, 1974).

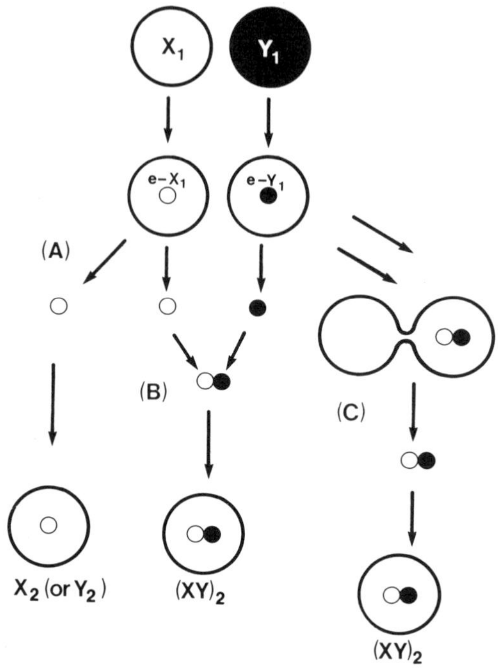

Fig. 5. Diagrammatic representation of the various fates of endoparticles analogous to different modes of reproduction.
(a) "Asexual" cycle
(b) "Sexual," external fertilization
(c) "Sexual," internal fertilization
X_1 and Y_1 = first generation microspheres
e-X_1 and e-Y_1 = endoparticles from X_1 and Y_1
X_2 and Y_2 = homogeneous second generation microspheres
$(XY)_2$ = heterogeneous second generation microspheres

internal syntheses can be developed through cycles of combinations and recombinations between and among microspheres of varied enzymic capabilities. Experiments demonstrating this potential have been reported (Hsu and Fox, 1976; Hsu, 1974).

As proteinoid polymers arrange themselves into a higher level of structural complexity, their inherent orderliness and functions make a simultaneous hierarchial jump from those typical of polyamino acid macromolecules to properties of cellular character.

The above observation has profound significance on the origin of life in that most of the basic essential components of organic evolution appeared simultaneously in one evolutionary event. Growth, inheritance, reproduction, and variability emerged, by a quantum jump, already integrated and functioning in a manner analogous to biological evolving systems.

Apparently a nucleic acid genetic code, an ATP energy trapping system and internal biosynthesis are not obligatory prerequisites of dynamic proto-evolution. Nevertheless, the acquisition of these facilities would mark the advent of life and bio-evolution as we understand it today. Indications of how this can be accomplished are provided by two types of research data. One type shows evidence for proteinoid mediated synthesis of ATP (Fox et al., 1978), peptides (Fox and Nakashima, 1980), and oligo and polynucleotides (Jungck and Fox, 1973; Waehneldt and Fox, 1968; Nakashima and Fox, 1972). In the latter case, the relationship can be codonic, anticodonic or neither according to conditions. Another type of study using microspheres as model protocells has shown that enzymic properties of different kinds of proteinoids can be combined in hybrid microspheres through reproductive processes inherent to them (Hsu and Fox, 1976; Hus, 1974; Fig. 4). Expanded functional capacity as well as synergetic interactions between components of microspheres occur readily. That the proteinoid microsphere can evolve towards a contemporary cell is thus inferred.

The model protocell bridges the gap between molecular evolution and protobiological evolution. It provides a unique opportunity to study evolutionary processes in general. We now have a scientifically-based overview of the sequential transitions whereby simple inorganic materials can develop, step by step, into dynamic protocells endowed with the potential for evolution toward contemporary life. This overview permits a larger and clearer view of two general principles of evolution: (a) the manner in which natural selection works, and (b) the time course of evolutionary change.

NATURAL SELECTION: A LARGER VIEW

Conventionally, natural selection is said to work by allowing the survival of the fittest (Darwin) by competitive exclusion of others (Gausse). This principle has been supported by many evolutionary scientists including Mayr and J. Huxley.

Observations in molecular and protobiological evolution, however, indicate that a more accurate view is that the environment permits the simultaneous survival of all those which are able. Coexisting and interacting variants can produce changes in many directions. For example, many kinds of building blocks of biomolecules appear simultaneously in one reaction. Given the same reactants, different experimental conditions favor the production of certain products but not to the exclusion of others. Thus, precursors have the opportunity to accumulate in the same place and can interact to form a variety of biomolecules such as different kinds of proteinoids.

Similarly, many types of protocells in the form of microspheres made of varied proteinoids can coexist in the same milieu and have been shown to interact with each other in as many ways as their constitution and opportunity allow. Such interactions are avenues for deriving new variants in numerous directions. Evolution progresses pluralistically, or linearly if circumstances are strictly limiting.

In the larger sense, natural selection and evolution is not necessarily restricted to particular materials or mechanisms. Life can conceivably originate on Earth without preformed DNA because functional polyamino acids offer a feasible alternative which can evolve towards a nucleic acid-based life. A larger view is especially important in searching for possible life in extraterrestial environments where materials and conditions may well support some type of evolution different from our DNA-based evolutionary processes.

It can be argued that natural selection in the Darwin-Gausse sense is intended to apply only to, and is still appropriate for, biological systems. There are now sufficient data from ecological studies to indicate that a different perspective called the "variable-environment" view must be considered as a primary mode of ecological interaction. As summarized by Schoener (1982), the variable-environment view says that populations are often well supported by their environments so that interspecific competition is weak or unnecessary. When there is an occasional period of stress, the existing phenotypic variability can easily cope with the hardship. It is inferred that selection on the basis of competition cannot fully explain genetically controlled characteristics in biological systems.

The variable-environment view agrees very well with the larger view of natural selection derived from studies in prebiological evolution in that the environment is seen to play a relatively permissive and passive role. The potential for variation is primarily set by the inherent nature of the interacting members and secondarily limited by the environment. The creative element in chemical and protobiological evolution is thus an intrinsic matter. Natural selection does not create, but channels the creativity into directional changes (evolution). What is learned of intrinsic sources of variability in prebiological systems may suggest to the contemporary evolutionist to look in the same direction to seek internal sources for variability in biological systems.

TIME-COURSE OF EVOLUTIONARY CHANGE

A comparative analysis of the different levels of evolution shows each to have its own character and, at the same time, share some fundamental features of all evolving systems.

Evolution at the molecular level is composed of series of reactions yielding products that are generally completely different from the precursors. All information directing their interactions are inherent in the structure and chemistry of the materials. Such information is non-segregatable and cannot be cycled in the absence of reproductive mechanisms. Without a cell structure for protection and preservation of materials, there are kinetic problems for interactions, and the system is very much vulnerable to environmental influence.

The accumulation of appropriate precursors in the right places, in adequate concentrations, may be tenuous. However, once conditions are favorable, such reactions require little time to complete. The time-course of such changes are depicted as very steep steps (mega-evolution) preceded by periods of quiescence (Fig. 6).

One such mega-step bridges the gap between molecular and protocellular evolution as demonstrated by the assembly of proteinoid polymers to form microspheres. Protocellular evolution resembles cellular evolution in that both take place in phase-separated cell structures which provide protection and independence from the environment.

By mechanisms phenomenologically analogous to combination and inheritance, protocellular variants are produced by transfers of actual portions of functional materials between protocells and between generations. Variants can acquire large (or small) amounts of new functional materials and the rate of change, in general, is somewhere intermediate between those of molecular and biological evolution.

Biological evolution is in a state of adaptedness closest to the optimal. Information is stored separately from functional materials in precise genetic codes. Large deviations usually have negative survival effects. The time-course of change is thus much finer when compared to either molecular or protocellular evolution. For the past four decades the predominant view has been that organic evolution exhibits a very slow, smooth time-course as proposed by J. Huxley in his theory of "Modern Synthesis". This view of organic evolution describes the process to be a smooth continuum; the only source of new genetic information is through point mutations (micro-evolution). Such a restricted view is now under challenge by today's evolutionary scientists (Stanley, 1979; Lewin, 1980). Although biological evolution does not occur by drastic leaps

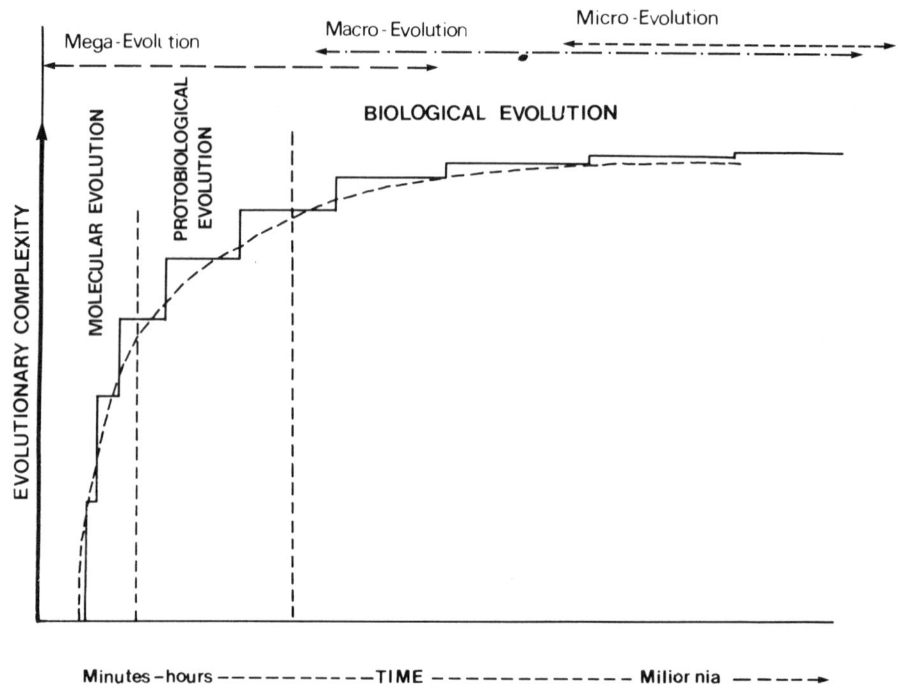

Fig. 6. The major evolutionary periods in the origin and development of life on a logrithmic time scale shows a changing time-course from very abrupt (molecular evolution) to much controlled (biological evolution). Changes occur in step-wise fashion; the degree of change generally decreases as evolution progresses. There are undetermined areas of overlap between mega-, macro-, and micro-evolutionary patterns.

(mega-evolution), its general step-wise, jerky mode of activity (macro-evolution) is consistent with the overall pattern of evolutionary change. Fig. 6 is a very rough sketch of the types of evolutionary patterns.

SUMMARY

Some salient features of prebiological evolution are recapitulated below, some of which may have relevance to biological evolution.

1. Natural selection usually supports the simultaneous existence of as many variants as are capable of existing, by whatever means available to them. Protoevolution is, in general, a pluralistic process.
2. Natural selection channels the directions of evolutionary changes, but the source of variability is intrinsic to evolving units.
3. Changes in materials have corresponding changes in functions. The properties of products of prebioloigcal evolution cannot be predicted by studying the precursors individually.
4. The time-course of protoevolutionary change is by quantum jumps (mega-evolution). There are sedentary intermissions between bursts of activities. At any time, larger or finer evolutionary changes may proceed, insofar as they are compatible with the intrinsic nature of the evolving units, are supported by the environment, and can produce heritable changes.
5. All components of functional protoproteins as well as those of evolvable protocells can emerge simultaneously in one evolutionary mega-step. A corollary to this observation is that protoevolution cannot be studied by discovering "missing links" alone, because there may have been none.

REFERENCES

Calvin, M., 1962, Adv. Biol. Med. Phys., 8:315.

Fakhrai, H., van Roode, J. H. G., Orgel, L. E., 1981, J. Mol. Evol., 17:295.

Fox, S. W., 1969, Naturwiss., 56:1.

Fox, S. W., 1974, Mol. Cell. Biochem., 3:129.

Fox, S. W., and Harada, J., 1958, Science, 128:1214.

Fox, S. W., Harada, K., and Kendrick, V., 1959, Science, 129:1221.

Fox, S. W., McCauley, R. J., and Wood, A., 1967, Comp. Biochem. Physiol., 20:773.

Fox, S. W., and Dose, K., 1972, "Molecular Evolution and the Origin of Life," Freeman, San Francisco.

Fox, S. W., Hsu, L. L., Brooke, S., Nakashima, T., and Lacey, Jr., J. C., 1972, Int. J. Neurosci., 3:182.

Fox, S. W., Adachi, T., Stillwell, W., Ishima, Y., and Baumann, G., 1978, in: "Light Transducting Membrances," D. W. Deamer, ed., Academic Press, New York.

Fox, S. W., and Nakashima, T., 1980, BioSystems, 12:155.

Hsu, L. L., 1972, in: "Molecular Evolution: Prebiological and Biological," D. L. Rohlfing, and A. I. Oparin, eds., Plenum, New York.

Hsu, L. L., 1974, Proteinoid Microspheres as Models for Primordial Organic Evolving Systems, Ph.D. dissertation, University of Miami, Coral Gables.

Hsu, L. L., Brooke, S., and Fox, S. W., 1971, Curr. Mod. Biol., 4:12.

Hsu, L. L., and Fox, S. W., 1976, *BioSystems*, 8:89.
Jungck, J. R., and Fox, S. W., 1973, *Naturwiss.*, 60:425.
Kerr, R. A., 1980, *Science*, 210:42.
Koesian, J., 1974, *in*: "The Origin of Life and Evolutionary Biochemistry," K. Dose, S. W. Fox, G. A. Deborin, and T. E. Pavlovskaya, eds., pp. 221,-231, Plenum, New York.
Lewin, R., 1980, *Science*, 210:883.
Morrison, P., 1973, *in*: "Molecules in the Galactic Environment," M. A. Gordon, and L. E. Snyder, eds., p. 443, Wiley-Interscience, New York.
Nakashima, T., and Fox, S. W., 1972, *Proc. Nat. Acad. Sci. U.S.*, 69:106.
Oparin, A. I., 1938, "The Origin of Life," 2nd edn., Macmillan, London.
Oparin, A. I., 1957, "The Origin of Life on the Earth," Academic Press, New York.
Schoener, T. W., 1982, *Amer. Sci.*, 70:586.
Stanley, S. M., 1979, "Macroevolution, Pattern and Process," W. H. Freeman, San Francisco.
Waehneldt, T. V., and Fox, S. W., 1968, *Biochem. Biophys. Acta*, 160:239.
Watson, J. D., and Crick, F. H. C., 1953, *Nature*, 171:964.

THE UROBOROS

Ronald F. Fox

Georgia Institute of Technology
Atlanta, Georgia 30332

INTRODUCTION

The Uroboros myth has a very long history which can be traced back to 16th century B.C. Egypt (Hopkins, 1967), and perhaps earlier. A strong parallel developed in 4th century B.C. Greece in Platonic-Aristotelian philosophy. By the 4th century A.D., the writings of Zosimus provided the foundation for alchemy (Berthelot, 1885), which would flower in various forms during the centuries which followed. Throughout this long history, one concept was recurrently central, the Uroboros. As will be delineated in the following, this concept acquires a whole new level of significance in the modern context of molecular biology.

The Uroboros symbol is that of a serpent apparently grasping its own tail in its mouth (Hopkins, 1967; Berthelot, 1885). This configuration was originally intended to convey the idea of the serpent begetting itself from its own mouth. Often, naive students of this symbolism have misconstrued its meaning in precisely the opposite sense, a serpent biting, or eating, its own tail. The word, Uroboros, is from the Greek word for self-begetting (Neumann, 1954).

The idea that alchemy is concerned with the transmutation of base elements into gold is really a debased interpretation of the significance of gold for the ancient Greek alchemists and philosophers. To the Greeks, gold was the highest on the scale of metals, parallel with fire on the scale of elements. It was to them the only substance to resist fire; it was yellow like fire and the sun, which symbolized gold in the ancient literature. The gold sought by the ancients was a gold-colored metal, alloy or bronze. It was the color which they sought both in Greece and earlier in Egypt.

The gold color was the ideal of perfection. Indeed, they called color a spirit. In the case of the highest color, gold, the spirit was endowed with the power to reproduce itself and transmute all metals into perfect gold. Much later, the alchemists of the last millenium became fixated on simply producing gold for its monetary value, and some truly mastered the art of using a little color to convert a white metal into a golden one, right before one's eyes.

To Plato, gold is perfection (Plato, 1957). To his pupil, Aristotle (Aristotle, 1941), in his theory of forms, organic nature was based on a series of cycles in which the goal was that stage which could reproduce itself just as gold could beget gold. It was this view which has the strongest analogue in modern molecular biology. The symbol of the Uroboros serpent repeats this view in the early technique of Mary the Jewess (Hopkins, 1967; Berthelot, 1885) in her device, the kerotakis, which has come down to this day as a reflux condenser. Through heating, a substance vaporized, the vapor rose in the container, condensed and fell to the bottom again, only to be recycled. This technique, without great elaboration, became the principal technique of all subsequent alchemical manipulation of metals, sulphur, mercury, and other ingredients. This is how one made the philosopher's stone (Hopkins, 1967; Berthelot, 1885). Zosimus is credited with having described the preparation of cinnabar with such methods.

The modern, molecular view of the Uroboros also involves an energy flux and cycles leading to an ultimate stage of self-replication. In the following sections, the problem of the origin of a molecular Uroboros will be discussed, as well as its subsequent evolution.

ELEMENTS AND MOLECULES

If one looks at the constituents of a simple organism, one finds that nearly all of the essential components are based on just 12 of the elements: H, C, N, O, Na, Mg, P, S, Cl, K, Ca, and Fe. One can enlarge this list some, if one prefers, but the essence of the following remarks will remain invariant to which set of elements one believes is truly significant. Under the ambient conditions of temperature and pressure, the lowest lying free energy combinations of these elements are a variety of small molecules such as water, carbon dioxide, methane, ammonia, hydrogen cyanide, sulfuric acid, phosphoric acid, formaldehyde, and the ions in water of Na, Mg, K, Ca, and Cl. With the input of any of a variety of energy sources, such as heat, electrical discharge, shock waves, radiation, etc., monomeric molecules are produced. These include amino acids, simple sugars, purines, pyrimidines, quinones, and some simple phosphates. Thus, three levels of hierarchy, elements, small molecules, and monomers, are easily discerned. It is no longer the case that the abiogenesis of monomers on the primitive earth is considered a difficulty (Fox and Dose, 1972; Calvin, 1969; Miller and Orgel, 1974). Many of the small molecules are found in interstellar clouds, and amino acids have been detected in samples of moon rocks and meteors. These findings attest to the relative ease and naturalness of the smaller constituents of living matter.

THE LIVING STATE OF ORGANIZATION

The remarkable feature of life is that from such humble constituents as the twelve elements listed above, a state of organization as complex as a living organism can emerge. This emergence is predicated on a source of energy which is required to energize the molecular machinery needed to generate and replicate organismic molecules, some of which are of great complexity. The organization is such that the living state begets itself, once it has commenced. For this reason, we may think of it as a molecular Uroboros.

The most difficult question posed by a molecular Uroboros is the question of its origin. This may be couched slightly differently as the question of how to initiate such a self-replicating molecular cycle. In most respects, modern molecular biology provides an explanation of how one cell gives rise to another. This explanation involves a detailed account of energy metabolism, monomer synthesis, and polymer synthesis. However, the question of how to initiate a cellular replication scheme before the first cell exists is not answered. In this chapter, the point of view to be presented which addresses this question is based on the idea of energy flow ordering; the establishment of molecular organization in response to an energy flux. Moreover, it will be argued that the impetus behind evolution at each of its hierarchical levels, from simple cells on up to socialized populations of metazoans, is energy flow ordering.

The Monomer to Polymer Transition (Fox, 1982)

The abiogenesis of monomers, such as amino acids, sugars, purines, and pyrimidines, has removed these molecules from the category of strictly biological molecules. The polymers, proteins, polynucleotides, and polysaccharides, however remain characteristically biological. These macromolecules are made from appropriate sets of monomers through dehydration linkages. Proteins are amino acids linked by peptide bonds; polynucleotides are nucleotides linked by phosphodiester bonds, and polysaccharides are sugars linked by glycosidic bonds. In each instance, the linkage is a dehydration linkage, which is why these polymers are unstable, or at best metastable, in aqueous milieu. Without an energy source to maintain the polymeric state of a cell, the polymers hydrolyze into their constituent monomers, on the way towards thermodynamic equilibrium and death. The apparatus employed by the cell to synthesize these polymers is the most complex in the cell. It is also the most energy intensive, with a great requirement for monomer activation energy, ultimately derived from the ubiquitous cellular energy currency of the molecule ATP.

DNA is perhaps the quintessential macromolecule from the viewpoint of the Uroboros. Its structural ability to be replicated makes it at first sight appear to be self-begetting, i.e., a molecular Uroboros. However, it needs many proteins to achieve this replicability, and it needs a lot of ATP to activate the precursor monomeric constituents, the nucleotide monophosphates. The requisite energy metabolism required to generate the ATP also requires hundreds of proteins as catalysts. The proteins, on the other hand, are not self-replicating. They are the product of protein biosynthesis which decodes the information of the DNA base sequence in the genes. This decoding requires polynucleotides: DNA genes, RNA messengers, ribosomal RNAs, transfer RNAs; and hundreds of special proteins, including ribosomal proteins, tRNA-synthetases, RNA polymerases, and many more needed for energy coupling. ATP is again needed to activate the precursor amino acids, and the entire energy metabolism apparatus is required for the generation of the ATP. Suddenly, we are no longer able to view DNA as a molecular Uroboros, but are forced to think simultaneously about protein biosynthesis and energy matabolism. Moreover, the energy requirement for all polymer biosynthesis is met by ATP, the generation of which depends upon the full panoply of polymers in order to establish the energy matabolism apparatus. We are forced to the view that the first molecular Uroboros emerges only at the level of organization recognized as the cell, and not at any macromolecular sublevel. This view has previously been espoused by Jacob (Fox, 1982; Jacob, 1973).

Rather than argue whether proteins or polynucleotides came first, the question appears to be how polymer production emerged at all. How did the monomer to polymer transition commence? Which came first: polymers or energy metabolism? This is the kind of question to ask. Nevertheless, the concept of the Uroboros teaches us that neither came first, but that instead the issue should be how energy matabolism and polymer synthesis co-evolved. A rather detailed scenario for the co-evolution of energy metabolism and polymer synthesis has been presented elsewhere by this author (Fox, 1982). The model provides a potential explanation for the origin of the genetic code. Primitive energy metabolism has much simpler requirements than for contemporary cells and pyrophosphate serves as the precursor to ATP. The primitive genetic apparatus is much simpler than the contemporary apparatus of cells, and does not yet require ribosomes, tRNAs, or tRNA-synthetases. It is based on a primitive RNA gene-messenger which is directly read into polypeptides through the intermediacy of 2'-OH amino-acyl ribose esters. The reader is referred to (Fox, 1982) for greater detail.

ENERGY FLOW ORDERING

The twelve elements at the beginning of this chapter do not make an organism. If they are properly organized and coupled to an energy flux, they behave collectively as a self-reproducing cell, a molecular Uroboros. The ability of energy to engender organization is exhibited in a tremendous variety of contexts. How does such organization develop in response to an energy flux is a problem which has not been studied as much as it deserves to be. In the case of a molecular Uroboros, the act of initiation is really the crucial step to explain.

In the physical world, energy can be used to make a fluid flow past a stationary object, thereby creating the beautiful von Karmen vortex trail (Prandtl and Tietjen, 1934); heat flux in a fluid will generate the elegant Benard cell structures (Chandrasekhar, 1961); chemical energy of oxidation-reduction will drive the propagating colored rings in the Belousov-Zhabotinski reaction (Winfree, 1974). Many other examples of energy flow ordering exist. When the energy source is depleted or cut off, the system relaxes back into its equilibrium state, a state lacking the organizational features of the driven state. In each of these examples, the coupling of the energy flux to the system occurs as soon as the system and the energy come together. There is no problem of initiating the coupling. An induction phase may be required (Belousov-Zhabotinski) or a threshold may have to be surpassed (Benard cells), but once the appropriate requirement is met, the coupling drives the production of dynamic organization.

In the biological world, the situation is fundamentally different. Take as the energy flux, sunlight, and take as the system the twelve basic elements for biology. A simple photosynthetic alga exhibits the effect of energy coupling. Indeed, it is an Uroboros! However, coupling is mediated by a complex molecule, chlorophyll, and by the attached energy metabolism apparatus which is highly concatenated. If you simply shine sunlight on the twelve elements, you do not get algae. Therefore, energy flow ordering by itself does not solve the problem of the Uroboros even if it does provide an explanation of many examples of remarkable organization in simple physical contexts.

The problem of biological energy flow ordering is tractable only if one takes an evolutionary point of view. The energy coupling has evolved through many intermediate levels of hierarchy. In our model (Fox, 1982) of protein biosynthesis evolution and the genetic code evolution, a simple precursor energy metabolism was envisioned and the apparatus for both primitive polymer synthesis and primitive energy metabolism was depicted in terms of much simpler structures than are used by contemporary cells. The problem is to devise, and test, simpler sequences which could subsequently evolve into the contemporary complexity. We believe that this objective has been made conceptually plausible in our earlier model (Fox, 1982). In the following we wish to adumbrate the point of view that the subsequent evolution of higher forms, including socialized metazoans, is also a consequence of critical stages in energy flow evolution.

THE IMPETUS TO EVOLUTION

In contrast to the contemporary concerns of human society over energy resources and efficient utilization, the energy needs of organisms throughout evolution have been abundantly fulfilled. Even the most advanced, contemporary photosynthetic organisms only use between one and five per cent of the available sunlight because the limiting constraint is carbon dioxide, not energy. Our thesis in the following is that at any particular level of evolution the ability of organisms to transduce energy in their particular manner became sufficiently refined that energy reached abundant levels, thereby triggering a transition to a new level of organization.

The model (Fox, 1982) for the origin of energy metabolism and protein biosynthesis assumes that the earliest stages of energy transduction involved geophysical energy fluxes which gave rise to abundant oxidation-reduction energy. This primitive redox energy was then transduced into the chemical energy of pyrophosphate. The model envisages a build-up of pyrophosphate energy of sufficient potenital to drive the primitive monomer to polymer transition. A primitive molecular genetics emerges at this stage, and we called this stage of organization the primitive Uroboros.

The primitive Uroboros refined its energy metabolism through its evolutionary capacity inherent in its primitive genetic mechanism. Eventually, both a primitive chemisomotic mechanism and the production of ATP evolved. With abundant ATP production, polymer synthesis could really become advanced, and the genetic mechanism and protein biosynthesis apparatus of contemporary prokaryotes evolved. This stage is called the contemporary Uroboros, but it is merely the prokaryotic stage.

Energy metabolism continued to evolve. The cytochrome chains developed in the cellular membranes and the energy metabolism pathways of glycolysis and the citric acid became fully refined. Energy organelles developed leading eventually to the potential for simple mesazoa. This stage involves the first truly multicellular Uroboros.

Further advances in energy transduction made increasingly larger multicellular forms stable, until true metazoans emerged. Energy became available for motility. The ability to change environments was a concommitant of this motility, which in turn made food more available, thereby making energy even more abundant. Primitive nervous systems could be energized. As ATP, calcium and phosphate became even more abundant, depositions of rigid, bony structures emerged. This is the stage of truly sentient, motile, organomorphs. The mollusks emerged in one lineage, and the fishes in another.

Eventually, limbs developed from fins, and terrestial forms evolved. These forms, of course, required a dramatic advance in energy transduction, the development of oxygen based energy metabolism. This development resulted in an explosion of terrestial varieties. Hierarchial food chains emerged and resulted in even greater variation.

At an earlier stage, energy was sufficiently abundant that the synthesis to polynucleotides in abundance was not a difficulty. This promoted both greater genetic diversity and the potential for sexuality and diploidy.

Energy again becomes abundant with the terrestials, and giant forms developed, culminating in the great age of the giant reptiles,, or dinosaurs. Bone and muscle become preeminent in biology. The calcium, phosphate, and ATP attest to the energy based heritage of these structures.

The Earth cools somewhat, and energy is required to generate heat. Again it is available and this development leads to mammalian evolution. Isothermy provides the condition for the refined development of higher forms of neurological structure. Brainy forms evolve.

The preceding eight paragraphs provide examples of how the view that energy is the impetus to evolution may be supported. Many other examples come to mind, and each should be analyzed in detail. The availability of energy and the type of energy determine the properties of the emergent organomorphic stages of evolution.

By looking at the recent history (last ten thousand years) of human society, we can identify a recapitulation of the evolutionary sequence of life itself in terms of the impetus of energy. It is easily appreciated that significant impact on social development followed a variety of purposeful uses of energy. Examples include the impact of fire on the development of metallurgy. This not only provided the transition from the stone age into the iron and then bronze ages, but it also gave rise to the earliest precursors to alchemy. The use of gravity in irrigation revolutionized society by making food abundant which permitted dense populations to exist. Explosives transformed warfare and created redistributions of populations. Sailing vessels, using wind energy, revolutionized commerce on a global scale. Steam power transformed transportation. Electricity transformed communications and provided a vast variety of work saving applications. In each case, once the form of energy is identified, it is straightforward to see how it caused the subsequent societal evolution. One cannot help but wonder how the story will read when the nuclear energy chapter is finally written!

CONCLUSION

In this chapter, we have described the concept of the Uroboros. Of particular interest is the problem of initiating a Uroboros. This problem has been discussed in greater detail elsewhere (Fox, 1982), and the central theme is the evolution of energy transduction to a degree sufficient to promote the monomer to polymer transition. In addition, we have argued that the impetus to evolution is energy. Only when energy considerations are included, is it possible to explain the physical basis for evolution. We have proposed that evolution has occurred by transitions between successive hierarchical levels of organization which became possible because the energy metabolism at the earlier level became refined enough to make energy especially abundant. The abundant energy provided the condition for emergence of novel organomorphic organization. Such an analysis not only provides a basis for explaining the origin of life itself (Fox, 1982), and the phylogenetic radiations therefrom, but it also provides of the ontogeny of human society. Thus, energy transduction is the key to a comprehensive, unified approach to the understanding of biological organization and evolution. It provides a molecular manifestation of the ancient concept of the Uroboros.

REFERENCES

Aristotle, 1941, "The Basic Works of Aristotle," Random House, New York.
Berthelot, M., 1885, "Les Origines De L'Alchimie," Steinheil, Paris.
Calvin, M., 1969, "Chemical Evolution," Oxford Univ., New York.
Chandrasekhar, S., 1961, Hydrodynamic and Hydromagnetic Stability," Clarendon Press, Oxford.
Fox, R. F., 1982, "Biological Energy Transduction: The Uroboros," Wiley, New York.
Fox, S. W., and Dose, K., 1972, "Molecular Evolution and the Origin of Life," W. H. Freeman, San Francisco.
Hopkins, A. J., 1967, "Alchemy, Child of Greek Philosophy," AMS Press, New York.
Jacob, F., 1973, "The Logic of Life," Pantheon, New York.
Miller, S. L., and Orgel, L. E., 1974, "The Origins of Life on the Earth," Prentice-Hall, Englewood Cliffs, New Jersey.
Neumann, E., 1954, "The Origins and History of Consciousness," Chpt. 1, Princeton Univ. Press, Princeton.
Plato, 1957, "Plato's Cosmology: The Timaeus of Plato," Liberal Arts Press, New York.
Prandtl, L., and Tietjen, O. G., 1934, "Fundamentals of Hydro- and Aero-mechanics," Dover, New York.
Winfree, A. T., 1974, Sci. Amer., 230:82.

RELATIONAL BIOLOGY AND THE ORIGIN OF LIFE

Robert Rosen

Department of Physiology and Biophysics
Dalhousie University
Halifax, Nova Scotia

INTRODUCTION

The science of biology rests upon the fact that we perceive the sharpest possible distinction between organisms and inanimate systems. Perhaps the basic task of biology is to isolate and characterize the nature of this distinction; to determine why certain systems are living and others are not.

Organisms, no less than inorganic systems, are material and hence equally subject to the laws which govern matter. But it is difficult to believe that the distinction between them can be drawn at this level; indeed, there is nothing in physical law which implies the existence of organisms. It is easy to image a physical universe devoid of organisms, though otherwise governed by the same general laws as our own.

The "origin of life" problem arises precisely because (a) we distinguish so sharply between the organic and the inorganic, and (b) we can imagine a time early enough so that there existed only inorganic systems. Furthermore, we know that (at least under present conditions, or indeed under any conditions we have been able to establish) omnia vivum ex vivo; the only organisms in the biosphere are those which have arisen from the reproduction of pre-existing ones. But since we also believe that nil ex nihilo, it follows that under some conditions "inorganic matter" becomes "living matter." The general problem of the origin of life is to characterize such a set of conditions; the more restricted problem of the origin of life is to determine those particular conditions prevailing on the early, abiotic earth which enabled life to appear there. The first (general) problem is a purely scientific one; the second (restricted) one is also historical.

There are two basic difficulties faced by those interested in the restricted origin-of-life problem. The first is common to all historical studies; namely, the information pertinent to the problem has been partially or totally obliterated. In all such cases, alternate sources of information msut be sought, and this is always done by embedding the problem into one or another specific context. For instance, in all evolutionary studies the context most frequently utilized is the character of the biosphere as it exists now; the nature of present organisms is utilized as an alternate information source about extinct ones. However, powerful objections can be raised against the use of such contextual information; indeed, a popular view at present is that historical processes are simply not governed by logical necessity and are hence exempt from science from the outset.

A second and more serious problem rests in the absence of any precise definition of what an organism is. It is one of the most astonishing and instructive facts of biology that no satisfactory characterization of living systems has been found. Thus, there is no clear, unambiguous point at which we can say that an investigation into the origin of life has succeeded, or that it has failed. The lack of a clearly defined end-point has been a major obstacle in all studies of origins.

The tacit characterization of living systems implicit in biology in general, and in origins studies in particular, revolves around the genotype-phenotype dualism. We shall have much more to say about this as we proceed. For the present, it is sufficient to recognize two distinct camps among those studying origins problems: (a) those who believe that genotype is primary and that the corresponding phenotype was somehow generated by or wrapped around a primordial genome; (b) those who believe that phenotype is primary and that somehow a genomic system became incorporated into it. The first camp is inhabited by those concerned with the "origin of the genetic code," who see the basic event as the emergence of replicating macromolecules which can also serve as templates. The second camp contains, among others, those who regard such things as vesicular organization as fundamental and who study e.g., coacervation (Oparin) or proteinoid condensation (Fox). Both camps are modern manifestations of a long tradition which used to be called "biomimesis," in which it was felt that a physical system manifesting "sufficiently many" of the properties of life would have to be called alive (much as Turing's Test, in artificial intelligence, asserts that a machine manifesting "sufficiently many" properties of mind would have to be called intelligent), even though neither life nor intelligence can be categorically defined.

Both of these camps are themselves nourished by the sea of investigators who find abiological ways to generate organic molecules. These proliferating researches show us only that organic molecules can be generated in abundance. This fact only serves to stress that the real origin-of-life problem lies elsewhere.

RELATIONAL BIOLOGY

Until well into the present century, attempts to understand biological processes in terms of underlying physicochemical mechanisms were few and far between. On the experimental side, the early biochemical investigations which ultimately culminated in molecular biology were disdained by most traditional biologists, who argued that the slushes and homogenates which were the starting-point for such investigations were abstract artifacts unrelated to the original organisms from which they came. Likewise, on the theoretical side, the mathematical tools which had served physics so well were rejected out of hand as inadmissible; inadequate in principle to deal with "biological reality."

The great pioneer in systematically applying the methods of mathematics and physics to biological phenomena was Nicolas Rashevsky. Starting in the late 1920s, he and his co-workers established the credibility of this approach, making major advances in understanding such diverse biological processes as nerve excitation, cytokinesis, cardiovascular dynamics, central nervous system function, and many others. Nowadays the techniques of mathematical modeling which these investigators championed have become commonplace, so that it is hard to recall how radical they appeared initially.

Rashevsky, as a physicist, began with an essentially reductionistic view of the relation of biology to chemistry and physics. However, his successes at modeling specific biological processes, instead of reinforcing his reductionism, began to worry him increasingly. As he said: "There is no record of a successful mathematical theory which would treat the integrated activities of the organism as a whole . . . this integrated activity of the organism is probably the most essential manifestation of life. . . . These fundamental manifestions of life drop out from all our present theories of mathematical biology. . . . We must look for a principle which connects the different physical phenomena involved and expresses the biological unity of the organism and of the organic world as a whole." (Rashevsky, 1954).

Rashevsky's basic idea was the following: Instead of starting with a host of reductionistic fragments and attempting to find some a posteriori way of fitting them together to generate integrated biological behavior, let us rather try to represent this integrated behavior from the outset. In this way, he was led to an abstract topological structure (the details need not concern us here) which serves as a kind of functional bauplan manifested by any problem we would want to call an organism. In other words, any specific organism would constitute a realization of this abstract structure; the manner in which particular organisms relate to (or map onto) the bauplan would establish their relations to each other.

Rashevsky coined the term relational biology to characterize this approach, as distinct from quantitative or metric biology. One of its crucial features is this: Experimental approaches (e.g., biochemistry or molecular biology) proceed by initially destroying all higher-level biological organization, leaving behind a purely physicochemical system to be studied entirely by physicochemical means. In other words, they proceed by abstracting away all organizational properties, hoping to recapture them later by synthetic arguments based on their analytic data. The relational approach, on the other hand, proceeds in an exactly inverse way; in effect, it initially abstracts away all purely physicochemical aspects, leaving behind a pure organization to be represented and studied entirely by mathematical means. Here, the detailed physics and chemistry of such a system are to be recaptured by a process of realization.

In particular, given such a relational structure, we may ask for its simplest realization in some appropriate physicochemical sense. This will by definition be a real physicochemical system, which also manifests the organization characteristic of organisms. Such a system will be an obvious candidate for a primordial organism. Thus, among many other things, the relational approach suggests a new way to attack the origin-of-life problem; a way which sidesteps the specific difficulties of historicity and undetermined end-point which have hindered all non-relational approaches. The remainder of our discussion will be devoted to exploring these ideas.

RELATIONAL CELL MODELS: THE (M, R)-SYSTEMS

For well over a century the cell has been regarded as the basic unit of biological organization, both structurally and functionally. Cells exist both as independent organisms and as constituents of organisms. In extant organisms, most new life begins as a single cell (zygote). Thus, it is natural to regard cells as being central to the origin-of-life problem; for many investigators, the origin of life translates into the origin of the cell.

Cells can exist in an unlimited variety of forms, from the simplest PPLO through the ciliated protozoa through the differentiated cells of higher organisms, and can manifest a corresponding variety of chemistries and morphologies. Thus the question arises: what is a cell? How is a cell to be recognized? One answer is in terms of conventional anatomical and physiological criteria; e.g., cells are membrane-bound and contain recognizable organelles such as nuclei, mitochondria, plastids, etc. But in fact, cells are as hard to define in these terms as is life itself.

Let us then try another approach; a relational approach. Viable cells, however they may differ from one another in structural terms, invariably manifest a characteristic partition of their activities into two classes. On the one hand, we have activities pertaining to the processing of materials drawn from the environment. On the other hand, we have activities connected with the generation and maintenance of the specific machinery involved in processing environmental material. The first class of activities roughly correspond to cytoplasmic function; the latter to nuclear. We shall employ more neutral, less structural terms; the first class will be called metabolic activities; the second class will be called repair. Any abstract system whose activities can be partitioned in this way will be called an (M, R)-system.

This partition, which is independent of any specific structural characteristics, provides the basis for a relational treatment of cellular organization. In the present section, we will briefly sketch some of the theory of such systems, both to see what it is like, and to introduce some of the central mathematical ideas which will be used later. For fuller details, the reader may consult the literature (cf. Rosen, 1973).

The basic unit in an (M, R)-system is called a component. Components are functional units which serve to transform or convert inputs into outputs. Mathematically, such a component is represented simply by a mapping

$$f : A \to B \tag{1}$$

where A is the set of inputs to the component f, and B is the set of outputs.

Such components may be composed, or arranged into networks, since their associated mappings can be composed. Any such network will be a candidate for the metabolic part of an (M, R)-system. For simplicity, we shall consider here only the simplest possible such network, which consists of a single mapping (1); our results can readily be generalized to arbitrary networks.

Now we turn to specifying the R-part of such a system. The metabolic machinery, represented here by the single mapping (1), is to be generated by a new kind of component. We require that the input to this new component be the output of the metabolic machinery; the output of the new component is to be a mapping from inputs A to outputs B. Hence the R-part of the system is itself represented by a mapping of the form

$$\Phi : B \to H(A, B) \tag{2}$$

where H(A, B) is the set of mappings from A to B. The entire (M, R)-system can thus be represented as a composite of the form

$$A \xrightarrow{f} B \xrightarrow{\Phi} H(A, B). \tag{3}$$

It is, in a sense, the simplest formal system which manifests the kind of organization which is characteristic of biological cells.

Let us look at one consequence of this picture. In biology, we know that the repair component Φ should not only produce copies of the metabolic machinery; it should also replicate. In the above picture, no replication mechanism is visible. However, we shall now conjure one, just from what we have already put into the (M, R)-system, with no further special assumptions.

Quite generally, let X, Y be sets, and let $\rho : X \to Y$ be a mapping drawn from a set H(X, Y) of maps. Traditionally, we interpret the expression $\rho(x)$ by assuming that the mapping ρ is held fixed in H(X, Y), while the argument x is allowed to vary in X. Suppose, however, that we hold x fixed in X, and allow ρ to vary in H(X, Y). Then x associates to each mapping ρ in H(X, Y) a corresponding value in Y. That is, x induces a mapping (sometimes called an "evaluation map")

$$\hat{x} : H(X, Y) \to Y$$

defined by

$$\hat{x}(\rho) = \rho(x).$$

If this mapping $\hat{x}$ should be invertible, then its inverse $\hat{x}^{-1}$ is a mapping from Y to H(X, Y).

The above considerations are perfectly general; they apply to any sets X, Y. In particular, let us return to the (M, R)-system (3), and let us choose

$$X = B, \quad Y = H(A, B), \quad \rho = \Phi$$

In this case, the mapping $\hat{b}^{-1}$, if it exists, is of the form

$$\hat{b}^{-1} : H(A, B) \to H(B, H(A, B)). \tag{4}$$

That is, the range of this induced map $\hat{b}^{-1}$ is precisely the set to which the repair map Φ belongs (cf. (2) above). It is easy to see that the effect of this induced map $\hat{b}^{-1}$ is to replicate the repair component Φ.

Now we observe that not every (M, R)-system will admit such an induced replication of Φ; there is a condition to be satisfied. That condition is

$$\hat{b}(\Phi_1) = \hat{b}(\Phi_2) \quad \text{implies } \Phi_1 = \Phi_2$$

or

$$\Phi_1(b) = \Phi_2(b) \quad \text{implies } \Phi_1 = \Phi_2$$

which assures the invertibility of $\hat{b}$. It says: If any two possible repair components in (3) agree on the environmental output b, then no replication map (4) exists.

A deeper investigation of this condition requires us to specify the nature of the _category_ from which our mappings are drawn; this is a most interesting exercise, but one which we cannot review here (cf. Rosen, 1973). The main conclusion to which we wish to draw attention is that _abstract cellular organization already implies the possibility of replication_, without any further hypotheses. However, such replication requires a rather stringent (invertibility) condition to be satisfied. In fact, this condition can be regarded as an abstract form of the familiar _one-gene-one-enzyme hypothesis_.

REALIZATIONS OF (M, R)-SYSTEMS

We pointed out earlier that one of the basic problems facing the origin-of-life researcher has been the absence of a clearly defined end-point for these studies. The relational approaches we have described above may be considered as providing such an end-point, and thus suggest a new strategy which can be applied to origins problems. Specifically, in the context of the (M, R)-system: Insofar as these systems accurately reflect the basic characteristic organization manifested by all cells, we may say that, whatever else a cell may be _structurally_ (i.e., in terms of its physics and chemistry) it must also be an (M, R)-system. That is, any specific cell must constitute a _physical realization_ of the formalism which represents an (M, R)-system. We may go further and argue plausibly that _any physical system which realizes this formalism is, to that extent, a cell_.

Thus, _any_ physical realization of any (M, R)-system is simultaneously, by definition, (a) a physical system and (b) possessed of the characteristic organization of cells. As a physical system, it has a description in terms of traditional laws; it has a state set and equations of motion which govern its behavior in time. Once a physical system has been characterized to this extent, it can (at least in principle) be _synthesized_; conditions leading to its preparation can be articulated and established. In other words, once such a system has been characterized, it can be built.

The synthesis of such a system would in effect provide a constructive solution to the general origin-of-life problem (which, we recall, asks only for the articulation of conditions under which we may pass from a situation in which biological organization is absent to one in which it is present, without the intervention of living agents). Indeed, if we succeed in constructing a physical realization of an (M, R)-system for which the replication mapping (4) above is also realizable, we have good grounds for asserting that we have thereby constructed a new, autonomous life form.

The restricted origin-of-life problem, which deals with ascertaining those specific conditions on the abiotic earth which led to the appearance of cellular life, can also be approached through the above strategy. We still seek to physically realize an (M, R)-system, but instead of dealing with any realization, we are interested only in those which satisfy certain structural boundary conditions. These boundary conditions involve such things as the presence of protein in the metabolic machinery and DNA in the repair components. Such boundary conditions restrict attention to a certain subclass of the class of all physical realizations of (M, R)-systems. Of course this subclass is still sizable, but for the study of origins we would wish to restrict attention further, to those systems which are in some sense simplest. Now "simplicity" itself imposes a further boundary condition in the form of a minimum principle; this is in fact a general procedure used in mathematics for extracting a single "canonical form" from an indefinitely large equivalence class. Thus, although the problem is technically much harder, the strategy we have sketched remains applicable to the restricted origin-of-life problem.

The key to the implementation of this strategy is, of course, to develop a procedure for physically realizing an abstract mathematical mapping in a manner appropriate for the (M, R)-system formalism. This is not as simple a matter as it looks, although a number of helpful hints are available; it may be helpful to discuss a few of them here.

The most important clue to this problem is the recognition that the laws which describe any physical system can be regarded as defining a mapping, or a set of mappings. In fact, when we do this, some provocative relationships between inorganic and organic concepts arise, almost of themselves. To see this, it is best to proceed by means of a specific example. Consider the familiar van der Waals equation, a thermodynamic equation of state characterizing the equilibria of certain non-ideal gases:

$$(p + a/v^2)(v - b) = rT$$

where p, v, T are the thermodynamic state variables; a, b, r are parameters which depend on the "species" of gas. Abstractly, the van der Waals equation is a functional relation involving six arguments:

$$F(p, v, T, a, b, r) = 0.$$

We can, however, rewrite this single relation as a three-parameter family of functions of three arguments:

$$F_{abr}(p, v, T) = 0. \tag{5}$$

This three-parameter family is in fact a stable family; an arbitrary change, or mutation, $(a, b, r) \mapsto (a', b', r')$ in the "species" of gas can be compensated by a simple scale transormation of the state variables; i.e., the equation of state (5) is invariant under the transformation

$$\begin{aligned} &(a, b, r) \mapsto (a', b', r'), \\ &(p, v, T) \mapsto (p', v', T') \end{aligned} \tag{6}$$

where p', v', T' respectively satisfy the relations

$$\begin{aligned} &p/(a/b^2) = p'/(a'/b'^2), \\ &v/b = v'/b' \\ &T/(a/rb) = T'/(a'/r'b'). \end{aligned} \tag{7}$$

This invariance expresses the familiar Law of Corresponding States. As we have explained in detail elsewhere (cf. Rosen, 1978) the above argument is a very special case of much more general considerations of similitude or similarity.

We can see this even more sharply by rewriting the relation (5) in the following way. The pressure p and temperature T of a gas in equilibrium are determined by external conditions (environment); the remaining quantity, the volume v, is then determined according to (5) by the "species" of the gas and by the environment. Thus, we can rewrite (5) as a three-parameter family of mappings

$$f_{abr} : (p, T) \mapsto (v) \tag{8}$$

The invariance of the van der Waals equation under the transformations (6), (7) is then expressed by the commutativity of the diagram

$$\begin{array}{ccc} (p, T) & \xrightarrow{f_{a,b,r}} & (v) \\ \alpha \Big\downarrow & & \Big\downarrow \beta \\ (p', T') & \xrightarrow[f_{a',b',r'}]{} & (v') \end{array}$$

where the maps α, β depend only on the "species" of the gases involved, and are given explicitly by (7) above.

In the context of the (M, R)-systems, the form (8) is clearly highly suggestive; it is precisely of the form (1) and represents the conversion of "inputs" determined by the environment into corresponding "outputs."

The basic ideas illustrated by the above example are in fact perfectly general. Given any equation of state, we can partition its arguments into a number of different classes; in one class are those arguments which are "species-determined," and which we may therefore call *genomic*; the remaining arguments can be further partitioned into those which are environmentally determined (*environment*); and those determined by genome and environment according to (8), which can thus be regarded as *phenotype*.

Thus we see that there is a precise sense in which the genotype-phenotype dualism arises in *any* class of systems governed by (i.e., realizing) a common equation of state; in other words, this dualism is in itself *not inherently biological*. Of course, the biological genome has certain unique characteristics not encompassed by the above characterization. In particular, it has none of the characteristics which we associated with "repair" in our discussion of the (M, R)-systems. But there is a definite sense in which *any* physical system (e.g., the microspheres of Fox) possesses a genotype, and a corresponding sense in which a "naked genotype" (e.g., a replicating template) already manifests a phenotype. These facts may be most helpful in assessing the biological relevance of such systems.

It might be thought that the realization problem for abstract mappings is completely solved along the lines of the above examples. This is not strictly the case, however, because the kinds of equations of state employed in physics are usually *not causal relations*; they are phenomenological expressions of linkages of observables (cf. Rosen, 1978). In philosophic language, the above examples regard "input" only as *formal cause* of output (i.e., as *information*), while in the abstract mapping processes in an (M, R)-system, we would require that "input" be *material cause* of output. These observations point up an unexpected difficulty we must confront when dealing with such realization problems; the same mathematical relation may arise as an abstraction of fundamentally different physical situations; in particular, it abstracts out all the underlying causal structure. This has led to an extraordinary amount of confusion, especially in biology where the distinction between formal cause (information), material cause and efficient cause is absolutely essential.

Nevertheless, we hope that the reader is persuaded that the approach to origin-of-life through the physical realization of relational structures represents a significant addition to the scientific armory, and that it opens new possibilities and perspectives for synthetic approaches to biological organization in general.

REFERENCES

Rashevsky, N., 1954, Bull. Math. Biophys., 16:317.

Rosen, R., 1973, in "Foundations of Mathematical Biology," R. Rosen, ed., Vol. II, Academic Press, New York.

Rosen, R., 1978, "Fundamentals of Measurement and the Representation of Natural Systems," Elsevier, New York.

PROTOBIOLOGY: A THEORETICAL SYNTHESIS*

Koichiro Matsuno

Technological University of Nagaoka
Nagaoka 949-54
Japan

INTRODUCTION

What is significant to protobiology (Fox, 1983; Dose, 1983)--the construction of the first biosystems from atoms and abiotic molecules--is the way the constituent elements are regulated and organized internally. The most fundamental physical attributes describing the construction process are both the number of constituent elements and the different types of these elements. The internal regulation that brings about material self-organization places a certain restriction upon the kinetic freedom the constituent elements can maintain. This implies that none of the constituent elements can move independently of each other. The number of degrees of internal freedom inherent to the aggregate of constituent elements is smaller than that conceived when all of these elements move independently. The number of degrees of internal freedom, that is, the number of independent variables, decreases as the internally regulated molecular construction proceeds. This aspect makes protobiology unique compared with most other physical phenomena, in the latter of which the number of degrees of internal freedom or, simply, the number of independent variables, is no more than a form of specification originating in the boundary condition applied externally through an external constraint. The decrease in the number of degrees of internal freedom induces a change in a physical boundary condition.

---

*Most of the work was done while the author was on leave at NFCR Laboratory at Institute for Molecular and Cellular Evolution, University of Miami, Coral Gables, Florida 33134.

If the decrease in the number of degrees of internal freedom was not reasoned upon a physical basis inasmuch as physics cannot reason the very existence of the boundary condition within which it works, one would have to ask for something beyond physics to reason the origin and evolution of biosystems (Polanyi, 1968). On the other hand, if physics still has in itself room for incorporating the process allowing for the decrease in the number of degrees of internal freedom, the underlying physical mechanism must be worked out (Pattee, 1977). Protobiology is itself distinguished from other branches of physical sciences in that it deals with physical processes that can change, especially to describe the number of degrees of internal freedom.

An approach in understanding the problem of decrease in the number of degrees of internal freedom in molecular dynamics is to specify the molecular elements and the way they interact among themselves unambiguously beforehand and then to contrive a possible set of architectures. This scheme of the axiomatic construction has been proven to be very successful in many branches of physical science under the name of structure reductionism. However, Gödel's incompleteness theorem reminds us that one cannot claim that all of the possible theorems derivable be consistent with the initial axioms (e.g., Lahiri, 1977). The possibility exists that there is a theorem which can neither be proved nor disproved within the given axiomatic theorization. In particular, if the theorem becomes too complex as in the case of the molecular description of complex biosystems, it would become technically difficult to construct such a theorem on a molecular basis. It would become even more difficult to examine its consistency with the axioms expressed in molecular terms. One alternative in avoiding the incompleteness associated with axiomatic construction is the experimental construction of protobiological structures and processes from atoms and abiotic molecules (Fox, 1956; Young, 1983; Cherkin, 1983), since in the latter we are freed from the burden of specifying the character of each molecular element beforehand. Nevertheless, experimental constructionism alone does not solve all the problems. A remaining problem concerns how to describe the process and the product that protobiogenesis provides. If one employs an axiomatic description which allows a set of immutable elements and well-defined relationships among the elements to be described, the same problem of incompleteness exists as with the axiomatic construction. What is needed in order to overcome the incompleteness of the axiomatic construction is the consistency between the descriptive elements and the derivable descriptions (T. O. Fox, 1983). This closure property (Pattee, 1982) is what is required for the description or, the theorization of the laboratory construction of protobiological structures and processes.

The decrease in the number of degrees of internal freedom associated with molecular construction in protobiogenesis leads to the fact that each of the molecular elements participating in the construction is under the influence of what it has made. Molecular construction is a process of material accumulation and is open to material flow. If molecular construction reaches a certain stationary state and if further changes are ruled out, the closure character between the constituent molecular elements and the macroscopic structure they make stands still. However, molecular evolution moves away from the stationary state. The consistency between the molecular elements and their construction must be understood as a process of on-going molecular construction. A material aggregate open to material flow is thus a carrier maintaining this consistency between the microscopic elements and the resulting macroscopic structure. This character makes the description of the evolutionary process very unique in the sense that one is urged to abandon the axiomatic theorization of the process based upon immutable fundamentals or primitives. Although such an axiomatic description might be expected if the evolutionary process were to approach an asymptotic stationary state, this is not the case in the real world.

A well-tested starting point of molecular construction of small molecules like amino acids is the synthesis from their constituent atoms (Choughuley, 1983; Harada, 1983b). Furthermore, thermal polymerization of amino acids would be highly conceivable during the earlier geological stage of the Earth because of the abundance of various energy sources, such as geological heat, electrical discharge and ultraviolet light irradiation (Kenyon, 1983). These facts provide us with sound reasoning for the simulation experiments in constructing higher molecular complexes with the aid of various energy sources (Harada, 1983a).

For instance, from an evolutionary perspective, a significant step in the making of molecular complexes is the formation of thermal proteins or proteinoids from amino acids (Fox, 1956; Fox and Harada, 1958; and Harada and Fox, 1958). Although the presence of other polymers such as polynucleotides, the making of which is conceivable during terrestrial evolution on the early Earth (Fakhari et al., 1981; Paecht-Horowitz, 1983), via catalysis using metallic ions or clays, thermal proteins figure prominent among many possible polymers (Orgel, 1968; Fox and Dose, 1977) due to their relative stability and ease of production. During the thermal polymerization of amino acids, the portion that has already been polymerized conditions the further incoming amino acids stereochemically and the internal state of the polymeric aggregate changes with time as the polymer grows. This indicates that the state of the molecular elements is always under the influence of the molecular complex they join. A similar closure property is expected also between proteinoids and their spherical aggregates formed on contact with water or, what we call proteinoid microspheres (Fox and Dose, 1977). A description of proteinoids in isolation differs from that in proteinoid microspheres, because the latter provides a new context within which proteinoids must be described. What experimental constructionism emphasizes is that amino acids in isolation differ from those in proteinoid molecules and that proteinoids in isolation differ from those composing proteinoid microspheres. If one takes these successive incompletenesses seriously, it might be said that an inevitable conclusion would be a kind of defeatism by which we mean an impossibility of describing the process of molecular construction in terms of immutable molecular elements. However, one can also expect the possibility that if only the consistency of description is emphasized, the molecular elements can adjust their contents in a consistent manner with the molecular construction.

A descriptive method which could be compatible with experimental constructionism is the one that emphasizes the consistent closure property between elements and their aggregated construction. Unless one has such a descriptive scheme with the closure property, one cannot successfully describe what experimental constructionism provides. The constructionistic sequence from amino acids to proteinoids and further to proteinoid microspheres suggests a need for a new descriptive scheme or a paradigm with a closure property, which is very much different from the scheme of the axiomatic construction from immutable elements that are common to structural reductionism.

The laboratory construction of molecular complexes may or may not accurately simulate terrestrial evolutionary processes. Undoubtedly, the evolution in the real world is an instance of an intricately tangled web of historical events. It is almost impossible to faithfully retrace terrestrial evolutionary processes in a laboratory setting. This might raise the question of relevancy in applying experimental constructionism in attempting to understand early evolution. In fact, if the particularity of the terrestrial boundary condition on both geological and cosmological scales are of prime importance in the evolutionary process taking place there, experimental constructionism could not claim its relevance to the real process unless such a terrestrial boundary condition be simulated faithfully in its fine details in the laboratory. This circumstance would indirectly force us to admit that laboratory construction of molecular complexes could not provide a dependable clue to terrestrial molecular evolution because of the impossibility of reconstructing the historical boundary condition of the early Earth in extremely small and simple laboratory settings.

On the other hand, however, the situation will be drastically changed if the particularity of the terrestrial boundary condition is of secondary importance compared with something else influencing terrestrial evolutionary processes. This alternative view is to emphasize matter itself over the external boundary condition that does indeed exert an influence upon matter. Matter itself assumes the role of a primary cause of evolution. It is this primacy of matter that could make the laboratory construction of molecular

complexes relevant to terrestrial molecular evolution, since one can expect to simulate the fundamentally essential process of terrestrial evolution in the laboratory. If one looks at the laboratory construction of molecular complexes especially with an emphasis on its external boundary condition, any experiment cannot remain invulnerable against the charge of its imperfect modelling of the terrestrial boundary condition (Kenyon, 1983). This is, however, not the point of concern here. What does count is to demonstrate the intrinsic evolutionary capability of matter that does not depend upon the particularity of the laboratory boundary condition employed for the laboratory construction. A crucial problem concerning the laboratory construction is to distinguish the processes and the products that are subject to laboratory boundary conditions from those not so subject. The intrinsic evolutionary capability of matter can be observed in the processes and in the products that remain unchanged even if laboratory boundary conditions are varied to a certain extent. It is this hope of being able to identify the intrinsic evolutionary capability that could make the experimental construction of molecular complexes relevant to terrestrial molecular evolution. The underlying theorization needs to have the consistent closure property between the descriptive elements and the derivable descriptions.

OPEN SYSTEMS AND THEIR ORIGINS

Any material aggregate, whether it be atom, molecule, or a complex of molecules, is open to material flow when it is in the process of increasing or decreasing the amount of material accumulation. The boundary separating the material aggregate from its exterior is self-generative in the sense that the boundary is formed through a self-consistent interaction between its interior and exterior. What is more, if a material aggregate thus formed can have any significance to the intrinsic evolutionary capability of matter at all, it cannot be a mere passive entity shaping its own boundary with the exterior only by responding to an externally controlled boundary condition. For instance, the honeycomb-like hexagonal cells due to the heat flow-convection flow instability of Benard (Paecht-Horowitz, 1983) shape their structures by responding simply to the contour of the container in which the fluid is placed. These Benard cells, although certainly open to material flow, are exogenous in their generation and do not maintain the internal capability of changing their own structures by themselves with time. A material aggregate which has an intrinsic evolutionary capability is the one that can successively change its interaction with the exterior not only by responding to the impetus from outside but also by influencing the exterior as a reaction to the former process in a recurrent manner. If the material aggregate fails to induce a change in the exterior that could again influence the aggregate successively, it would remain simply as a passive entity to the exterior. The evolutionary cause would have to be sought in the external boundary condition.

A principal reason why open material aggregates maintain their intrinsic evolutionary capability is that these aggregates not only react to the impetus from their exterior but also induce some change in their surroundings in response to the preceding impetus. During the formation of a polymer from monomers, for instance, it can change the way of interaction with its exterior as the elongation progresses because the then aggregated monomers altogether specify how and in which way the next monomer attaches. Consequently, what is crucial to the intrinsic evolutionary capability of matter is identified as its possession of internal freedom that remains intact without being exhaustively controlled by external means or agents. The internal freedom is simply the capability of matter to change its own internal state while exerting an influence upon its surroundings. At this point, it is important to recognize that the influence of external sources acting upon matter and the subsequent reaction of matter upon its surroundings never occur in a simultaneous manner. Instantaneous propagation of interaction change in matter is physically unacceptable. Because of this sequential nature of the reaction, a material aggregate changes its internal character by itself although initiated by an impetus from the surroundings. A consequence is a subsequent variation in the surroundings caused by the change in interaction with the aggregate. This change in the surroundings can again exert a new influence upon the same material aggregate. The successive generation of variation and reaction between matter and its surroundings can thus proceed indefinitely unless prohibitive boundary conditions are applied externally.

In terms of evolution the idea of a primordial soup of atoms and abiotic molecules (Florkin, 1975) provides the most significant model of an open material aggregate. The combined process of material association and dissociation can give rise to material units which are open to material flow, provided the material aggregates which are formed can participate in further association with other elements during their inevitable partial degradation. Open material aggregates can materialize if the three energies of material association, dissociation, and thermal agitation exist within a certain range. Material association could proceed if the energy for material association does not much exceed the thermal agitation energy. Furthermore, material aggregates, once formed, would soon disintegrate if the energy for material dissociation is much less than the thermal agitation energy. These aggregates, on the other hand, would almost avoid their disintegration if the energy for material dissociation far exceeds the thermal agitation energy. One could expect open material aggregates only when the association energy does not much exceed the thermal agitation energy and when the dissociation energy is of the order of the thermal agitation energy (Matsuno, 1983).

Any open material aggregate specifies both its own association energies with other molecular units and its dissociation energies with smaller units. But, both the association and dissociation energies vary as the underlying material aggregate changes itself. These changes in the association and dissociation energies are due to the reaction of the material aggregate to its surroundings as responding to the preceding molecular association or dissociation. The molecular association energy changes as the association or dissociation proceeds. Similarly, the molecular dissociation energy changes as the association or dissociation proceeds. Open material aggregates thus have the capability of changing their own states through the change in the molecular association and dissociation energies which respond to the preceding association or dissociation. Such material aggregares are more than those that remain passive toward the impetus from the surroundings. The change in the molecular association and dissociation energies is a form of change in response to the interaction with the surroundings. This is the consequence of the open material aggregate. Thus, open material aggregates formed in the primordial soup are intrinsically active toward their surroundings in the sense that they successively change their interaction with the surroundings in response to the preceding molecular association or dissociation or both. Those material aggregates, however, cease to be active toward the surroundings if the three energies of molecular association, dissociation, and thermal agitation are not found inside the energy window within which open material aggregates could materialize. Open material aggregates endogenously generate and vary their boundaries with the surroundings through the successive changes in the energies of material association and dissociation.

One form of generating open material aggregates is thermal polymerization of amino acids (Fox, 1956; Rohlfing, 1983). During the process of polymerization, the growing thermal polyamino acid is in a state of open material aggregation. These thermal polyamino acids or proteinoids (Kokufuta and Harada, 1983) can in turn serve as the constituent elements of protocellular microparticles or proteinoid microspheres (Fox and Dose, 1977). The material association process differs between the thermal polymerization of amino acids and the formation of microspheres from proteinoids in aqueous melieu. The peptide bond synthesis during the thermal polymerization is due to covalent-bond interaction, whereas the material accretion underlying the production of microspheres includes the noncovalent-type interaction that is anisotropic (Wolken, 1983). The change in the character of interaction during material association is thus endogenous between the growing open material aggregate and its surroundings. In particular, the proteinoids having lipid-like properties, including anisotropy, are responsible for forming membrane and spheres (Przybylski et al., 1982; Stratten, 1983). One kind of evidence suggests that the resulting layer structure is seen as membrane excitability which is associated with transmembrane charge disequilibrium (Przybylski, 1983).

FLOW EQUILIBRATION

Open material aggregates continue to change the energies of material association and dissociation as they interact with their surroundings endogenously and successively. The interaction between an open material aggregate and its surroundings is reciprocal in the sense that one of the two always causes a change in the other in a successively cyclic manner. This is because the boundary of the open material aggregate is generated endogenously through the interaction with its surroundings. What is significant to open material aggregates that are generated endogenously, is their capability of influencing their surroundings, in addition to being influenced by the same surroundings. Matter, when it is open to material flow, is an active agent influencing its surroundings. This observation naturally raises the question as to what extent matter is an active agent. We shall argue that the extent to which open material aggregates are active agents is certainly acceptable in the realm of physical science.

One of the physical restraints to which any endogenously generated open system must be subject is the fulfillment of the continuity of material flow. What is needed subsequently, especially from the theoretical viewpoint, is the observation of how the continuity of material flow materializes. For this purpose, let us consider a material aggregate open to material flow. The aggregate itself is in a state of material-flow-inside-the-material-aggregate since any material element within the aggregate can be either interchanged with or influenced by others from the exterior (Matsuno, 1982a). The material flow entering the aggregate equilibrates with the flow-inside-the-material-aggregate, which in turn equilibrates with the flow out of the aggregate. The continuity of material flow must be satisfied at both interfaces. The interaction between the material aggregate and the exterior is responsible for the maintenance of the continuity of material flow. However, it is important to notice that the material aggregate maintains continuity by itself by adjusting its interaction with the exterior.

Regulation of the interaction in an open system is self-generative (Matsuno, 1983). Suppose that the continuity of material flow is first satisfied at both interfaces, and then the flow entering the aggregate happens to vary due to some change in the interaction with its exterior. Unless a proper adjustment is made, continuity between the flow into the aggregate and the flow-inside-the-material-aggregate would be destroyed. Material flow equilibration at the interface forces the flow-inside-the-material-aggregate to change so as to preserve continuity with the incoming flow. The change in the flow-inside-the-material-aggregate propagates toward the interface with the outgoing flow. Material flow equilibration at the interface now induces a change in the outgoing flow so as to recover the continuity by adjusting the interaction with the surroundings. The change in the interaction between the material aggregate and its surroundings accompanying the flow equilibration at the interface with the outgoing flow again induces a variation in the incoming flow as the interaction change propagates in the medium.

Most fundamental to the maintenance of the flow continuity through a material aggregate is that equilibration at the interface between the incoming flow and the flow-inside-the-material-aggregate does not occur concurrently with similar equilibration at the interface between the flow-inside-the-material-aggregate and the outgoing flow. This is because the change in the medium is propagated with a finite velocity. An immediate consequence is that material flow equilibration at one interface always becomes a cause for similar equilibration at other interfaces in a successive manner. Material flow equilibration in open material aggregates continues indefinitely once it has begun. The cause of material flow equilibration in an open material aggregate is always the preceding equilibrations.

The interaction change underlying successive material flow equilibration is identified with those changes in the energies of material association and dissociation. As material accretion or depletion proceeds in an open material aggregate, the energies of material association and dissociation change successively. For instance, the material accumulation mediated by the association energy fixed at a preceding instant serves as a cause for the new change in the association energy for further material accumulation. Similar successive changes also apply to material dissociation. The endogenous and successive changes in the energies of material association and dissociation in open material aggregates are thus under the restraint of material flow equilibration. The extent to which open material aggregates are active agents is: they influence their surroundings while complying with the continuity of material flow.

Material aggregates under the influence of material flow equilibration induce adjustments of interaction with the surroundings. The preceding adjustments incessantly supply endogenous causes for the later ones in a successive manner. Successive adjustments of interaction for material flow equilibration are manifest as a form of variation imposed upon the configuration of interaction among material elements constituting the open material aggregate. Although one could imagine arbitrary variations in the configuration of interaction, which in turn influences the energies of material association and dissociation, only those fulfilling the conditions for material flow equilibration are realizable in a material aggregate open to material flow. Random variation means that every conceivable configuration of interaction between atoms and molecules could appear with equal likelihood. Yet the real variation is never random. Configurations of interaction which violate material flow equilibration are internally excluded. Endogenously generated variations in open material aggregates are already constrained compared with hypothetically random ones, and constrained variations provide an impetus to the genesis of the succeeding constrained variations.

Material flow equilibrations which always generate subsequent equilibration of similar character are agents for raising constrained nonrandom variations. A material aggregate open to material flow equips itself with the regulative capability of limiting allowed variations only to those which are nonrandom and constrained by material flow equilibration. Both the successive genesis of variations and the constraint acting upon them originate in material flow equilibration. What appears as vitalistic in the successive genesis of constrained variations is based firmly upon the established physical process, that is to say, there is a propagation of interaction change in the medium with a finite velocity. For instance, material accumulation and the associated change in the energy of material association are sequential and do not occur simultaneously. The result of material flow equilibration is a fulfillment of the continuity of material flow. However, material flow equilibration is more than the continuity of material flow. It incorporates into itself the action for the continuity of material flow.

Experimental evidence supports the successive genesis of constrained variations due to material flow equilibration. Evidence for this is a tripeptide synthesis by thermal polymerization of glutamic acid, glycine, and tyrosine (Nakashima et al., 1977; Hartmann et al., 1981). It has been observed that the two occurring tripeptides that result are pyroglutamyltyrosylglycine (<glu-tyr-gly) and pyroglutamylglycyltyrosine (<glu-gly-tyr). No other theoretical tripeptide isomers have been found. This experimental observation suggests that theoretical tripeptide isomers other than the two sequences <glu-tyr-gly and <glu-gly-tyr fail to survive material flow equilibration during their formative periods. This is because if a tripeptide other than <glu-tyr-gly and <glu-gly-tyr survived material flow equilibration during the formative period, there would be no argument against the presence of such an isomer. Any polymeric species which cannot exist as a stable entity fails to maintain itself as a stable material aggregate (Hsu, 1983). Any material aggregate which fails to remain as a stable entity cannot survive as a stable carrier for material flow equilibration. For instance, even if one could imagine a hypothetical isomer such as <gly-tyr-tyr, it fails to materialize because it failed to become a stable carrier of material flow equilibration. The self-limiting capability of amino acid sequences during thermal polymerization is an indication of successive endogenous constraints working during the same formative period of polymers (Melius, 1983). These constraints originate in material flow equilibration. What material flow equilibration makes explicit is that the operation of equilibration is internally selective and that the resulting products which survive the flow equilibration have already been constrained.

Material flow equilibrations having self-constraining capability successively limit the range of possibilities over which open material aggregates survive. This aspect of constraining the range of possibilities is found to share similarity to a certain extent with the information that can constrain the range of ambiguity on the part of the receiver or observer. This similarity refers to the range of possibilities of surviving material aggregates on the one hand, and to the range of ambiguity to the observer on the other. However, an important difference is that although the information specific to information theory always refers to how and to what extent the observer is informed, the self-constraint pertinent to material flow equilibration refers to how and to what extent matter informs matter itself. One must distinguish between the information to the observer and the information to matter. The information to the observer that information theory deals with takes it for granted that if the observer receives a message from a source matrix of information, he will accordingly be informed and decrease the ambiguity about the source matrix. It is not the subject matter of information theory to scrutinize how the observer becomes informed when he receives a message. It is also beyond information theory to reason how the source matrix of information comes into existence. The source matrix of information is simply a boundary condition that makes information theory workable.

On the other hand matter, as open material aggregate, behaves as both the source matrix of information and the information receiver. The underlying process governs how matter as a source matrix of information or as an agent of constraint changes itself and, at the same time, how matter becomes informed or constrained. Protobiological information as a specific form of information to matter thus goes beyond the realm set by information theory. What we have observed is that material flow equilibration of open material aggregates underlies an origin and evolution of protobiological and biological information.

SYMMETRY AND SYMMETRY-BREAKING

Open material aggregates maintain a certain active capability of changing the energies of material association and dissociation in response to the preceding material accumulation or disintegration. The extent of open material aggregates being active agents is, however, limited within the fulfillment of material flow equilibration. Furthermore, the successive endogenous operations of material flow equilibration continue to generate constrained nonrandom variations. What thus remains to be seen is the implication of the endogenous constraint due to material flow equilibration to protobiological and biological information.

One of the reliable starting points for investigating the origin and evolution of protobiological information is the primordial soup, where experimental simulation is repeatedly verifiable (Florkin, 1975). In particular, a mixture of free amino acids serves for this purpose because of its conceptual clearness. Although it might be argued that a mixture including monomers other than amino acids could also be considered (Orgel, 1968; Cairns-Smith, 1982), this would cause a further complication of the simpler one. The choice of the mixture of free amino acids should be judged according to what kind of results it can provide. If any result of protobiological significance were not available, the previous choice of the primordial soup would have to be abandoned and an alternative tried. On the other hand, if the result turns out to be relevant to protobiological information, as we shall actually recognize, attempts to simulate a more complex and more realistic primordial soup can be tried by continuously transforming the mixture of free amino acids into the more complex one through, for instance, adding monomers other than amino acids.

Open material aggregates expected during thermal polymerization in the mixture of free amino acids are thermal proteins, i.e., proteinoids. Endogenous constraint due to material flow equilibration applies to the amino acid sequence of the synthesized polypeptides. If we consider the tripeptide synthesis (Nakashima et al., 1977; Hartmann et al., 1981) due to thermal polymerization of the three different amino acids, glutamic acid, glycine and tyrosine as discussed in the previous section, the observable sequence turns out to be restricted only to <glu-tyr-gly and <glu-gly-tyr. The hypothetical tripeptide tyr-glu-gly resulting from glu-tyr-gly by the permutational change between glutamic acid and tyrosine, for instance, is not observed in the products of thermal polymerization. On the other hand, the mixture of free amino acids of the three different kinds, especially before the stage of thermal polymerization, does not change its character of thermal molecular motions even if any pair of, let us say, glutamic acid and tyrosine interchange their positions. These artificial positional interchanges are not distinguishable from the similar interchanges originating in the thermal motion inherent to the mixture of free amino acids. This indicates a change in the symmetry property during the thermal polymerization. A permutation symmetry between an arbitrary pair of different types of amino acid, that was preserved before thermal polymerization, is broken after the polymerization. The active agent responsible for polymerization is material flow equilibration of open material aggregates. It is thus found that material flow equilibration underlies the breaking of the permutation symmetry that has been observed in the mixture of free amino acids.

The symmetry-breaking is a result of applying an extra-constraint to a system which would otherwise not lose the symmetry property. Consequently, material flow equilibration as a generator of successive constraints applied to open material aggregates incessantly induces symmetry-breakings. The decrease in the number of degrees of internal freedom is associated with the symmetry breaking because of the increase in the number of additional constraints that the material system must comply with.

One more aspect of the symmetry breaking is the dissolution of the previous degeneracy of internal states observed at a higher symmetry stage. The degeneracy means the multitude of indistinguishable internal states. For instance, let us again consider the mixture of free amino acids comprising the three different amino acids of glutamic acid, glycine, and tyrosine. The mixture of free amino acids does not change the character of internal states even if single glycine and tyrosine residues interchange their positions. This is because the two hypothetical mixtures, one of which differs from the other with regard to the interchange between glycine and tyrosine, are indistinguishable as states of thermal motion and thus degenerate. On the other hand, the mixture of free amino acids changes its internal character after thermal polymerization. The tripeptide <glu-tyr-gly certainly differs from <glu-gly-tyr. This marked difference indicates that the internal states of thermal proteins made from a mixture of free amino acids are not degenerate. The number of distinguishable internal states increases as the mixture of free amino acids transforms itself into thermal proteins. The previous degenerate internal state loses its degeneracy and dissolves into more than one state. The increase in the number of distinguishable internal states is associated with the symmetry breaking that originates in the decrease in the number of degrees of internal freedom.

The number of degrees of internal freedom points to the number of independent variables, and the number of internal states refers to the value range that each variable can take. The number of independent variables available to the mixture of free amino acids is much greater than the anologous number available to thermal proteins because the amino acids in a given peptide sequence have lost the independent kinetic freedom they had when they were in the mixture of free amino acids. The agent that induces the decrease in the number of independent variables is the constraint due to material flow equilibration. The number of internal states, on the other hand, increases with the decrease in the number of independent variables because of the dissolution of the degeneracy observed at the stage of higher symmetry. Still, the reason why the higher symmetry can be available as the number of independent variables is greater must be worked out.

The conceptual relationship between the number of independent variables and the number of distinguishable internal states can be seen when one examines the number of discernible elements in each case. The mixture of free amino acids comprising the three different amino acids, glutamic acid, glycine, and tyrosine, for instance, carries the number of independent variables of the order of the total number of amino acid molecules. In contrast, the number of discernible amino acid types is only three in the present case. This imbalance between the number of independent variables and the number of types of distinguishable elements is a cause of the degeneracy in the internal states. Even if the number of independent variables is great, the distinguishable internal states would be extremely restricted if the number of types of distinguishable elements is limited. If one can imagine or prepare such a system in which the number of independent variables far exceeds the number of types of distinguishable elements like the evolutionary matrix (e.g., a primordial soup, either natural or simulated), it can be said to be in a state of a high symmetry.

The degeneracy of internal states can be dissolved if the number of types of distinguishable elements increases, like the formation of the tripeptides <glu-tyr-gly and <glu-gly-tyr from glutamic acid, glycine and tyrosine. The tripeptide <glu-tyr-gly, for instance, constitutes a unique element distinguishable from all of the three different amino acids. A cause of generating distinguishable elements or of dissolving the degeneracy of internal states is the addition of a new constraint with which all of the elements that have survived must comply. One consequence of complying with the new constraint is a generation of further discernible types made of the then surviving elements. The additional constraint can limit the extent to which some of the former elements move independently of each other and can raise a new correlated aggregate out of the elements thus displaying the physical realization of the constraint. The agent responsible for raising the successive constraints is material flow equilibration. The effect of material flow equilibration is seen as the decrease in the number of degrees of internal freedom on the one hand and as the increase in the number of discernible internal states on the other.

The symmetry-breaking due to the decrease in the number of degrees of internal freedom and the accompanying dissolution of degenerate internal states are the two fundamental characters of evolving open material aggregates. The evolutionary capability of matter is found in the higher symmetry with the greater degeneracy of internal states. The arrow of evolutionary time is in the direction along which the symmetry breaks and at the same time the degeneracy of internal states dissolves. Unless there is the symmetry to be broken and the degenerate internal states to be dissolved, the evolution of matter could not be reasoned. However, the primordial soup makes itself highly degenerate with regard to its internal states because of its extremely small ratio of the number of types of distinguishable elements to the number of total elements inherent to the soup. The symmetry-breaking and the dissolution of degenerate internal states are common to the evolution of any open material aggregate.

The increase in the number of types of distinguishable elements or types in the course of material evolution sheds some light on the problem of how the diversity in the different types of protein molecules has been promulgated with evolutionary time. The earlier material system comprising a smaller number of different types of protein is the source matrix of the later one having a greater number of different types. The precursor proteins are limited in their types. Still, they are highly degenerate in the sense that the number of total proteins far exceeds the number of their different types. It is this degeneracy that can yield the wider variety of different types of protein in the later stage of evolution. Unless there is such a degeneracy in the internal states, the symmetry-breaking due to material flow equilibration cannot bring in further new types. The evolutionary process would meet a dead-end there. However, the degeneracy of the internal states, specifically, with regard to terrestrial proteins is not exhausted as far as the number of the total protein molecules is concerned for they are far greater in number than the number of different types of protein molecules. The multitude of the precursor proteins of the same type is the source of the evolutionary diversification in the types of successor proteins. The evolutionary amenability (Conrad, 1982) of protein molecules is thus sought in the degeneracy of, that is to say, the multitude of the precursor proteins of the same type.

The multitude of the elements of the same types signifies itself as a potential capability of matter that can constrain or inform matter itself. The informed result is a genesis of different material types, and the multitude of the same types produced accordingly again makes it possible for matter to inform matter itself in a recursive manner. What is unique about the protobiological and biological information of matter is its inseparability between the process of informing and that of being informed; or equivalently, between the constraining and matter being constrained. The information available to the observer within the realm of information theory, on the other hand, works only for the purpose of informing the observer or constraining the extent of ambiguity displayed to the observer. The source matrix of information does not change its character during informing the observer. This aspect exemplifies the marked difference between the protobiological and biological information of matter and the information available to the observer. The information which is carried by matter is changed during, and also by, the very process of informing other matter.

The symmetry-breaking due to material flow equilibration of open material aggregates points to how the information of matter changes itself as it informs matter. On the other hand, the informed result expressed upon matter is seen in the dissolution of the previous degenerate internal states. The presence of degenerate internal states is prerequisite to the following symmetry-breaking that can be identified as the evolution of matter, or as the genesis of discernible new types.

PROTOREPRODUCTION AND REPRODUCTIVE FIDELITY

One mode of symmetry-breaking in open material aggregates is the increase in material accumulation of the aggregates. The material elements participating in the material accumulation have lost the kinetic freedom that they had before. The decrease in the number of degrees of kinetic freedom indicates a dissolution of the previously degenerate internal states associated with many material elements of the same type moving independently of each other. Open material aggregates increase the material accumulation, thereby successively changing their interaction with the surroundings, in which successive interaction changes originate in the preceding ones. If the material accumulation exceeds a certain threshold beyond which the interaction change with the surroundings may originate principally in the exterior, the underlying open material aggregate cannot maintain itself as an organized entity toward its surroundings because of the fact that the time interval for successive changes internally grows with the size of the material aggregate. This is because the extent to which any open material aggregate organizes itself is its realization of successive interaction changes with the surroundings induced internally. If an open material aggregate becomes large enough to exceed such a threshold beyond which externally induced interaction changes may be predominant over internally caused successive ones, it can segment itself into the smaller ones, each of which can change its interaction with the surroundings successively in an endogenous manner. The partition of a large material aggregate into smaller ones and the subsequent material accumulation in the latter can be seen as an indication of a prototype of reproduction, or protoreproduction.

Experimental observations suggest that protoreproduction of microsystems was abundant (Fox et al., 1967). Budding of smaller protocells and the generation of endoparticles inside a parental protocell (Brooke and Fox, 1977) are common in experiments with proteinoid microspheres (Fox et al., 1967; Fox and Dose, 1977). However, it is important to notice that such protoreproduction differs from hypothetically perfect reproduction. The difference is in the reproductive fidelity of microsystem. The protoreproduction does not assume perfect fidelity between the successive generations, whereas perfect cellular reproduction does. The fidelity of systemic or cellular protoreproduction is determined through symmetry-breaking due to material flow equilibration of open material aggregates and through the associated dissolution of degenerated internal states. Later protoreproductive units are less degenerate internally compared with the earlier ones. This implies that the earlier protoreproductive units are more inferior in their cellular reproductive fidelity in the sense that they are more degenerate internally compared with the later ones. The cellular reproductive fidelity increases as the number of degrees of internal freedom, specifying the range of available protoreproduction, decreases. When a degenerate protoreproductive state segments itself into more than two of less-degenerate protoreproductive states, the earlier state gains its cellular reproductive fidelity by transforming itself into one of its less-degenerate protoreproductive states, because of the fewer degrees of internal freedom in the later new state.

The protoreproduction proceeds through the form of open material aggregates. This character of protoreproduction, due to openness, sheds some light on how the replication of macromolecules sets in and proceeds inside open material aggregates. The replication of polymers always assumes, within of course a limited extent of molecular reproductive fidelity, a cyclic reaction process in which the product of one stage acts as a template or an enzyme for the polymerization process of the next stage while the product at a certain later stage happens to coincide with that of an earlier stage. A new problem of the interplay between cellular and molecular reproductive fidelities thus arises.

Let us imagine that there exists a reaction cycle having N steps of polymerization and that the product of the N-th step is identical to the template or the enzyme for the polymerization at the first step. Each step needs an inflow of energized monomers for polymerization. If the inflow of energized monomers is not conditioned nor influenced by the N-step reaction cycle to survive, then the probability p_i that the externally-controlled inflow of monomers as a resource of the i-th step polymerization provides the right amount required for the polymerization over a unit time interval is always less than unity, $p_i<1$. If the kinetic and internal energies of the incoming monomers are too large or too small for the required polymerization, they cannot participate in the reaction cycle. The interaction among energized monomers always induces in themselves a widespread of energy distribution. Since our assumption has been that the reaction cycle itself cannot condition nor influence the amount of energy which the inflow of monomers carries with itself at each reaction step, the probability that the inflow carries the right amount of energy for a given polymerization process is always less than unity. Furthermore, if the amount of incoming monomer of each kind exceeds or falls below the amount required, the proper functioning of the reaction cycle will be disturbed.

The probability that the reaction cycle maintains itself over a unit time interval is $p=p_1p_2 \ldots p_N(<1)$. Consequently, the probability that the reaction cycle survives over the time interval

T, p^T, tends to vanish as T increases. Unless the inflow of monomers is properly constrained by the reaction cycle to survive, the cycle itself cannot survive for a long time. The reaction cycle which fails to influence the entering material resources cannot maintain the replication process after it exhausts the supplied resources and it fails to survive even for a while if the necessary resources are not supplied. What is significant to an evolutionary precursor of the biological machinery of polynucleotide and polypeptide replication is its own capability of influencing the interaction with the surroundings. Successive operation of reaction cycles is crucial to the protobiological and biological replication of polynucleotide and polypeptide.

The successive functioning of the reaction cycle does require a process that adjusts the interaction with the surroundings and allows only monomers that fulfill the condition of maintaining the reaction cycle. One such agent that can constrain the mode of interaction with the surroundings is an endogenously generated material aggregate open to material flow. This aggregate generates the boundary which separates it from the surroundings via the interaction with its surroundings. The necessary condition for the successive replication of macromolecules is the capacity of the underlying reaction cycle to influence the entering monomers. Open material aggregates have this capacity (Baltscheffsky and Jurka, 1983). Furthermore, open material aggregates are protoreproductive by themselves in addition to their potential capacity of housing reaction cycles in themselves.

Protoreproductive open material aggregates enhance their cellular reproductive fidelity by each generation due to the endogenous symmetry-breaking and the associated dissolution of degenerate internal states. The reaction cycle which allows a product of the previous reaction to serve as a template or an enzyme of the next reaction in a successive manner, once initiated, increases the products to be accumulated at least initially inside the underlying material aggregate exponentially with time. This indicates the loss of the number of degrees of kinetic freedom which the monomers, to be polymerized, originally had. The onset of a reaction cycle inside an open material aggregate is in accord with the symmetry-breaking due to the decrease in the number of degrees of internal freedom originating in material flow equilibration. A consequence of enhancing the cellular reproductive fidelity associated with the protoreproductive open material aggregates is an incorporation of a reaction cycle allowing replicating polymers. The molecular reproductive fidelity of replicating polymers is also enhanced through the endogenous operation of symmetry-breaking because of the successive dissolution of degeneracy in the internal states, and because of the successive enhancement of constraint.

The protoreproduction of open material aggregates is a result of both the ease with which open material aggregates are formed under a suitable protobiological condition and the specificity due to the ongoing symmetry-breaking accompanying the dissolution of degenerated internal states. The metabolism and the maintenance of a specific function processing matter is a necessary condition for protocellular reproduction (Rosen, 1983). Material flow equilibration of open material aggregates is certainly a physical embodiment of both the metabolism and the specificity associated with material flow. The protoreproductive cellular units precede the reaction cycle of replicating polymers.

One of the intriguing questions about the reaction cycle is how it evolves with time. Although it is possible to describe a form of arbitrary reaction cycle in terms of so-called reaction-kinetic dynamics, the crucial part of this question is how the descriptive form of the reaction cycle changes itself with time (Rössler, 1983). In this regard, open material aggregates carrying the reaction cycle provide some answer. The boundary condition of the reaction cycle is specified by the underlying open material aggregate. What is more, the boundary condition itself changes through material flow equilibration. This implies that open material aggregates are prerequisite for both the emergence and the evolution of the reaction cycle. A more satisfactory description of the reaction cycle must be accompanied by the dynamics of an open material aggregate which serves as the carrier of the cycle. Open material aggregates constitute the environment under which the reaction cycle materializes. The change in the reaction cycle that has been initiated by the change in interaction between the supporting material aggregate and its exterior can again influence the supporting carrier. This is because the change in interaction propagates in the medium with a finite velocity. The emergence and the evolution of the reaction cycle is thus inseparable from the supporting open material aggregates. The connecting thread is material flow equilibration.

The symmetry-breaking accompanied by the dissolution of degenerate internal states is also pivotal in organizing the reaction cycles appearing in an open material aggregate. The intricacy of interlocking reaction cycles is possible only when these are incorporated in an open material aggregate (R. F. Fox, 1983). The symmetry-breaking due to material flow equilibration can cause independent reaction cycles, if any, to make out of themselves an interlocked larger cycle. The interlocking is a physical realization of the symmetry-breaking that deprives those independent cycles of the degeneracy in their mutual disposition inside the open material aggregate. The reaction cycles, once interlocked, cannot operate independently of each other and thus lose the previous degeneracy observed in their arbitrary mutual disposition. A typical reaction cycle is those of polypeptides and polynucleotides. Open material aggregates can initiate the process of polypeptide replication. The plausibility of abiotic peptide synthesis cycle has already been suggested (Fox, 1981). Likewise, the residue-by-residue replication of polynucleotide is another possibility of the reaction cycle. An open material aggregate can make cause the interlocked replication of polypeptides and polynucleotides with each other. An interlocking of small reaction cycles can give rise to a larger one. The emergence of such a hierarchical reaction cycle is a consequence of the working of the symmetry-breaking intrinsic to open material aggregates. An example of the hierarchical reaction cycle is Eigen's hypercycle between polypeptides and polynucleotides, modelled after the contemporary coded protein synthesis (Eigen, 1971). An evolutionary precursor to this hypercycle is also proposed as a protohypercycle (Matsuno, 1982b). What is significant at this point is the potential capability of open material aggregates to interrelate both polypeptide and polynucleotide replications in a correlated cyclic manner.

The hierarchical reaction cycle also induces an enhancement of the molecular reproductive fidelity of polypeptides and polynucleotides when compared with those found in each of the uncorrelated reaction cycles. The enhancement came from the dissolution of degenerate internal states due to the symmetry-breaking as a result of forming a hierarchy of reaction cycles. The molecular reproductive fidelity of polymers thus observed differs from the cellular reproductive fidelity of open material aggregates. However, both originate in the same process of material flow equilibration inherent to open material aggregates. Protocellular processes are prerequisite to the onset and evolution of macromolecular multiplication. Enhancement of the molecular reproductive fidelity presumes an increase in the cellular reproductive fidelity of open material aggregates.

THE ORIGIN OF THE GENETIC CODE

Protoreproductive cellular units can eventually cause macromolecular multiplication in their interior if resources are available as discussed in the previous section. The underlying process is the symmetry-breaking originating in material flow equilibration of open material aggregates. An onset of macromolecular multiplication can take advantage of protoreproductive units, if available. Under these circumstances there are two possibilities of multiplication, one for polypeptides and the other for polynucleotides. What laboratory experiments have suggested (Fox and Dose, 1977) is that polypeptides synthesized thermally can form protoreproductive cellular units in aqueous milieu. On the other hand, however, it still remains to be seen whether polynucleotides synthesized with or without templates could form protoreproductive cellular units in the absence of polypeptides (Orgel,

1979). This does not, however, exclude the likelihood that polynucleotides could enter the protoreproductive cellular units made of polypeptides at a later evolutionary stage.

Thermal proteins originate in the self-sequencing of amino acids. If the polymerization of amino acids proceeds only in the form of their self-sequencing, the protocellular reproduction of open material aggregates made of thermal proteins would be at most heterotrophic because of the fact that the formation of polypeptides takes place independent of the protoreproductive cellular units. The heterotrophic reproductive units enhance the material accumulation by means of the accretion of polypeptides made elsewhere. However, the heterotrophic protoreproduction is not the only process. The side chain specificity of thermal proteins provides a possibility that the thermal proteins can work as catalysts for further polypeptide synthesis (Fox, 1981). A reaction cycle of polypeptide synthesis can be expected to follow accordingly. The protoreproductive cellular units incorporating a cyclic polypeptide synthesis in themselves is now autotrophic because they can synthesize their own polypeptides. Since the autotrophic reproductive units are autocatalytic in their polypeptide synthesis, the material accumulation is greatly enhanced compared with the case of non-autocatalytic synthesis. The heterotrophic reproductive units can alternate with the autotrophic ones because of the greater rate of material accumulation of the latter. The symmetry-breaking due to material flow equilibration acts so as to maintain consecutive accumulation of matter in open material aggregates. Consequently, only those aggregates with the greater accumulation rate survive. The alternation of the heterotrophic reproductive units with the autotrophic ones is just an exemplification of the symmetry-breaking associated with open material aggregates.

The energetic aspect of material aggregates (Lipmann, 1983) related to their openness suggests the preference of the autotrophic reproductive units over the heterotrophic ones. These energetics also influence the informational aspect of polypeptides found in the protoreproductive units. Open material aggregates have the informational capacity to inform or to constrain matter within themselves. The underlying physical process is the symmetry-breaking originating in material flow equilibration. Still, the constrained matter is informational in the sense that matter is in a state of being informed. An expression of the informed state of matter is the amino acid sequences in the polypeptides. It is, however, crucial to realize that the informational aspect of polypeptides is more than what has been expressed, for instance, as in the form of the amino acid sequences. The capacity of polypeptides to inform matter, or more precisely, the synthetic production process aided by the polypeptides must also be taken into account (Chernavskaya and Chernavskii, 1983). If one concentrates only on the expressed information or the products already synthesized, the informational capacity to constrain matter further would be lost from the expression. In order to retain the capacity of matter to inform itself, one must also take the physical process of constraint acting upon matter into account (J. L. Fox, 1983) when the expressed information is examined. This necessarily implies that the informational capacity of matter is always beyond what has already been expressed.

The informational capacity of thermal proteins is more than just the amino acid sequences expressed upon themselves. Thermal proteins can instruct the further polypeptide synthesis with the aid of their own side chain specificity. The extent to which the expressed information cannot cope with the intrinsic informational capacity is the change in the production process of polypeptide synthesis. Thermal proteins due to thermal polymerization of amino acids are different from those polypeptides synthesized by the thermal proteins. The informational capacity of the polypeptides synthesized by thermal proteins is more constrained. In other words, thermal proteins are informationally more inclusive than the polypeptides they synthesize. Thermal proteins themselves have a potential capability for changing the mechanism of polypeptide synthesis. The evolutionarily new mechanism emerges through the symmetry-breaking due to material flow equilibration. More precisely, the emergence of the new mechanisms of synthesis accompanies the greater rate of material accumulation of open material aggregates. The information to instruct a change in the peptide synthesis mechanism is already latent in the aggregate of the precursor polypeptides. The characters specific to this information of instruction are the successive endogenous constraints acting upon themselves.

A further change in the peptide synthesis is expected if polynecleotides participate in the process. There exists an affinity between amino acids or polyamino acids and polynucleotides (Weber and Lacey, 1978). The attractive interaction between some amino acids and their anticodonic homopolynucleotides is an indication of this affinity (Lacey, 1983). If nucleotides or polynucleotides are available, the likelihood will be greatly enhanced that the then polypeptide synthesis mechanism can incorporate into itself polynucleotides (Ponnamperuma and Hobish, 1983). The synthesis of polynucleotides from nucleotides proceeds with the aid of polypeptides working as catalysts. At the same time, the synthesis of polypeptides from amino acids is also under the influence of those polynucleotides synthesized by polypeptides, because of their mutual affinity (Nakashima, 1983). One can thus visualize an emergence of a reaction cycle of polynucleotides synthesized by polypeptides if a polypeptide synthesized under the influence of a certain polynucleotide can catalyze the synthesis of the polynucleotide of the same type directly or indirectly with the aid of some of the derivative polypeptides. This will give an interdigitation of both polypeptide and polynucleotide synthesis cycles.

A simple polypeptide synthesis cycle without an interdigitation of polynucleotides indicates that one type of polypeptide can work as a catalyst for the synthesis of another different type of polypeptide and that all the related polypeptides constitute a catalytic reaction cycle. In this case, the average number of polypeptides working as a catalyst for making an arbitrary polypeptide is always unity. On the other hand, however, if the two reaction cycles of polypeptides and polynucleotides are interdigitated with each other, the average number of polypeptides working as catalysts for making one polypeptide is always greater than unity. This is because the synthesis of one polypeptide requires both a polypeptide working as a catalyst and the intervening polynucleotides, the latter of which also requires polypeptides as their catalysts. If one calls the average number of catalytic polypeptide molecules required for the synthesis of one polypeptide as the order of autocatalysis, the interdigitated reaction cycle of polypeptides and polynucleotides thus turns out to maintain the order of autocatalysis greater than unity. Furthermore, the material accumulation of an open material aggregate in which polypeptide synthesis proceeds will be enhanced if the order of autocatalysis increases. The reason behind this enhancement of material accumulation is that the production rate of polypeptides is proportional to the product present raised to the power of the order of autocatalysis. The interdigitated reaction cycle with the greater order of autocatalysis in polypeptide production dominates over the other cycles with the smaller order of autocatalysis.

The interdigitated reaction cycle of polypeptides and polynucleotides requires more than one polypeptide molecule to produce or to instruct the making of one polypeptide. This exhibits a marked contrast compared with the simple cycle of polypeptides only, in the latter one polypeptide molecule instructs the making of one polypeptide. More specifically, the interdigitated reaction cycle requires more than one amino acid residue to instruct one residue in the production process of polypeptides, whereas the simple cycle of polypeptides hypothetically requires only one amino acid residue to instruct one residue (Fox and Matsuno, 1983). One thus observes that the interdigitated reaction cycle can have greater information content per each amino acid residue compared with the similar information content associated with the simple cycle of polypeptides only. The origin of the superficial increase in the information content per each amino acid residue is undoubtedly internal. That is to say, the information for instruction is already latent and comes to materialize at a later stage through an internally constrained process. The utilization of more than one amino residue in instructing one amino acid residue in polypeptide synthesis is a consequence of the endogenous symmetry-breaking originating in material flow equilibration that intrinsically favors open material aggregates with the greater accumulation rate.

The interdigitated reaction cycle of polypeptides and polynucleotides suggests that both polypeptides and polynucleotides can work as information carriers to instruct further polymeric synthesis. Then, the problem arises regarding which is informationally more inclusive, polypeptides or polynucleotides. One can estimate the average number of polypeptide molecules required for instructing the making of one polypeptide, since the contribution from polynucleotides can be equivalently translated in terms of polypeptides because of the cyclic nature of the polymeric synthesis. The same estimation can also apply to polynucleotides. If the average number of polypeptide molecules required for the instruction of making one polypeptide is greater than the average number of polypeptides required for instructing the making of one polynucleotide molecule, one can say that the polypeptide is informationally more inclusive than the polynucleotide because of the participation of the greater number of polypeptide molecules in the synthesis of one polymer. In this case, the polypeptide will turn out to be a more inclusive information carrier than the polynucleotide. The original polymeric information translation thus proceeded from polypeptide to polynucleotide, not vice versa. On the other hand, if the average number of polypeptides required for instructing the making of one polynucleotide molecule is greater than the similar number for making one polypeptide, the polynucleotide will now become informationally more inclusive. The polymeric information translation will proceed from polynucleotide to polypeptide.

The polymeric information translation from polypeptide to polynucleotide could be reversed if the process is accompanied by an increase in the otder to autocatalysis in peptide synthesis. In fact, a theoretically proposed protohypercycle which is autocatalytic to the order 4/3 in polypeptide is an example of the interdigitated reaction cycle of polypeptide-polynucleotide complex with the translation from polypeptide to polynucleotide (Matsuno, 1982b). One more theoretical model of the interdigitated reaction cycle with the translation from polynucleotide to polypeptide is Eigen's hypercycle (Eigen and Schuster, 1979), with its order of autocatalysis 3/2 in polypeptide (Matsuno, 1982b). As far as these theoretical models are concerned, the protohypercycle would seem to serve as a precursor of the hypercycle during the evolutionary development of the interdigitated reaction cycle of polypeptides and polynucleotides. The informational capacity of the instruction for making the hypercycle is already latent in the protohypercycles. The latent information can be specified explicitly only after it has been expressed in the form of a hypercycle. The carrier of this latent information is an open material aggregate that endogenously undergoes the symmetry-breaking originating in material flow equilibration.

The evolutionary significance of the mechanisms involved with the genetic code requires scrutiny of the origins of polynucleotides such as DNAs, messenger RNAs, ribosomal RNAs and transfer RNAs;

and polypeptides such as ribosomal proteins, tRNA-synthetases and RNA polymerases (Oshima, 1983). One of the integrated views contributing to the solution of this problem is to start from the polymerization of amino acids energized by various means such as heat. The resulting proteins, e.g., thermal proteins, can make out of themselves open material aggregates or protocells in the aqueous melieu. An open material aggregate is a carrier of the endogenous symmetry-breaking originating in material flow equilibration. One of the consequences of the symmetry-breaking inherent to open material aggregates is an increase in the material accumulation rate with evolutionary time. This can be responsible for the start of the reaction cycle of polypeptide synthesis as well as its maintenance. The symmetry-breaking can also enhance the interdigitation of various reaction cycles that could otherwise exist independently of each other. Accordingly, if nucleotides or polynucleotides are further available, the interdigitation of reaction cycles of both polypeptide and polynucleotide syntheses could be followed because of the very nature of the symmetry-breaking. One can thus visualize a materialization of the interdigitated reaction cycle of polypeptides and polynucleotides as a candidate for an evolutionary precursor of the genetic code apparatus. The transition from the direct interaction between amino acids and their homopolynucleotides (Lacey, 1983) to the indirect interaction between amino acids and their anticodons (Dillon, 1983) is in fact an indication of enhancing the interdigitation of further reaction cycles by incorporating the intervening adaptors.

A significant aspect of the present viewpoint is that the evolutionary precursors always equip themselves with the latent information to instruct the making of the evolutionary successors (Jungck, 1983). This latent information is maintained through the process of material flow equilibration intrinsic to open material aggregates. The extent to which the expressed information is available is what has already been constrained by the sequence of the preceding symmetry-breaking. What remains to be specified is the latent information. This inference tells us that the informational aspect of open material aggregates must be accompanied by both the then expressed information in an explicit manner and the instruction for making the succeeding products, including evolutionary novelties. The instruction for evolutionary novelties cannot be expressed in advance in an exhaustive manner. Nevertheless, the informational description of an open material aggregate cannot be satisfactory unless it is accompanied by the inevitable latent information, even if the content remains indefinite.

The evolutionary shceme starting from polypeptides conforms with the very basic character of open material aggregates, implying that the evolutionary informational capacity is already latent in the evolutionary precursors. What looks like an emergence of evolutionarily new information, such as de novo structure and function, is a result of applying a constraint to the latent information. The constraint originates in material flow equilibration of open material aggregates. On the other hand, another evolutionary scheme starting from the synthesis or the replication of polynucleotides in the absence of polypeotides, although hypothetically, must acquire the necessary evolutionary information from an external agent. This is because open material aggregates as a vehicle of applying endogenous constraints to the latent information could not be available from polynucleotides alone without the intervention of polypeptides (Orgel, 1979). Experimental evidence suggests that open material aggregates require polypeptides in their making irrespective of whether polynucleotides are also present (Fox and Dose, 1977). Evolution of polynucleotides alone without an intervention of polypeptides must rely solely upon the impetus from the environment, no matter how small its likelihood, because of the absence of open material aggregates that can influence their exterior (Orgel, 1968). In contrast, if polypeptides are available at an early evolutionary stage as suggested by geological reasonings (Kenyon, 1983) and also by simulated experiments (Fox and Dose, 1977), the presence of open material aggregates working as protocells would be highly conceivable. Once open material aggregates are formed, the evolution can proceed through the successive endogenous constraints acting upon the latent information. Cells processing material intake and outflow are responsible for providing endogenous boundary conditions that constrain the molecular multiplication proceeding inside these cells.

One cannot deny, on a purely logical basis, the possibility of successive replication of polynucleotides, without being accompanied by polypeptides and by the associated open material aggregates, simply because of its probabilistic unlikelihood (Eigen and Schuster, 1982). However, the conceptually difficult part of this possibility lies in the fact that the mechanism providing evolutionary novelties is not worked out. The evolutionary scheme starting from polypeptides and their associated open material aggregates, on the other hand, makes clear the process underlying the emergence of evolutionary novelties. The emergence and the evolution of the genetic code can also be conceptualized in terms of the successive constraints acting upon the latent information present within matter itself. It is protocellular reproduction which causes the molecular replication that eventually leads to the genesis of the genetic code.

NATURAL SELECTION IN PROTOBIOLOGY

Evolutionary complexity and its coordination in biology needs as an operational basis natural selection (Mayr, 1982). This observation naturally raises the question of how the process of natural selection emerged. Phenomenologically, natural selection is understood to operate through differential reproduction. The premise is the presence of reproductive or replicative units, cellular or molecular or both. This way of thinking leads us to pay attention to the origin of reproductive units for the functional origin of natural selection (Morgan, 1932). The problem now reduces to how reproductive units, either molecular or cellular, could emerge. What is required is a spontaneous emergence of a certain reproductive unit with a reasonably confined reproductive fidelity. Unless the reproductive fidelity is confined within a certain range, the reproduction in its literal sense could not proceed. The mechanism responsible for maintaining the reproductive fidelity within a fine-tuned range is prerequisite to the survival of reproductive units. Further, how both molecular and cellular reproduction could come to be integrated together must be elucidated unless both could appear in a simultaneous manner. The fine-tuning of reproductive fidelity is needed for both molecular and cellular processes.

One possibility, if molecular reproduction is viewed as preceding a cellular one, is the process of inorganic crystallization of clays (Cairns-Smith, 1982). The formation of the crystalline structure in clays is repetitive with regard to its layering even if crystalline defects are present. Those crystalline defects propagate through the repetitive layers. This indicates both the process of replication and the preservation of molecular reproductive fidelity to a certain extent. However, the functional underpinning of the molecular reproductive fidelity cannot be found anywhere other than in the process of crystallization. An immediate consequence of the inseparability between the very process of molecular reproduction and the preservation of molecular reproductive fidelity is that any crystalline defect, occurring through an arbitrary external means, cannot be removed from the process of reproduction. This is due to the fact that the maintenance of molecular reproductive fidelity is subject to the process of molecular reproduction. Reproductive fidelity is vulnerable to external disturbances. The content of molecular reproductive fidelity spontaneously changes due to those external disturbances. The cumulative enhancement of molecular reproductive fidelity, that is ubiquitously common to biological evolution, is thus ruled out from the system which lacks its own dynamics of maintaining and enhancing reproductive fidelity. A principal difficulty with the scenario claiming clay molecules as the precursor of biologically reproductive molecules lies in the inseparability between molecular reproduction and molecular reproductive fidelity. This difficulty is more fundamental than the takeover of clay molecules by carbon-dominant ones, since what is required in the transition from non-biological to biological reproductive units is the functional separation between molecular reproduction and molecular reproductive fidelity, i.e., an establishment of dynamics of maintaining and enhancing molecular reproductive fidelity.

One more possibility of molecular reproduction preceding a cellular one is the residue-by-residue replication of polynucleotide with the aid of base complementarity (Orgel, 1968). Contemporary polynecleotides are replicated by residue-by-residue processes. However, the molecular reproductive fidelity of the contemporary process seeks its functional underpinning not only upon the residue-by-residue replication, but also upon the entire machinery of biosynthesis (Commoner, 1968; Kornberg, 1980). The residue-by-residue replication of polynucleotide without any help from proteinous enzymes cannot separate the process of reproduction from its reproductive fidelity, since any replication error of exogenous origin propagates through the succeeding process. The inseparability between molecular reproduction and molecular reproductive fidelity is inevitable as far as only the residue-by-residue replication of polynucleotide is concerned. Thus, one recognizes that the real evolutionary problem of reproductive units is whether the residue-by-residue replication of polynucleotide with its functional separation between molecular reproduction and molecular reproductive fidelity could be preceded by the analogous residue-by-residue replication in which both molecular reproduction and reproductive fidelity are inseparable. The problem is again how the separability between molecular reproduction and molecular reproductive fidelity could materialize during evolution. The answer to this question forces us to distinguish between the origin of reproductive units and the functional origin of reproductive fidelity. Even if the origin of reproductive units such as enzyme-free replicating polynucleotides can be simulated, it could not have a direct relevance to the emergence of biologically reproductive units unless the process of maintaining the molecular reproductive fidelity is separated from the very process of reproduction. There are abundant possibilities of hypothesizing the replication process of abiotic molecules. However, what matters is how and where the process of maintaining the reproductive fidelity finds its own underpinning. If the functional basis of maintaining molecular reproductive fidelity happens to be identical with that of molecular reproduction, it could still remain irrelevant to the biological origin of replicative molecular units.

The biological reproductive fidelity presumes that a reproductive unit carries within it some internal capacity of responding against those excitations and disturbances coming from its exterior. Reproductive protocellular units are open to their surroundings. This observation indicates to us the possibility that open systems, or open material aggregates, can carry to a certain extent within them the capacity of working as active agents. In fact, material flow equilibration of open material aggregates, the internal capacity of fulfilling the continuity of material flow, is an indication of the fact that open material aggregates are active toward their surroundings irrespective of whether they are reproductive or not. We have already noted that open material aggregates can be protoreproductive because of the symmetry-breaking due to material flow equilibration and that the reproductive fidelity of open material aggregates increases with time because of the enhancement of internal constraint originating in the equilibration. Material flow equilibration underlying the enhancement of cellular reproductive fidelity is responsible for causing molecular reproductive units in open material aggregates. An onset of molecular replication in open material aggregates is a consequence of enhancement of the internal constraint due to material flow equilibration.

Natural selection is understood in terms of the differential reproductivity of reproductive units. However, the consequence of natural selection and differential reproductivity are tautological in their implications. The functional underpinning of natural selection is the process that gives differential reproductivity. This process is more than a mere passive one. The self-constraining capability of matter as exemplified in material flow equilibration of open material aggregates underlies the process known as natural selection, making the event of reproduction and the process of maintaining the reproductive fidelity separable. The discrete aspect of the self-constraining capability points to the occurrence of reproduction as an event along the time axis. In contrast, its continuous aspect is seen in the enhancement of the reproductive fidelity.

In summary, there exist two types of reproductive units. One is cellular, and the other molecular. Molecular reproduction presumes cellar reproduction, not vice versa. The enhancement of molecular reproductive fidelity follows an increase in cellular reproductive fidelity originating in the ongoing self-constraint inherent to cellular units.

An experimental indication of the self-constraining capability of open material aggregates is observed in the process of self-sequencing of amino acids during their thermal polymerization. The direct progenitor-progeny relationship is not yet available at the stage of self-sequencing of amino acids. An onset of protoreproductive units is postponed until material aggregation progresses further. Protocellular reproductive units in the form of proteinoid microspheres, for instance, can appear only after the self-constraining of matter expressed as an increase of material aggregation exceeds a certain threshold. The resulting reproductive selection among those protocellular units can be summarized in terms of their differential reproductivity. However, the process underlying the realization of ex post differential reproductivity is in the successive self-constraint inherent to open material aggregates.

The self-constraining capability of matter is endogenous and internal to matter itself. It must be sharply distinguished from the exogenous constraint acting upon matter. In contrast to the self-constraint that can operate indefinitely through material flow equilibration of open material aggregates, the exogenous constraint requires its operational basis totally in an external agent. Even if a certain molecular reproductive unit is caused by an external agent, the process of maintaining the molecular reproductive fidelity would also remain exogenous unless a cellular process is accompanied. The exogenous molecular reproduction remains passive toward the exogenous constraint. Exogenously controlled reproduction cannot serve as a mechanism of yielding endogenous reproductive units. What distinguishes the endogenous self-constraining of matter from the exogenous type is that the endogenous process is equipped with the internal tendency of enhancing the reproductive fidelity, in contrast to the exogenous one whose reproductive fidelity is totally up to external agents. Exogenously controlled reproduction is foreign to protobiological and biological processes.

The functional origin of natural selection in protobiology rests upon the process of self-constraining in cellular units, but not upon a mere emergence of reproductive units. The self-constraining capability is inherent to cellular aggregates open to material flow. Differential reproduction referred to as natural selection is a consequence of the interplay between cellular and molecular reproductive units. Between these two types of reproductive units, the cellular ones are more fundamental evolutionarily as a motor of endogenous self-constraint. It is this cellular process of self-constraint which enhances both cellular and molecular reproductive fidelities.

REFERENCES

Baltscheffsky, H., and Jurka, J., 1983, in this volume.
Brooke, S., and Fox, S. W., 1977, BioSystems, 9:1.
Cairns-Smith, A. G., 1982, "Genetic Takeover," Cambridge University Press, New York.
Cherkin, A., 1983, in this volume.
Chernavskaya, N. M., and Chernavskii, D. S., 1983, in this volume.
Choughuley, A. S. U., 1983, in this volume.
Commoner, B., 1968, Nature, 220:334.
Commoner, B., 1983, in this volume.
Conrad, M., 1982, BioSystems, 15:209.
Dillon, L. S., 1983, in this volume.
Dose, K., 1983, in this volume.
Eigen, M., 1971, Naturwiss., 58:465.
Eigen, M., and Schuster, P., 1979, "The Hypercycle," Springer-Verlag, Berlin.
Fakhari, H., van Roode, H. H. G., and Orgel, L. E., 1981, J. Mol. Evol., 17:295.

Florkin, M., 1975, in: "Comprehensive Biochemistry," M. Florkin, and E. H. Stotz, eds., Vol. 29B, p. 231, Elsevier, Amsterdam.
Fox, S. W., 1981, Int. J. Quantum Chem., QBS8:441.
Fox, S. W., 1983, in: "Beyond Neo-Darwinism," M. W. Ho, and P. T. Saunders, eds., Academic Press, London.
Fox, S. W., and Harada, K., 1958, Science, 128:1214.
Fox, S. W., McCauley, R. J., and Wood, A., 1967, Comp. Biochem. Physiol., 20:773.
Fox, S. W., and Dose, K., 1977, "Molecular Evolution and the Origin of Life," 2nd edn., Marcel Dekker, New York.
Fox, S. W., and Nakashima, T., 1980, BioSystems, 12:155.
Fox, S. W., and Matsuno, K., 1983, J. Theor. Biol., 101:321.
Fox, T. O., 1983, in this volume.
Harada, K., 1983a, in this volume: chapter on "Historical Aspects."
Harada, K., 1983b, in this volume: chapter on "Bioorganic Compounds."
Harada, K., and Fox, S. W., 1958, J. Amer. Chem. Soc., 80:2694.
Hartmann, J., Brand, M. C., and Dose, K., 1981, BioSystems, 13:141.
Hsu, L. L., 1983, in this volume.
Jungck, J. R., 1983, in this volume.
Kenyon, D. H., 1983, in this volume.
Kokufuta, E., and Harada, K., 1983, in this volume.
Kornberg, A., 1980, "DNA Replication," W. H. Freeman, San Francisco.
Lacey, Jr., J. C., 1983, in this volume.
Lahiri, A., 1977, BioSystems, 9:57.
Lipmann, F., 1983, in this volume.
Matsuno, K., 1982a, Int. J. Quantum Chem., QBS9:181.
Matsuno, K., 1982b, BioSystems, 15:1.
Matsuno, K., 1983, in: "Beyond Neo-Darwinism," M. W. Ho, and P. T. Saunders, eds., Academic Press, London.
Mayr, E., 1982, "The Growth of Biological Thought: Diversity, Evolution and Inheritance," Belknap Press, Cambridge, Massachusetts.
Melius, P., 1982, BioSystems, 15:275.
Melius, P., 1983, in this volume.
Morgan, T. H., 1932, "The Scientific Basis of Evolution," W. W. Norton & Co., New York.
Nakashima, T., 1983, in this volume.
Nakashima, T., Jungck, J. R., Fox, S. W., Lederer, E., and Das, B. C., 1977, Int. J. Quantum Chem., QBS4:65.
Nakashima, T., and Fox, S. W., 1980, J. Mol. Evol., 15:161.
Nakashima, T., and Fox, S. W., 1981, BioSystems, 14:151.
Orgel, L. E., 1968, J. Mol. Biol., 38:381.
Orgel, L. E., 1979, Proc. Roy. Soc. London, B205:435.
Oshima, T., 1983, in this volume.
Paecht-Horowitz, M., 1983, in this volume.
Pattee, H. H., 1977, Int. J. Gen. Syst., 3:259.
Pattee, H. H., 1979, BioSystems, 11:217.
Pattee, H. H., 1982, Cognition and Brain Theory, 5:325.
Polanyi, M., 1968, Science, 160:1308.

Ponnamperuma, C., and Hobish, M. K., 1983, in this volume.
Przybylski, A. T., 1983, in this volume.
Przybylski, A. T., Stratten, W. P., Syren, R. M., and Fox, S. W., 1982, Naturwiss., 69:561.
Rohlfing, D. L., 1983, in this volume.
Rosen, R., 1983, in this volume.
Rössler, O. E., 1983, in this volume.
Stratten, W. P., 1983, in this volume.
Weber, A., and Lacey, Jr., J. C., 1978, J. Mol. Evol., 11:199.
Wolken, J. J., 1983, in this volume.
Young, R. S., 1983, in this volume.

INDEX